Production in Forest Nurseries and Field Performance of Seedlings

Production in Forest Nurseries and Field Performance of Seedlings

Editors

Mohammed S. Lamhamedi
Damase P. Khasa
Steeve Pepin

MDPI

Basel • Beijing • Wuhan • Barcelona • Belgrade • Novi Sad • Cluj • Manchester

Editors

Mohammed S. Lamhamedi
Ministry of Natural
Ressources and Forests
Québec
Canada

Damase P. Khasa
Université Laval
Québec
Canada

Steeve Pepin
Université Laval
Québec
Canada

Editorial Office
MDPI
St. Alban-Anlage 66
4052 Basel, Switzerland

This is a reprint of articles from the Special Issue published online in the open access journal *Forests* (ISSN 1999-4907) (available at: https://www.mdpi.com/journal/forests/special_issues/Field_Performance).

For citation purposes, cite each article independently as indicated on the article page online and as indicated below:

Lastname, A.A.; Lastname, B.B. Article Title. *Journal Name* **Year**, *Volume Number*, Page Range.

ISBN 978-3-0365-8768-4 (Hbk)
ISBN 978-3-0365-8769-1 (PDF)
doi.org/10.3390/books978-3-0365-8769-1

Cover image courtesy of Mohammed S. Lamhamedi

Contents

About the Editors

Mohammed S. Lamhamedi

Dr. Mohammed S. Lamhamedi is a researcher emeritus of the Quebec government and an associate professor at Laval University (Canada). He is an expert with more than thirty years of experience on ecophysiology applied to the seedling production chain in forest nurseries, plantations, and international forestry. From 1985 to 1998, he was a professor-researcher at the Institut Agronomique et Vétérinaire Hassan II (Morocco), a postdoctoral researcher at the Institut de recherche en biologie végétale of the University of Montréal, a visiting researcher at the Canadian Forest Service, a scientific director (World Bank Project on modernization of forest nurseries, Tunisia) and a researcher at Laval University. He served as associate editor at the Canadian Journal of Forest Research (2006 – 2012) and has been a guest editor or reviewer for several scientific journals (Forests, etc.). Dr. Lamhamedi is the author of 112 refereed publications, two books, and 210 technical, popularized and professional publications. He has presented more than 80 conferences as a guest speaker. His research programs have addressed scientific and operational issues on the optimization of cultural practices in forest nurseries, seed technology, composting, ecophysiology and environmental stresses, the genetics of ectomycorrhizal fungi, clonal variability, somatic embryogenesis, cuttings, assisted migration in relation to climate change, and the restoration of mining sites. Dr. Lamhamedi was also responsible for the support, transfer of expertise, knowledge and know-how to forest nurseries in Canada and other countries. His major contributions to various international projects have earned him a worldwide renown status as a scientist and practitioner in the modernization of forest and agroforestry nurseries within the framework of reforestation, restoration, agroforestry programs, and the fight against desertification in different countries (Tunisia, Ghana, Nicaragua, Morocco, Algeria, etc.).

Damase P. Khasa

Prof Damase Khasa is an expert in agroforestry and international forestry. He has over 20 years of experience in agroforestry, plant symbioses, molecular ecology and eco-genomics, and restoration ecology. Over the past six years, 84 highly qualified personnel (HQP) of all levels from 21 countries have been trained or are in training in his lab, 77.3% of whom are international students, and 38.7% of whom are female. He is the associate editor of three peer-reviewed journals (Agroforestry Systems, Canadian Journal of Forest Research and Revue Scientifique et Technique Forêt et Environnement du Bassin du Congo) and has produced over 180 refereed publications (+4634 citations, h: 40, i10 = 101, in Google Scholar, 2023-08-15). He was the Program Director of postgraduate studies in agroforestry for 15 years (2004–2019). He was also the Director of the Natural Resource Management Training Programme (2008-2022) funded by Global Affairs Canada and the Congo Basin Forest Fund, which has trained 25 PhD and MSc students at Laval and several hundreds in central African universities. Outputs: Winner, 2011 Canadian Institute of Forestry International Forestry Achievement Award; the Central African Forest Commission 2022 Wangari Maathai Prize for the Congo Basin Forests; 2022 Trees Of Lives award offered by the company Viridis Terra International (Québec). Khasa is also Water Management Theme Lead (2018-2024) for the India-Canada (IC-IMPACTS) Network Centre of Excellence. Outputs since 2013: 78 funded projects, 1,331 HQP trained, 1,431 scientific publications, 382 partners and 32 patents.

Steeve Pepin

Dr Steeve Pepin has been a professor of environmental plant physiology in the Department of Soils and Agri-Food Engineering at University Laval since 2004. He specializes in the whole-plant and leaf-level gas exchange of forestry and horticultural species, as well as in environmental biophysics, crop irrigation and peat substrate properties. He has made important contributions in ecophysiology through his work on hydraulic constraints on plant gas exchange and on the responses of mature deciduous trees to elevated CO2. He is an associate editor of the peer-reviewed journal Botany and has published 65 scientific papers, two book chapters and one patent (2871 citations, h: 25, i10 = 47, in Google Scholar, 2023-08-21). Since 2005, he has supervised or co-supervised two postdocs, 14 PhD and 25 MSc students.

forests

MDPI

Editorial

The Production Chain of Tree Seedlings, from Seeds to Sustainable Plantations: An Essential Link for the Success of Reforestation and Restoration Programs in the Context of Climate Change

Mohammed S. Lamhamedi [1,*], Steeve Pepin [2] and Damase Khasa [3]

1 Ministère des Ressources Naturelles et des Forêts du Québec, Direction de la Recherche Forestière, 2700 Rue Einstein, Québec, QC G1P 3W8, Canada
2 Centre de Recherche et d'Innovation sur les Végétaux, Département des Sols et de Génie Agroalimentaire, Pavillon de l'Envirotron, Université Laval, 2480 boul. Hochelaga, Québec, QC G1V 0A6, Canada; steeve.pepin@fsaa.ulaval.ca
3 Centre for Forest Research and Institute for Integrative Systems Biology, Université Laval, Quebec, QC G1V 0A6, Canada; damase.khasa@ibis.ulaval.ca
* Correspondence: mohammed.lamhamedi@mffp.gouv.qc.ca

Abstract: Although the evolution of principles, procedures, and predictive abilities related to seedling quality throughout the plant production chain (i.e., from seeds to sustainable plantations) has been reviewed over the past decades in various technical and scientific publications, there is still a need to develop and integrate new and efficient practices in forest nurseries and at planting sites, in order to improve the morphophysiological quality of seedlings and saplings, and their survival and growth under different site and environmental conditions in the context of climate change. We have grouped together different scientific articles in this Special Issue of *Forests*, entitled "Production in Forest Nurseries and Field Performance of Seedlings". They cover different topics relating to the seedling production chain in different countries and continents, from growing media to planting performance related to reforestation, restoration, and agroforestry programs.

Keywords: forest nursery; cultural practices; microbial inoculation; planting performance; environmental stress

Citation: Lamhamedi, M.S.; Pepin, S.; Khasa, D. The Production Chain of Tree Seedlings, from Seeds to Sustainable Plantations: An Essential Link for the Success of Reforestation and Restoration Programs in the Context of Climate Change. *Forests* **2023**, *14*, 1693. https://doi.org/10.3390/f14091693

Received: 14 August 2023
Accepted: 19 August 2023
Published: 22 August 2023

1. Introduction

Across the globe, the successes of various reforestation programs constitute a major challenge for the forest managers and practitioners responsible for these programs. This success is based on the use and production of seeds and seedlings of high morphophysiological quality in forest nurseries, and the monitoring and maintenance of this quality throughout the plant production chain, i.e., from seeds to sustainable plantations (Figure 1). However, across continent and country, the constraints, issues, and technological innovations concerning different components of the seedling production chain are dissimilar. The efforts made by each country to improve the components of the reforestation sector are closely linked to the objectives and priorities of their programs (reforestation, restoration, conservation of genetic diversity, sustainability of ecosystems, increases in forest productivity and wood quality, clonal forestry, mitigation of climate change, integration of innovations and forest biotechnologies, fighting against erosion and desertification, improvement of the incomes of local populations, agroforestry, etc.), and the availability of financial resources expertise (scientific, professional, practical, and technical) in the different areas of the plant production chain.

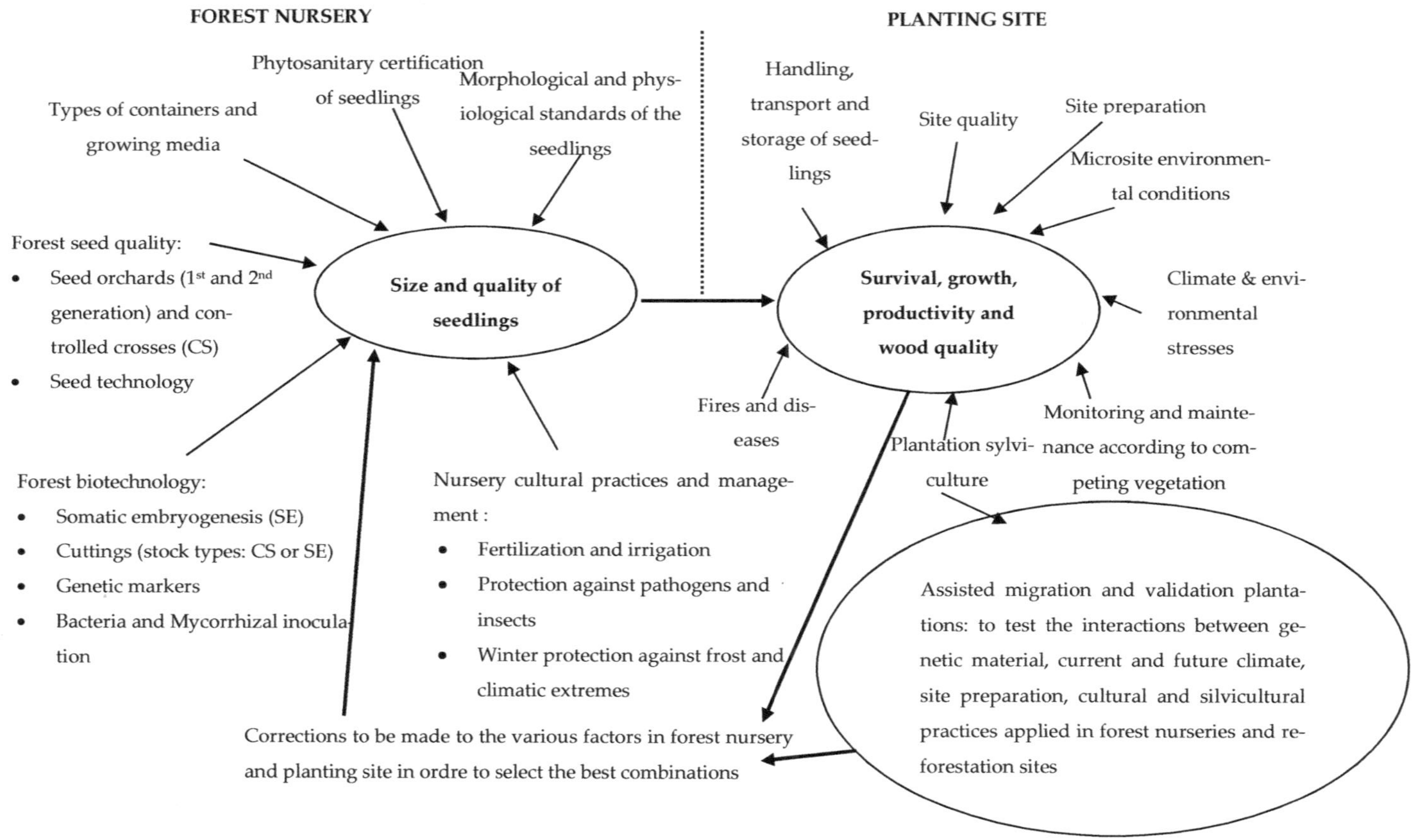

Figure 1. The main factors and components affecting the morphophysiological qualities of seeds and seedlings in forest nurseries, and the processes for monitoring and maintenance of these qualities throughout the plant production chain, i.e., from seeds to sustainable plantations (figure adapted from Lamhamedi and Fortin [1]).

2. Perpetual Readjustment of the Plant Production Chain of Tree Seedlings in the Face of the Diversified Challenges of Tomorrow

Resorting to the use of seedlings of high morpho-physiological quality allows forest managers to easily achieve the objectives of their reforestation and restoration programs, in terms of survival, growth, and sustainability, while facing the combination of an increased frequency of extreme droughts, heat waves, severe winter frosts, pests and diseases, invasive alien species, etc. Otherwise, replanting with additional seedlings will increase the cost of plantations, which increases the time to reach maturity and harvesting of tree plantations. The evolution of principles, procedures, and predictive abilities related to seedling quality throughout the plant production chain has been reviewed over the past decades in various technical and scientific publications [2–12].

Despite the many scientific and technical publications related to the plant production chain, we are aware that several facets of the state of the art in this field continue to advance and readjust rapidly in the contexts of climate change and the supply of substitute substrates for peat. For example, efforts are being made at the international scale to develop (*i*) a sound scientific basis for the assisted migration of tree species; (*ii*) genetically diverse seed sources with ecophysiological mechanisms to adapt to future climate conditions; (*iii*) substitute substrates for peat due to political and environmental pressures, increasing transportation costs, and the decreasing availability of imported peat in many countries; and (*iv*) new cultural practices in seed technology, the growing of seedlings in forest nurseries, seedling ecophysiology, and handling and planting techniques to improve survival and growth under multiple stress site conditions.

To this end, we have grouped together an Editorial and 12 scientific articles in this Special Issue of *Forests*, entitled "Production in Forest Nurseries and Field Performance of Seedlings". These 12 articles were peer reviewed confidentially and anonymously, and subsequently published. They cover different topics related to the seedling production chain in different countries and continents from growing media to planting performance.

Boudreault et al. [13] compared the hydraulic and aeration properties of peat substrates used to produce containerized white spruce seedlings (1 + 0) in eight forest nurseries. Their results suggest that the air-filled porosity at the container capacity of forest nursery substrates should be between 0.03 and 0.10 cm^3 cm^{-3}, in order to improve the root and shoot growth of white spruce seedlings (1 + 0). Moreover, a substrate bulk density between 0.07 and 0.115 g cm^{-3} had a positive effect on root biomass. However, the extraction of peat and its decomposition when used as growing media in forest and horticulture nurseries constitutes a source of greenhouse gas emissions [14]. Because of the depletion of peat resources, their impacts on climate change and certain countries' prohibition on importing peat, several countries have developed national strategies to reduce peat and to find renewable alternatives for peat substitution [14–18]. The challenge is to find alternatives to peat whose physico-chemical properties remain relatively stable, both throughout the growing season and from year to year under the cultural practices applied under forest nursery conditions. Du et al. [19] tested the effects of compost tea applications on tree growth and root mycorrhizal colonization for five common urban tree species (*Acer negundo, Corymbia maculata, Ficus platypoda, Hymenosporum flavum, Jacaranda mimosifolia*) over six months. In another study, Asmara et al. [20] examined a mixture of woody and herbaceous plant species with the introduction of microsymbionts through inoculation, and the application of biochar amendments for accelerating the post-mining restoration. Their results suggest that the restoration processes are improved with combined factors, including microbial inoculation, biochar amendment, crop mixture, plant density, and their direct effects on microclimate improvements.

Trujillo-Elisea et al. [21] evaluated the effect of 10 rhizobacteria strains, commonly named plant growth promoting rhizobacteria (PGPR), during the early stages of production of *Swietenia macrophylla* King [Meliaceae]) under forest nursery conditions. Their results support the advantages of using PGPRs in commercial tropical tree production, because they significantly increase the growth of seedlings. This timber species is of significant ecological and economic importance in the Neotropics.

Speetjens and Jacobs [22] investigated the effects of controlled-release fertilizer (CRF), chelated Fe treatments, and two pot host species (*Acacia koa* and *Dodonaea viscosa*) on the seedling development of Hawaiian sandalwood (*Santalum paniculatum*). They showed that high-quality *S. paniculatum* seedlings can be grown in containers by providing adequate nutrition, and that *S. paniculatum* may benefit from chelated iron fertilizers in a nutrient-limiting growing environment.

With climate change becoming a reality, Lamhamedi et al. [23] simulated different periods of warm weather at the beginning and end of winter, and evaluated their effects on the dehardening and growth of boreal forest seedlings (*Picea mariana* and *Picea glauca)* in response to different freezing temperatures in northern forest nurseries. In winter, regardless of the warming treatment, the seedlings of the two species tolerated the different freezing temperatures without any apparent damage. However, at the end of winter, and in the absence of snow cover, the seedlings did not show frost tolerance at −20 °C. On the other hand, the seedlings showed normal growth after undergoing frosts at −4 °C and −12 °C, comparable to that observed for control seedlings. Different cultural practices and protection strategies are proposed to improve frost tolerance and reduce the winter loss of seedlings.

Landhäusser et al. [24] investigated how initial differences in size, biomass allocation, and non-structural carbohydrate (NSC) storage affect the subsequent partitioning of new biomass, growth potential, and drought response in seedlings of a broad-leaved deciduous tree species (*Populus tremuloides* Michx.) and an evergreen coniferous species (*Pinus banksiana* Lamb).

The results of this study highlight (1) the complexity of how differences in biomass allocation and changes in seedling size may alter storage and the response of species to droughts, and (2) the importance of accounting for initial seedling characteristics (both morphological and physiological) when predicting seedling growth and the impacts of environmental stressors.

Harayama et al. [25] evaluated the effects of seedling size, seven stock types, and mechanical site preparation methods on the initial survival and growth of Japanese Larch (*Larix kaempferi*) seedlings during four consecutive growing seasons. Seedling type (bare-root versus container) had no effect on seedling height during the four growing seasons after planting. Their findings suggest that seedlings with a sufficiently large root-collar diameter and a young age, regardless of seedling type, can grow taller than the surrounding vegetation more quickly. In a literature review, South et al. [26] reviewed and discussed the various factors that govern the successes and failures of reforestation and restoration programs with healthy pine seedlings after leaving the nursery. With a focus on pine seedlings planted in the southern United States, the authors also listed non-nursery factors that have killed pine seedlings in North America, Africa, and Europe.

With an increase in the population, the quantity of wastewater and the scarcity of water resources due to a significant drop in rainfall and long periods of severe drought, especially in semi-arid areas, several countries are beginning to move towards recycling and decontaminating wastewater, using tree seedlings as a biological means. In this Special Issue, Bousbih et al. [27] investigated the potential use of two forest species (*Salix alba* and *Casuarina glauca*) in the rhizofiltration of heavy-metal-contaminated industrial wastewater. *S. alba* exhibited a greater removal capacity for heavy metal ions, and it could be effective as a phytoremediation species for toxic-metal-polluted industrial effluent water.

To ensure the sustainability of arid ecosystems, improve people's incomes, and fight their rural exodus, rural development approaches have focused on optimizing the management of high value-added agroforestry species. In this Special Issue, there are articles focused on assessing the quality of argan oil extracted from natural stands and urban plantations [28], and the evaluation of arbuscular mycorrhizal fungi (AMF) in different argan stands representative of the distribution and genetic diversity of this species [29], with the aim of selecting AMF isolates and inoculating seedlings in the forest nursery, in order to confer them increased tolerance to drought in different plantation sites.

The ecosystems of the Moroccan argan forest constitute an original socio-agrisylvopastoral system which is unique in the world, and a bulwark against the advancement of desertification. It is estimated that nearly 1.3 million people derive their incomes from the valorization of various products of high added value from argan trees; in particular, the extraction of argan oil and the synthesis of cosmetic and medicinal products, whose demand continues to increase on national and international markets. This oil has become one of the most expensive oils in the world. With multiple uses for argan trees, the growing needs, and the severity of the climate, the areas covered by argan trees regress by 2000 ha/year on average. To ensure its sustainability, in 2014, the United Nations Educational, Scientific and Cultural Organization (UNESCO) granted the Moroccan argan grove a biosphere reserve status (intangible cultural heritage of humanity). The Arganeraie Biosphere Reserve was recently created on "International Day of the Argan Tree", by a declaration by the United Nations General Assembly on 10 May 2022.

Sabiri et al. [28] showed that the chemical characteristics of argan oil extracted from the plantations are similar to the oil from the two natural stands of argan trees. These results suggest that it is possible to implement an assisted migration of this tree species outside its natural area into sites exposed to sea spray, without affecting the quality of its edible argan oil. Because of its great resistance to drought, the enhancement and added value of each constituent of the tree (foliage, fruit, wood, etc.), and because of the severity of the droughts that have raged in recent decades across the globe, the argan tree has become a species of first choice for reforestation outside of its natural range; for example, in several countries in Europe (France, Portugal, Spain), in Latin and North America (Chile, Argentina and USA), in Africa (Tunisia, Algeria, Egypt, Sudan, South Africa, etc.), in the Middle East (Syria, Israel, etc.), and in Australia [30–32].

3. Conclusions

With an unprecedented increase in the frequency of fires, and the millions of hectares of forests burnt on a planetary scale in recent years, several programs of reforestation, restoration, and fighting against desertification are underway in different countries. After being planted in several regions of the world, the seedlings are subjected to severe droughts and extremely high temperatures. Faced with these major constraints and challenges, it is not enough to simply plant millions of seedlings, but also to ensure the selection of suitable genetic sources, the morphophysiological quality of the seedlings, and the optimization of cultural practices (e.g., from seed to planting) and silvicultural techniques, in order to achieve the objectives of the various programs (reforestation and wood production, restoration, prevention of desertification, agroforestry, etc.) in terms of survival and growth.

Author Contributions: Conceptualization, M.S.L.; writing—original draft preparation, M.S.L.; writing—review and editing, M.S.L., S.P. and D.K. All authors have read and agreed to the published version of the manuscript.

Funding: Funding for this research project was provided by the Direction de la recherche forestière (DRF)/ Forest Research Branch of the Ministry of Natural Resources and Forests of the Government of Quebec as part of two projects: (1) no. 112310038: Determination of the frost tolerance thresholds of seedlings in winter in relation to climatic extremes and development of a model for predicting the evolution of the hardening state of seedlings in forest nurseries, and (2) no. 142332093. Optimization of the cultural practices affecting root insufficiency and the morpho-physiological quality of plants produced in forest nurseries (holder: Mohammed S. Lamhamedi).

Data Availability Statement: Not applicable.

Acknowledgments: We sincerely thank all authors for their work in producing these high-quality articles. We also thank the many reviewers for their constructive comments, which improved the clarity of the content of each article, the scientific rigor, and the operational scope of the results.

Conflicts of Interest: The authors declare no conflict of interest.

References

1. Lamhamedi, M.S.; Fortin, J.A. La qualité des plants forestiers. In Proceedings of the Actes de la Première Journée Nationale Sur les Plants Forestiers, Rabat, Maroc, 3 June 1992; Abourouh, M., Ed.; Division de Recherches et D'Expérimentations Forestières: Rabat, Maroc, 1994; pp. 35–49.
2. Sutton, R.F. Planting stock quality and grading. *For. Ecol. Manag.* **1979**, *2*, 123–132. [CrossRef]
3. Chavasse, C.G.R. Planting stock quality: A review of factors affecting performance. *N. Z. J. For. Sci.* **1980**, *25*, 144–171.
4. Burdett, A.N. Quality control in the production of forest planting stock. *For. Chron.* **1983**, *59*, 132–138. [CrossRef]
5. Duryea, M.L. *Evaluating Seedling Quality: Principles, Procedures, and Predictive Abilities of Major Tests*; Forest Research Laboratory, Oregon State University: Corvallis, OR, USA, 1985.
6. Rose, R.; Campbell, S.J.; Landis, T.D. *Target Seedling Symposium: Proceedings, Combined Meeting of the Western Forest Nursery Associations, August 13–17, 1990, Roseburg, Oregon*; RM-GTR-200; U.S. Department of Agriculture, Forest Service: Fort Collins, CO, USA, 1990.
7. Grossnickle, S.C. *Ecophysiology of Northern Spruce Species: The Performance of Planted Seedlings*; NRC Research Press: Ottawa, ON, Canada, 2000.
8. Landis, T.D.; Dumroese, R.K.; Haase, D.L. *The Container Tree Nursery Manual. Seedling Processing, Storage and Outplanting*; Agricultural Handbook 674; U.S. Department of Agriculture, Forest Service: Washington, DC, USA, 2010; Volume 7, 199p.
9. Dumroese, R.K.; Landis, T.D.; Pinto, J.R.; Haase, D.L.; Wilkinson, K.W.; Davis, A.S. Meeting Forest Restoration Challenges: Using the Target Plant Concept. *Reforesta* **2016**, *1*, 37–52. [CrossRef]
10. Grossnickle, S.C. Seedling establishment on a forest restoration site—An ecophysiological perspective. *Reforesta* **2018**, *6*, 110–139. [CrossRef]
11. Grossnickle, S.C.; MacDonald, J.E. Seedling Quality: History, Application, and Plant Attributes. *Forests* **2018**, *9*, 283. [CrossRef]
12. Riikonen, J.; Luoranen, J. Seedling Production and the Field Performance of Seedlings. *Forests* **2018**, *9*, 740. [CrossRef]
13. Boudreault, S.; Caron, J.; Lamhamedi, M.S.; Pepin, S. Comparison of Hydraulic and Aeration Properties of Peat Substrates Used to Produce Containerized White Spruce Seedlings (1 + 0) in Forest Nurseries. *Forests* **2023**, *14*, 858. [CrossRef]
14. Hirschler, O.; Osterburg, B.; Weimar, H.; Glasenapp, S.; Ohmes, M.-F. *Peat Replacement in Horticultural Growing Media: Availability of Bio-Based Alternative Materials*; Working Paper 190; Johann Heinrich von Thünen-Institut: Braunschweig, Germany, 2022; pp. 1–64. [CrossRef]
15. Stoffella, P.J.; Kahn, B.A. *Compost Utilization in Horticultural Cropping Systems*; Lewis Publishers: New York, NY, USA, 2001; pp. 1–414.
16. Raviv, M.; Lieth, J.H. *Soiless Culture: Theory and Practice*, 1st ed.; Elsevier: London, UK, 2008; pp. 1–587.
17. Landis, T.D.; Tinus, R.W.; McDonald, S.E.; Barnett, J.P. The Container Tree Nursery Manual. Vol. 3: Containers and Growing Media. In *Handbook Agric*; n 674, U.S.D.A.; Forest Service: Washington, DC, USA, 1990; 87p.
18. Lamhamedi, M.S.; Abourouh, M.; Fortin, J.A. Technological transfer: The use of ectomycorrhizal fungi in conventional and modern forest tree nurseries in northern Africa. In *Advances in Mycorrhizal Science and Technology*; Khasa, D., Piché, Y., Coughlan, A.P., Eds.; NRC Research Press: Ottawa, ON, Canada, 2009; pp. 139–152.
19. Du, D.; Livesley, S.J.; Arndt, S.K.; Truong, C.; Miller, R.E. The Use of Compost Tea in a Containerized Urban Tree Nursery Shows No Evident Benefits to Tree Growth or Mycorrhizal Colonization. *Forests* **2023**, *14*, 1195. [CrossRef]
20. Asmara, D.H.; Allaire, S.; van Noordwijk, M.; Khasa, D.P. The Effect of Biochar Amendment, Microbiome Inoculation, Crop Mixture and Planting Density on Post-Mining Restoration. *Forests* **2023**, *14*, 856. [CrossRef]
21. Trujillo-Elisea, F.I.; Labrín-Sotomayor, N.Y.; Becerra-Lucio, P.A.; Becerra-Lucio, A.A.; Martínez-Heredia, J.E.; Chávez-Bárcenas, A.T.; Peña-Ramírez, Y.J. Plant Growth and Microbiota Structural Effects of Rhizobacteria Inoculation on Mahogany (Swietenia macrophylla King [Meliaceae]) under Nursery Conditions. *Forests* **2022**, *13*, 1742. [CrossRef]
22. Speetjens, T.M.; Jacobs, D.F. Nutrient Availability Has a Greater Influence than Pot Host on Seedling Development of Hemiparasitic Hawaiian Sandalwood (*Santalum paniculatum* Hook. and Arn.). *Forests* **2023**, *14*, 458. [CrossRef]
23. Lamhamedi, M.S.; Lambert, M.-C.; Renaud, M. Simulation of Episodic Winter Warming on Dehardening of Boreal Forest Seedlings in Northern Forest Nurseries. *Forests* **2022**, *13*, 1975. [CrossRef]
24. Landhäusser, S.M.; Wiley, E.T.; Solarik, K.A.; Kulbaba, S.P.; Goeppel, A.E. The Importance of Initial Seedling Characteristics in Controlling Allocation to Growth and Reserves under Different Soil Moisture Conditions. *Forests* **2023**, *14*, 796. [CrossRef]
25. Harayama, H.; Tsuyama, I.; Kitao, M.; Yamada, T.; Furuya, N.; Utsugi, H.; Sasaki, S. Effects of Seedling Size, Stock Type, and Mechanical Site Preparation Method on Initial Survival and Growth of Japanese Larch (*Larix kaempferi*) Seedlings. *Forests* **2023**, *14*, 784. [CrossRef]
26. South, D.B.; Starkey, T.E.; Lyons, A. Why Healthy Pine Seedlings Die after They Leave the Nursery. *Forests* **2023**, *14*, 645. [CrossRef]
27. Bousbih, M.; Lamhamedi, M.S.; Abassi, M.; Khasa, D.P.; Béjaoui, Z. Potential Use of Two Forest Species (*Salix alba* and *Casuarina glauca*) in the Rhizofiltration of Heavy-Metal-Contaminated Industrial Wastewater. *Forests* **2023**, *14*, 654. [CrossRef]
28. Sabiri, C.; Tazi, B.; Maata, N.; Rahim, S.; Taki, H.; Bennamara, A.; Saad, L.; Derouiche, A. Nutritional Assessment and Comparison of the Composition of Oil Extracted from Argan Nuts Collected from a Plantation and Two Natural Forest Stands of ARGAN Trees. *Forests* **2023**, *14*, 180. [CrossRef]

29. Ganoudi, M.; Ouallal, I.; El Mekkaoui, A.; Majid, M.; Ibriz, M.; Iraqi, D. Diversity of Endomycorrhizal Fungi in Argan Forest Stands: Implications for the success of reforestation programs. *Forests* **2023**, *14*, 1649. [CrossRef]
30. Nouaim, R. *L'Arganier Au Maroc: Entre Mythes et Réalités*, 1st ed.; l'Hartmattan: Paris, France, 2005; pp. 1–227.
31. Falasca, S.L.; Pitta-Alvarez, S.; Ulberich, A. The Potential Growing Areas for *Argania spinosa* (L.) Skeels (Sapotaceae) in Argentinean Drylands. *Int. J. Agr.* **2018**, *2018*, 9262659. [CrossRef]
32. Benomar, L.; Elferjani, R.; Hamilton, J.; O'Neill, G.A.; Echchakoui, S.; Bergeron, Y.; Lamara, M. Bibliometric Analysis of the Structure and Evolution of Research on Assisted Migration. *Curr. For. Rep.* **2022**, *8*, 199–213. [CrossRef]

forests

MDPI

Article

Comparison of Hydraulic and Aeration Properties of Peat Substrates Used to Produce Containerized White Spruce Seedlings (1+0) in Forest Nurseries

Simon Boudreault [1], Jean Caron [1], Mohammed S. Lamhamedi [2] and Steeve Pepin [1,*]

[1] Centre de Recherche et d'Innovation sur les Végétaux, Département des Sols et de Génie Agroalimentaire, Pavillon de l'Envirotron, Université Laval, 2480 boul. Hochelaga, Québec, QC G1V 0A6, Canada; sboudreault@activaenviro.ca (S.B.); jean.caron@fsaa.ulaval.ca (J.C.)

[2] Ministère des Ressources Naturelles et des Forêts du Québec, Direction de la Recherche Forestière, 2700 Rue Einstein, Québec, QC G1P 3W8, Canada; mohammed.lamhamedi@mffp.gouv.qc.ca

* Correspondence: steeve.pepin@fsaa.ulaval.ca

Abstract: The physical properties of peat substrates from eight tree nurseries were characterized to determine bulk density, air-filled porosity, saturated hydraulic conductivity, pore effectiveness, relative gas diffusivity and chemical properties. There were significant variations among nurseries both in growth of white spruce [*Picea glauca* (Moench) Voss] seedlings (1+0) and substrate properties. Shoot dry mass and root collar diameter were negatively correlated with air-filled porosity and saturated hydraulic conductivity, whereas root dry mass was positively correlated with bulk density. Seedling growth increased with increasing substrate bulk density up to ~0.11 g cm^{-3}, above which value conditions may become limiting to plant performance. Our results suggest that there was no growth limitation due to restricted aeration ($D_s\ D_0^{-1} > 0.005\ m^2\ s^{-1}\ m^{-2}$ s for all substrates except one) and that over-aeration reduced seedling growth under dry irrigation management.

Keywords: air content; bulk density; gas diffusivity; growing media; *Picea glauca*; tree nursery; container; peat moss

Citation: Boudreault, S.; Caron, J.; Lamhamedi, M.S.; Pepin, S. Comparison of Hydraulic and Aeration Properties of Peat Substrates Used to Produce Containerized White Spruce Seedlings (1+0) in Forest Nurseries. *Forests* **2023**, *14*, 858. https://doi.org/10.3390/f14040858

Academic Editor: Antonio Montagnoli

Received: 13 March 2023
Revised: 6 April 2023
Accepted: 17 April 2023
Published: 21 April 2023

1. Introduction

Nearly 147 million forest seedlings are produced annually in the 18 forest nurseries (6 government and 12 private) in Quebec, Canada [1]. In 2021, white spruce [*Picea glauca* (Moench) Voss] accounted for 22.6% of the seedlings produced in Quebec, i.e., nearly 30 million seedlings [2]. Before planting, the seedlings are evaluated according to morpho-physiological quality criteria and standards established by the Ministry of Natural Resources and Forests (QC, Canada) [3]. Among these criteria, root insufficiency is one of the main causes of plant rejection [4]. In the case of root insufficiency, the plant will be rejected if the root plug partially or completely breaks after extraction from the cavity, shows distinct portions bound by an undamaged root system with more than 5 mm of discontinuity between the portions or if more than 33% of the roots located at the periphery are dead or necrotic [2]. These discarded seedlings obviously have a negative impact on the profitability and long-term viability of nurseries. Various studies have focused on the control and optimization of the main cultural practices, in particular irrigation, fertilization and the model of the container (shape and volume of the cavity, number of seedlings m^{-2}) to improve growth and root architecture [5–8]. However, few studies have focused on the evaluation of the physicochemical properties of substrates in relation to the growth of forest seedlings at an operational scale.

Extreme conditions of substrate water content such as excess water and water stress negatively influence plant physiology, root growth and mineral nutrition [6,9–12]. In addition, net photosynthesis decreases with increasing substrate water content when aeration becomes a limiting factor [13]. Optimizing the physicochemical and aeration properties of the substrate is a critical factor to further improve root growth.

A good artificial growing substrate should have good aeration, high water retention properties, relatively high cation exchange capacity (CEC), be easy to handle and available at low cost [14]. The substrates used in forest nurseries in Quebec are essentially composed of peat and a small proportion of vermiculite and/or perlite. It is common practice in forest nurseries in Quebec to use blond peat with little humification characterized by a high water retention capacity and air content. It has a low pH and a relatively high CEC, two chemical properties ideal for the production and growth of containerized forest seedlings [15]. However, the physicochemical properties of peat can vary depending on its origin and degree of humification [11,16]. In addition, the supply of low-humified peat tends to become difficult with the increase in the duration of peatland exploitation and the scarcity of those exploited near nurseries [17]. The different stages of peat handling, from harvesting to potting, can lead to changes in its physical properties. For example, the type of mixer used to prepare the substrate and a long mixing time can increase the proportion of fine particles [13]. Inappropriate humification and handling of the peat will have the effect of increasing the water retention capacity of the substrate and reducing its air-filled porosity (θ_a), thus increasing the risk of root asphyxiation and waterlogging [11]. Moreover, the height of the saturated zone at the bottom of the root plug after irrigation or rainfall events increases in fine-textured peat substrates, which may further limit gas exchange in the rhizosphere [15].

The physical properties of the substrate can influence the growth of plants [14,18] and forest seedlings [9,13,19,20], in particular air-filled porosity (θ_a), which is largely used as an index of aeration in peat substrates. However, gas exchange between substrates and ambient air also depends on pores' connection in the growing medium; consequently, high θ_a values are not always indicative of adequate substrate aeration. The relative diffusivity of gases (D_s $D_0{}^{-1}$; estimated from measurements of θ_a, pore tortuosity and saturated hydraulic conductivity) has been shown to be a better parameter than air-filled porosity to investigate how substrate aeration properties affect the development and growth of plant species [21,22]. Our previous study showed that the relative gas diffusivity of peat substrates should not be lower than 0.003 to 0.005 cm^2 s cm^{-2} s^{-1} for growth of white spruce seedlings [20]. However, to our knowledge, no study has verified the effect of D_s $D_0{}^{-1}$ in peat–vermiculite potting substrates on the growth of roots and shoots of forest seedlings produced in containers under tunnels during the first growing season in several tree nurseries.

The general objective of this study consists in characterizing the substrates used in the forest nurseries of Quebec using the index of their relative diffusivity of gases and their physicochemical properties in relation to the growth of white spruce seedlings (1+0) during the first growing season under forest nursery conditions. Physical properties were determined directly in the containers to minimize the impact of handling on the fragile structure of the artificial substrates.

2. Materials and Methods

2.1. Substrate Samples from Forest Nurseries

Peat substrates were collected at the beginning and end of the growing season from 8 out of 18 tree nurseries that produce containerized white spruce seedlings in Quebec (Figure 1). These nurseries were located in different ecological regions (bioclimatic domains), between 46°04′–48°49′ N lat. and 65°51′–74°38′ W long. (Figure 1). Accordingly, the number of growing degree-day units (mean air temperature above 5 °C) during the vegetative season varied between 1250 and 2000 among the different nurseries. For confidentiality purposes, we assigned a number from 1 to 8 to each forest nursery included in the study (see Section 3.2).

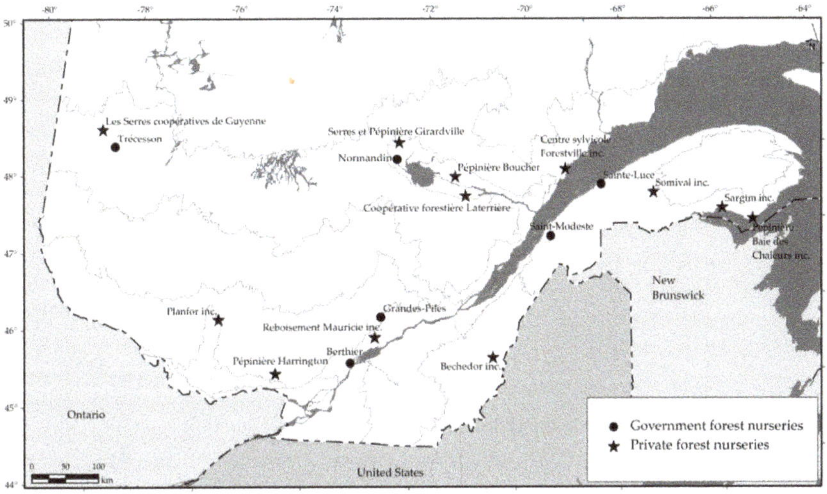

Figure 1. Locations of private and government forest nurseries in the province of Québec (adapted from Lamhamedi et al. [23]).

2.2. Production of White Spruce Seedlings (1+0)

White spruce seeds were sown into containers (model IPL 25-310: 25 cavities per container, 310 cm^3 per cavity, each 12 cm in height with top and base diameters of 6.5 cm and 5.2 cm; IPL®, Saint-Damien-de-Buckland, QC, Canada) between 7 and 20 May 2008, depending on the ecological region. Containers were filled with substrates composed of sphagnum peat moss of various grades (specific to each nursery) provided by Quebec and New Brunswick peat suppliers. Horticultural grade 3 or 4 vermiculite was added to all substrates in proportions that varied among nurseries from 9 to 40% ($v\,v^{-1}$); only the substrate of nursery 8 included perlite (9%, $v\,v^{-1}$). White spruce seedlings were cultivated under unheated polyethylene tunnels, with irrigation and fertilization management practices tailored to the operational conditions of each tree nursery. The amount of nitrogen (N), phosphorous (P), potassium (K) and oligo-nutrients that seedlings received during the growing season (see Table 1) were based on the fertility of peat substrates and seedlings' developmental stage and nutritional needs in order to meet the required growth standards for white spruce (1+0) stock [24,25]. The irrigation was adjusted by nursery managers according to seedling size; the volumetric water content (θ_v) of substrates was determined using time domain reflectometry [6,26] or gravimetric methods [27]. Typically, irrigation management under tunnel resulted in substrate water contents of about 0.35 to 0.45 $cm^3\ cm^{-3}$.

Table 1. Composition, chemical properties, and initial fertility of substrates after potting (n = 5), and quantities of nitrogen (N), phosphorous (P), and potassium (K) applied per seedling of white spruce (1+0) during the growing season in sampled forest nurseries.

No. [1]	Peat	Vermi [2]	Perlite	pH	EC [3] (mS m^{-1})	CEC_{eff} [4] (meq L^{-1})	Initial Fertility (µg 100 cm^{-3})						Applied Fertilizer (mg $Plant^{-1}$)		
		(%, $v\,v^{-1}$)					N_NH$_4$	N_NO$_{2,3}$	P	K	Ca	Mg	N_{total}	P	K
1	75	25	0	4.20	116.7	26.3	14	2	16	57	4	27	86	–	–
2	75	25	0	3.81	209.0	22.4	13	6	18	60	7	35	34	17	21
3	80	20	0	4.10	126.0	21.6	32	5	51	52	5	21	106	24	38
4	86	15	0	4.14	119.7	26.3	14	19	17	37	4	19	83	19	30
5	75	25	0	4.20	166.0	32.2	28	2	90	90	12	39	82	27	49
6	80	20	0	4.17	213.3	33.9	17	10	53	157	15	5	51	12	26
7	60	40	0	4.00	161.3	24.4	18	7	7	40	4	26	51	9	21
8	82	9	9	3.99	143.6	25.0	29	9	8	19	5	29	48	13	31

[1] Forest nursery number. [2] Vermi = vermiculite. [3] EC = electrical conductivity. [4] CEC_{eff} = effective cation exchange capacity.

2.3. Physical Properties of Peat Substrates

One week after potting and prior to seeding, five containers were randomly selected in each tree nursery and brought back to the laboratory to determine the physical properties of the substrates in situ (i.e., on root plugs in containers). Bulk density (ρ_b) was evaluated for three individual cavities per container by measuring the volume and dry weight (oven-dried at 105 °C for 24 h) of each root plug. Total porosity (θ_t) was measured on the same three samples by determining the organic matter content after calcination at 550 °C [28]. Particle size distribution was measured for 650 mL air-dried composite substrate samples (n = 3 root plugs) sieved mechanically (shaker model RX-29, W. S. Tylers®, Mentor, OH, USA) for 3 min on a nest of sieves (0.106, 0.25, 0.5, 1, 2, 4 and 8 mm stacked over a pan). The mean weight diameter (MWD) of particles was calculated as described by Kemper and Roseneau [29]:

$$\text{MWD} = \sum_{i=1}^{n} x_i f_i \quad (1)$$

where x_i is the weight fraction retained on the ith sieve divided by the total substrate sample mass, f_i is the average particle size on the ith sieve, and n is the number of size classes, here equal to 8.

Containers with remaining root plugs were slowly saturated by gradually increasing the water level at a rate of 0.5 cm h^{-1} for 24 h; substrates were then drained freely for 2 h to determine the volumetric water content at container capacity (θ_c, which corresponds to a matric potential of −0.6 kPa, i.e., at mid-height of the IPL 25–310 container). Substrate water content was measured using a 10.5 cm long three-rod (0.2 cm in diameter, spaced at 1 cm apart) TDR probe inserted vertically within a cavity and a cable tester (model 1502B, Tektronix Inc., Beaverton, OR, USA) (Figure 2). For all θ_v determinations, dielectric constants (K_a) between 5 and 50.8 were converted to θ_v using the K_a–θ_v equation established by Paquet et al. [30], which was found appropriate for these substrates [20]. Values of K_a between 50.8 and 81 were linearly interpolated from θ_v at K_a = 50.8 to θ_v = 1.0 cm^3 cm^{-3} at K_a = 81. The air-filled porosity after drainage (θ_a) was calculated as

$$\theta_a = \theta_t - \theta_c \quad (2)$$

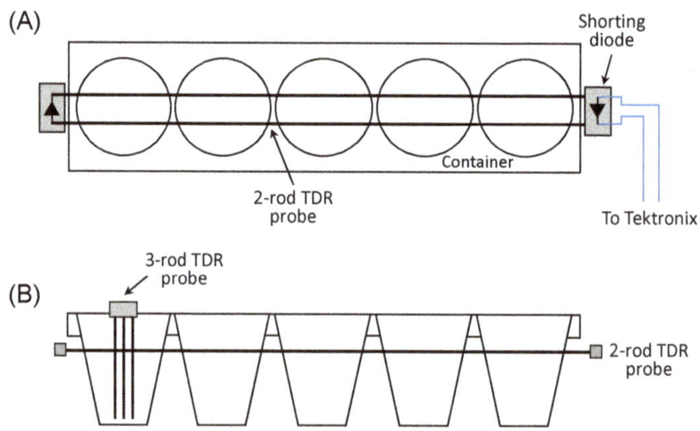

Figure 2. Schematic of the linear five-cavity container (aerial (**A**) and lateral (**B**) views) including a 10.5 cm long three-rod TDR probe inserted vertically within a cavity and a 36 cm long two-rod TDR probe with shorting diodes inserted horizontally across the five cavities at a height of 8 cm from the bottom of the container.

Five substrate root plugs were carefully extracted from each nursery container and placed in a linear five-cavity container, with the lower end of the 310 cm^3 cavities covered with mosquito netting. The five-cavity containers were slowly saturated as described above and then placed on a tension table under a plastic cover. Substrate moisture content was determined using a 36 cm long two-rod (1.5 cm spacing) TDR probe with shorting diodes at the beginning and end of the probe, inserted horizontally into the container across the five cavities at a height of 8 cm from the bottom of the cavities [26]. A correction factor was used for the calculation of K_a to account for propagation time that was not related to the substrate within the five cavities [26].

To determine the matric potential at which air first enters the substrate when going from saturated to unsaturated conditions (ψ_a, the point of air entry; Nemati et al. [31]), the five-cavity containers were rewetted from the bottom up to full saturation overnight and then placed on a tension table with a water potential of 0.6 kPa (i.e., a water head of 6 cm above the surface of the tension table). The matric potential (ψ) at mid-container height was reduced from 0 to −0.8 kPa by decreasing the water level in 0.2 kPa steps and the substrate θ_v was measured via TDR at each ψ step following a 2 h equilibrium period. From these ψ–θ_v curves, ψ_a was estimated visually and corresponded to the matric potential when a significant drop in θ_v was observed. If air entry had not occurred at −0.8 kPa, substrate ψ was further reduced from −0.8 to −1.2 kPa. For these measurements, matric potentials were evaluated from the height of the water level above the surface of the tension table, from which 0.125 kPa was subtracted to account for the zone of influence of the TDR probe (i.e., the distance required to detect a change in θ_v; see Caron et al. [32] and Boudreault et al. [20]). Afterward, substrates were covered by a plastic sheet to prevent evaporation and their water release curve was established by measuring θ_v at matric potentials of −1.4, −1.6, −1.8, −3, −5 and −10 kPa. The available (AW) and easily available (EAW) water content in the substrates were estimated from θ_c (ψ = −0.6 kPa) and values of θ_v at ψ = −10 kPa and −5 kPa, respectively [33].

$$\mathrm{EAW} = \theta_c - \theta_{-5\mathrm{kPa}} \tag{3}$$

$$\mathrm{AW} = \theta_c - \theta_{-10\mathrm{kPa}} \tag{4}$$

Saturated hydraulic conductivity (K_s) was determined on three individual cavities per container per nursery using an infiltrometer that provided a constant flow of water at the surface of the sample [28]. Substrates were saturated as described above; once steady-state conditions of water flow through the substrate were achieved, the pressure gradient and the flow of water were measured and K_s estimated using Darcy's law. A correction factor of 1.22 was used to account for the slightly conical shape of the cavity and narrowing of the wall near the bottom of the cavity (see Figure 1 in Boudreault et al. [20]). The aeration properties of the substrates were then determined using measured values of ψ_a and K_s and data from the water-release curve based on the multiple-point method described by Allaire et al. [18] and Caron and Nkongolo [22]. Briefly, this method consists in determining the parameters of the water desorption curve using a nonlinear, five-parameter function adapted to artificial substrates [34]:

$$\theta = \theta_r + \frac{\theta - \theta_r}{\left[1 + (a\psi)^n\right]^b} \tag{5}$$

where θ is the substrate volumetric water content (cm^3 cm^{-3}), ψ is the matric potential (kPa), θ_r is the residual water content (cm^3 cm^{-3}) and a, b and n are empirical parameters determined by minimizing the error between estimates of Equation (5) and points of the water release curve using Mathcad 7. Once established, the function was used with K_s values to determine the coefficient of pore effectiveness (γ) following

$$\gamma = \left\{ \frac{0.00028\rho g}{\eta K_s} \int_{\theta_r}^{\theta_{\psi_a}} a^2 \left[\left(\frac{\theta_s - \theta_r}{\theta - \theta_r} \right)^{-\frac{1}{b}} - 1 \right]^{-\frac{2}{n}} d\theta \right\}^{-1} \tag{6}$$

where $\theta_{\psi a}$ is the water content at air entry value ($cm^3\ cm^{-3}$), θ_s is the water content at saturation ($cm^3\ cm^{-3}$), ρ is water density ($g\ cm^{-3}$), g is the gravitational acceleration ($m\ s^{-2}$) and η is the viscosity of water (0.001 Pa s at 20°C). Because g represents the tortuosity, constriction and continuity of pores in the substrate, $D_s\ D_0{}^{-1}$ can be estimated using the equation from King and Smith [35], generalized for artificial media:

$$\frac{D_s}{D_0} = \gamma\, \theta_a \tag{7}$$

Equation (7) provides an estimate of relative gas diffusivity at an average ψ of −0.6 kPa found in the 25–310 nursery containers when substrates are at container capacity.

2.4. Chemical Properties of Peat Substrates

Mineral nutrients, effective cation exchange capacity (CEC_{eff}), pH and electrical conductivity (EC) were determined for substrates sampled from the eight tree nurseries prior to seeding. The substrate from three cavities per container was air-dried in the laboratory and combined to form a composite sample (*n* = 5 per nursery). Major cations were extracted from each composite sample with a 1 mol L^{-1} solution of NH_4Cl and CECeff was calculated as the sum of exchangeable bases plus actual acidity [36]. The nutrient (cation) composition of each extract was then measured using plasma atomic emission spectrometry (Model ICAP 9000, Thermo Instruments, Franklin, MA, USA) [36] The concentrations of NH_4–N, NO_2–N, NO_3–N, P, K, Ca and Mg in the substrates were measured in saturated aqueous extracts [36,37]. After extraction, the mineral N concentration was determined by colorimetry using a continuous flow spectrophotometer (QuickChem 8000, Lachat Instruments, Loveland, CO, USA), while P, K, Ca and Mg concentrations were determined by atomic emission spectrometry with plasma (ICAP 61E, Thermo Instruments, Franklin, MA, USA) [36,37]. Substrate pH and EC were measured directly on the aqueous extracts without reporting the results on a dry basis [38]. Mineral nutrient analyses of substrates were determined at the organic and inorganic chemistry laboratory of the Quebec Forest Research Branch.

2.5. Growth and Nutrition of White Spruce Seedlings (1+0) and Physical Properties of Substrates after One Growing Season

In spring 2009, three containers were randomly selected prior to bud burst in each of the eight tree nurseries to determine the growth parameters and nutrient content of white spruce seedlings (1+0). Seedling height and root collar diameter were measured in the laboratory on ten seedlings randomly selected per container (*n* = 30 per nursery). Seedlings were then severed at the root collar and their root system was carefully washed by hand. The root and shoot dry masses of these seedlings were measured after oven-drying at 70 °C for 48 h. Nutrient concentrations in root and shoot tissues were determined for three composite samples per tree nursery, each consisting of the ten seedlings per container used to determine growth parameters. After tissue grinding and acid digestion, P, K, Ca and Mg concentrations were measured by atomic emission spectrometry with plasma (ICAP 61E, Thermo Instruments), whereas N concentration was determined using the Kjeldahl method (flow injection colorimetry; QuickChem 8000, Lachat Instruments). All mineral nutrient analyses were performed at the organic and inorganic chemistry laboratory of the Quebec Forest Research Branch. Additionally, the bulk density and saturated hydraulic conductivity of substrates collected in spring 2009 were each measured on three different cavities per container. Volumetric water content at saturation (i.e., total porosity, θ_t) and container capacity (θ_c) were also determined to estimate air-filled porosity after drainage (θ_a) using Equation (2).

2.6. Statistical Analyses

An analysis of variance followed by a protected LSD test were used to compare substrate physical properties, seedling growth parameters and nutrient concentrations among tree nurseries (with significance determined at $p < 0.05$). Pearson's correlation coefficients and linear regressions were computed to examine the relationships between growth parameters of white spruce and the physical properties (ρ_b, θ_a, EAW, AW, K_s, MWD, γ and $D_s\ D_0^{-1}$) of substrates after potting. A similar analysis was performed with values of ρ_b, θ_a and K_s obtained after one year of growth (i.e., on substrates sampled in spring 2009). All statistical analyses were performed using SAS 9.1.3 (SAS Institute Inc., Cary, NC, USA).

3. Results

3.1. Physical Properties of Forest Nursery Substrates

The mean weight diameter (MWD) of particles varied between 1.17 and 3.06 mm after potting, with a mean (±SD) value of 1.92 ± 0.57 mm across all nurseries (Figure 3). Whereas the fraction of particles with a diameter of 1 to 4 mm varied from 25% in the substrate from nursery 1 to 52% in those from nurseries 5 and 6, there was a five-fold increase in the fraction of fine particles (<0.5 mm) in substrates from nursery 6 compared to those from nursery 1 (Figure 3). A greater water retention was observed in the substrates from nursery 1 and 5 over most of the measured range of matric potentials, while substrates from nurseries 2, 3, 4, 6, 7 and 8 had similar water release curves (Figure 4).

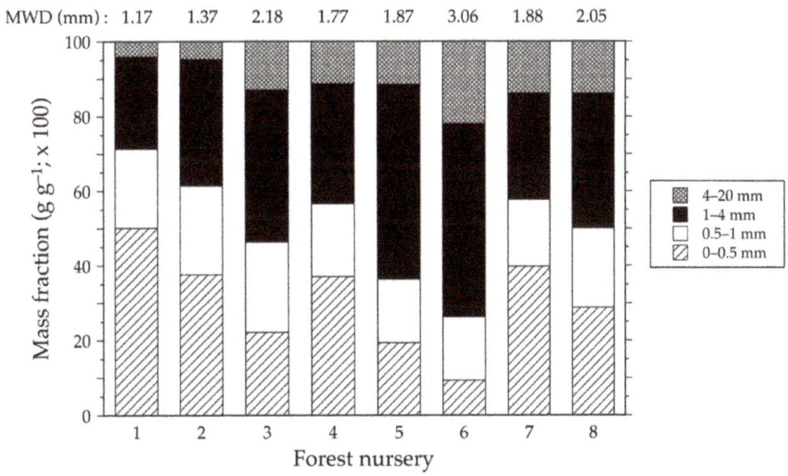

Figure 3. Mean weight diameter (MWD; above each stack column) and the proportion of peat particles with a diameter < 0.5 mm and between 0.5 and 1 mm, 1 and 4 mm and 4 and 20 mm in nursery substrates after potting.

The average saturated hydraulic conductivity of substrates after potting was 0.06 ± 0.01 cm s^{-1}, with values ranging from 0.02 (nursery 1) to 0.11 cm s^{-1} (nursery 6) (Table 2). Substrate bulk density, which varied between 0.076 and 0.105 g cm^{-3}, differed significantly ($p < 0.001$) among forest nurseries, those of nursery 1 and 5 having the highest ρ_b values (Table 2). There was little variation in total porosity of the substrates sampled after potting (0.944 ± 0.001 cm^3 cm^{-3}), yet the air-filled porosity (θ_a) was three times higher in substrates from nurseries 3 and 7 (>0.14 cm^3 cm^{-3}) compared to that of nursery 1 (0.04 cm^3 cm^{-3}) (Table 2). While no significant difference was observed in the easily available water (EAW) content of substrates ($p = 0.09$), values of AW differed significantly among nursery substrates ($p < 0.001$). EAW and AW averaged over all substrates after potting were 0.31 ± 0.04 cm^3 cm^{-3} and 0.42 ± 0.02 cm^3 cm^{-3}, respectively. There was no significant difference in pore effectiveness coefficient (γ) or relative gas diffusivity ($D_s\ D_0^{-1}$) among the eight substrates (Table 2). Values of γ changed relatively little across the different substrates, ranging from 6.6% in nursery 2 to 11.8% in nursery 6. The substrate from nursery 1 had the lowest $D_s D_0^{-1}$ value (0.0041 $m^2 s^{-1} m^{-2} s$), whereas the highest relative gas diffusivity was found in the substrate from nursery 3 (0.0171 $m^2 s^{-1} m^{-2} s$), in accordance with its high air-filled porosity.

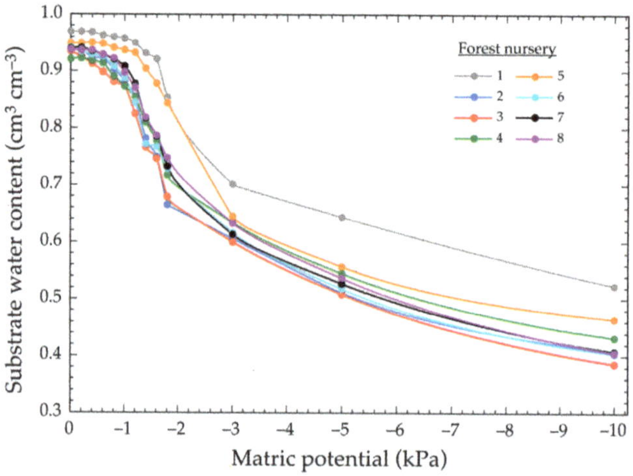

Figure 4. Water release curves of substrates from sampled forest nurseries after potting.

Table 2. Physical properties of substrates from eight forest nurseries after potting.

No. [1]	K_s	ρ_b	θ_t	θ_a	EAW	AW	γ	$D_s D_0^{-1}$
	($cm\ s^{-1}$)	($g\ cm^{-3}$)	($cm^3\ cm^{-3}$)	($cm^3\ cm^{-3}$)	($cm^3\ cm^{-3}$)	($cm^3\ cm^{-3}$)	($m\ m^{-1}$)	($m^2\ s^{-1}\ m^{-2}\ s$)
1	0.02 e	0.095 ab	0.943 bc	0.04 d	0.26 a	0.38 bc	0.093 a	0.0041 a
2	0.04 d	0.089 c	0.945 bc	0.11 bcd	0.29 a	0.44 ab	0.066 a	0.0070 a
3	0.09 bc	0.076 d	0.954 a	0.17 a	0.28 a	0.40 a	0.096 a	0.0171 a
4	0.08 bc	0.094 ab	0.943 bc	0.09 c	0.34 a	0.41 a	0.108 a	0.0088 a
5	0.03 e	0.105 a	0.938 d	0.11 bc	0.26 a	0.39 c	0.105 a	0.0119 a
6	0.11 a	0.099 a	0.942 bc	0.08 cd	0.30 a	0.46 a	0.118 a	0.0091 a
7	0.05 d	0.088 c	0.945 b	0.14 ab	0.28 a	0.40 a	0.075 a	0.0104 a
8	0.05 cd	0.092 bc	0.944 bc	0.08 cd	0.33 a	0.46 a	0.081 a	0.0061 a
Mean	0.06	0.091	0.944	0.11	0.31	0.42	0.087	0.0090
SD	0.01	0.002	0.001	0.02	0.04	0.02	0.020	0.0100

[1] No. = forest nursery number, K_s = saturated hydraulic conductivity, ρ_b = bulk density, θ_t = total porosity, θ_a = air-filled porosity at container capacity, EAW = easily available water content, AW = available water content, γ = pore effectiveness coefficient and $D_s D_0^{-1}$ = relative gas diffusivity. SD = standard deviation. Means followed by the same letter are not significantly different at $p = 0.05$ according to a protected LSD test.

3.2. Growth of White Spruce Seedlings (1+0)

The provenance of white spruce seedlings had a significant effect on measured growth parameters (Table 3). For instance, there was a 78% difference in shoot dry mass of seedlings between nursery 1 (1244 mg) and nursery 8 (277 mg), while root dry mass differed by 68% among sampled forest nurseries (Table 3). Average seedling height at the end of the growing season varied between 6.4 and 12.9 cm, whereas mean root collar diameter ranged from 1.36 to 3.39 mm. Furthermore, nutrient concentrations in the shoot tissues of white spruce seedlings were significantly different among nursery substrates ($p < 0.05$; Table 3).

3.3. Correlations between Seedling Growth Parameters and Substrate Physicochemical Properties

There were no correlations between morphological variables (height, root collar diameter, shoot dry mass, root dry mass) of white spruce seedlings and the 10 physical properties of substrates from eight forest nurseries after potting (Table 4). However, significant correlations were found between growth parameters and substrate physical properties measured at the end of the growing season (K_{s_1}, ρ_{b_1} and θ_{a_1}). A significant linear relationship was observed between saturated hydraulic conductivity (K_{s_1}) and shoot dry weight ($r = -0.53$, $p = 0.008$), root collar diameter ($r = -0.42$, $p = 0.039$) and height/diameter ratio ($r = 0.41$, $p = 0.042$). Air-filled porosity (θ_{a_1}) was significantly negatively correlated with shoot dry weight ($r = -0.53$, $p = 0.008$; Figure 5) and collar diameter ($r = -0.51$, $p = 0.01$). Additionally, there was a significant positive correlation between the bulk density of substrates (ρ_{b_1}) and root dry weight ($r = 0.63$, $p < 0.001$; Table 4 and Figure 6). Morphological variables were also significantly correlated with the quantity of nitrogen (N_{total}), phosphorous (P) and potassium (K) applied per seedling. A positive linear relation was observed between the dose of applied P and shoot dry weight ($r = 0.84$, $p = 0.019$) as well as seedling height ($r = 0.82$, $p = 0.021$). Root collar diameter was correlated with the doses of applied N ($r = 0.80$, $p = 0.030$) and P ($r = 0.86$, $p = 0.011$), whereas the H/D ratio was negatively related to applied N ($r = -0.81$, $p = 0.026$).

Table 3. Morphological variables of white spruce seedlings (1+0) and nutrient concentration of N, P, K, Ca and Mg in shoot tissues ($n = 3$).

No. [1]	Shoot DW (mg)	Root DW (mg)	H (cm)	D (mm)	H/D (cm mm^{-1})	N (g kg^{-1})	P (g kg^{-1})	K (g kg^{-1})	Ca (g kg^{-1})	Mg (g kg^{-1})
1	1244 a	520 a	9.3 b	2.9 b	3.2 e,f	26.8 a,b	3.1 b,c,d	5.5 d,e	1.4 c	1.3 d
2	1190 a,b	552 a	10.0 b	3.4 a	3.0 f	24.1 b,c	3.2 b	6.7 c,d	1.9 a,b	1.8 a
3	997 b,c	370 b	9.9 b	2.7 b	3.7 d,e	23.1 b,c,d	2.6 c,d	5.4 e	1.7 b	1.5 b,c,d
4	896 c,d	355 b	9.9 b	2.4 c	4.2 c	23.6 b,c,d	2.6 c,d	8.1 b	1.7 b	1.7 a,b
5	812 c,d	371 b	12.9 a	2.4 c	5.3 a	21.3 c,d	3.1 b,c	18.6 a	2.1 a	1.7 a,b,c
6	734 d,e	506 a	8.2 c	2.3 c	3.6 d,e	19.9 d	2.6 d	6.4 b,c	1.3 c	1.4 c,d
7	511 e,f	274 b,c	6.9 d	1.8 d	3.8 c,d	30.7 a	4.2 a	7.2 c,d,e	2.1 a	1.9 a
8	277 f	177 c	6.4 d	1.4 e	4.7 b	25.1 b,c	3.4 b	6.2 c,d,e	1.6 b	1.5 b,c,d

[1] No. = forest nursery number, DW = dry weight, H = seedling height, D = stem diameter, H/D = height–diameter ratio. Forest nurseries were sorted from 1 to 8 in tables and figures based on shoot dry weight of white spruce seedlings (1+0) at the end of the growing season. Means followed by the same letter are not significantly different at $p = 0.05$ according to a protected LSD test.

Table 4. Pearson correlation coefficients [1] between morphological variables and physical and chemical properties of forest nursery substrates after potting, quantities of nitrogen (N_{total}), phosphorous (P) and potassium (K) applied per seedling of white spruce (1+0) during the growing season and physical properties determined at the end of the growing season.

	Shoot DW [2]	Root DW	H	D	H/D
Physical properties after potting					
K_s	0.02	0.28	−0.18	0.17	−0.41
ρ_b	−0.03	0.28	0.31	0.00	0.34
θ_t	0.09	−0.19	−0.24	0.06	−0.35
θ_a	−0.18	−0.39	0.11	−0.10	0.17
MWD	−0.62	−0.36	0.07	−0.58	0.77
<0.5 mm	0.30	0.06	−0.29	0.17	−0.50
1_4 mm	−0.24	0.06	0.10	−0.16	0.24
EAW	−0.25	0.11	−0.44	−0.08	−0.27
AW	−0.45	−0.08	−0.50	−0.30	−0.05
γ	0.34	0.66	0.35	0.47	−0.21
$D_s D_0^{-1}$	0.02	−0.10	0.33	0.12	0.12
Chemical properties after potting					
pH	−0.11	−0.07	−0.30	−0.32	−0.06
CE	−0.30	−0.03	−0.39	−0.10	−0.23
CEC_{eff}	−0.41	−0.08	−0.47	−0.45	−0.03
NH_4	−0.13	−0.06	0.32	−0.09	0.37
$NO_{2\text{-}3}$	−0.20	−0.44	0.14	−0.23	0.40
CEC_{eff}	−0.41	−0.08	−0.47	−0.45	−0.03
NH_4	−0.13	−0.06	0.32	−0.09	0.37
$NO_{2\text{-}3}$	−0.20	−0.44	0.14	−0.23	0.40
Ca	−0.32	0.02	−0.43	−0.23	−0.24
Mg	0.32	0.61	0.30	0.41	−0.13

Table 4. *Cont.*

	Shoot DW [2]	Root DW	H	D	H/D
Applied fertilizer					
N_{total}	0.72	0.63	0.52	**0.80** *	**−0.81** *
P	**0.84** *	0.56	**0.82** *	**0.86** *	−0.56
K	0.39	0.37	0.27	0.52	−0.43
Physical properties at end of growing season					
K_{s_1}	**−0.53** **	−0.29	−0.09	**−0.42** *	**0.41** *
ρ_{b_1}	0.33	**0.63** **	0.28	0.35	−0.14
θ_{a_1}	**−0.53** **	−0.37	−0.28	**−0.51** *	0.23

[1] The smallest significant coefficient was 0.75 for variables measured after potting ($p < 0.05$; $n = 8$) and 0.40 for physical properties measured after one growing season ($p < 0.05$; $n = 24$). * Significant at $p < 0.05$, ** Significant at $p < 0.01$. [2] DW = dry weight, H = seedling height, D = root collar diameter and H/D = height–diameter ratio. K_s = saturated hydraulic conductivity, ρ_b = bulk density, θ_t = total porosity, θ_a = air-filled porosity at container capacity, MWD = mean weight diameter, <0.5 mm = fraction of peat particles with a diameter < 0.5 mm, 1–4 mm = proportion of peat particles with a diameter between 1 and 4 mm, EAW = easily available water content, AW = available water content, γ = pore effectiveness coefficient, $D_s\ D_0^{-1}$ = relative gas diffusivity, EC = electrical conductivity and CEC_{eff} = effective cation exchange capacity.

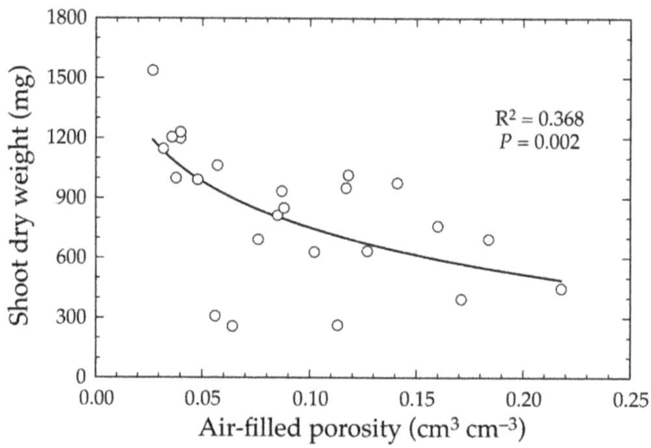

Figure 5. Relationship between shoot dry weight of white spruce seedlings (1+0) and the air-filled porosity at container capacity of substrates at the end of the growing season ($n = 24$).

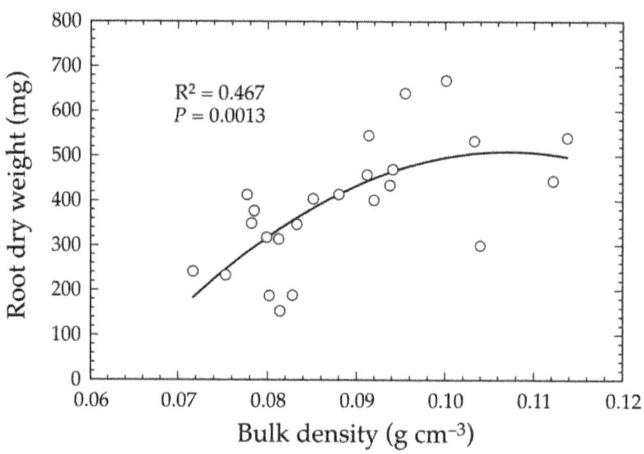

Figure 6. Relationship between root dry weight of white spruce seedlings (1+0) and bulk density of substrates at the end of the growing season ($n = 24$).

4. Discussion

Significant differences were observed between the physical properties of the substrates from the eight forest nurseries, despite the similarity of the different components of the substrates used to produce forest seedlings in Quebec [39]. The exception was the substrate from nursery 5, which contained a little more than 45% of particles larger than 1.7 mm (Figure 3). None of the substrates had a coarse (>2 mm)-to-fine particle ratio greater than 50%, as proposed as an acceptable standard for container forest seedling production [40]. It has been shown that the growth of white spruce seedlings produced in greenhouses was greater during the first year of production when the substrate was composed of fine particles (<1.3 mm) [9]. However, no significant relationship could be observed between MWD and growth variables during our study.

The θ_a of the substrates at container capacity measured just after potting varied between 0.04 and 0.17 $cm^3\ cm^{-3}$ (Table 2) and was similar to θ_a values reported in the literature for this type of substrate [19,41,42]. Except for air-filled porosity observed in the substrate from nursery 1, these values were near the lowest limit for acceptable gas diffusion (i.e., θ_a~0.10 $cm^3\ cm^{-3}$) [43] as well as adequate root growth and respiration (i.e., θ_a~0.10–0.15 $cm^3\ cm^{-3}$) [44] and should likely have limited plant growth under a wet irrigation management. The significant negative correlations observed between θ_a and K_s measured after one growing season and the growth variables instead indicated that low aeration of the substrate did not limit seedling growth, but instead suggested that too much aeration decreases shoot growth (Table 4, Figure 5). This would also be consistent with water stress affecting seedling growth. Indeed, these results are very similar to those obtained in a companion study for one-year growth (year 2008 in [20]). For aeration, this could be explained by the low need of conifers for aeration [45], especially during the first growing season, when the cavity volume is not very limiting for root growth. Moreover, for the correlation, θ_a was measured at container capacity (ψ = −0.6 kPa), a matric potential that is rarely maintained for long periods after irrigation [42] as the irrigation threshold was most likely less than −5 kPa based on the recommended threshold of 0.30–0.45 $cm^3\ cm^{-3}$ substrate water content to initiate irrigation.

Water stress appeared most likely to explain these negative correlations for the first year of growth, as previously reported [20]. Indeed, the irrigation system used in forest nurseries tends to maintain a volumetric water content between 0.30 and 0.45 $cm^3\ cm^{-3}$. The desorption curves presented in Figure 4 well describe the reduction of water content in substrates and its replacement by air under a matric potential of −0.6 kPa, and indicate threshold water potentials to trigger irrigation far below −5 kPa and even −10 kPa for most substrates, threshold values below which growth limitations were reported for *Prunus* × *cystena* [46] and attributed to too-low values of saturated hydraulic conductivity. Results were also consistent with a decrease in shoot growth with decreasing water potential in our parallel study for the first year (2008 in [20]). Under a dry irrigation management, at water potentials below −2 kPa, θ_a was high (>0.15 $cm^3\ cm^{-3}$) for most substrates from the eight nurseries. Values of θ_a at ψ = −0.6 kPa (see Figure 4) were then likely uncommon in substrates during the growing season under forest nursery conditions, which greatly facilitated the presence of air-filled pores [6,7].

Aeration limitations in substrates also appeared very unlikely based on relative gas diffusivity measurements, even if wet conditions may have prevailed occasionally in the container between irrigation events. $D_s\ D_0^{-1}$ values obtained in this study for peat substrates used in eight forest nurseries (0.0041–0.0171 $m^2\ s^{-1}\ m^{-2}$ s) were similar to those observed by Boudreault et al. [20] ($D_s\ D_0^{-1}$ = 0.0026 to 0.0105 $m^2\ s^{-1}\ m^{-2}$ s), Caron et al. ([47]; $D_s\ D_0^{-1}$ = 0.0026 to 0.0105 $m^2\ s^{-1}\ m^{-2}$ s) and Caron and Nkongolo ([22]; $D_s\ D_0^{-1}$ = 0.0067 to 0.0230 $m^2\ s^{-1}\ m^{-2}$ s) using the same method. Allaire et al. [18] reported $D_s\ D_0^{-1}$ values that ranged between 0.006 and 0.018 $m^2\ s^{-1}\ m^{-2}$ s for horticultural substrates. Pore effectiveness and $D_s\ D_0^{-1}$ measured at container capacity did not show any significant correlation with seedling growth variables (Table 4) contrary to several studies where positive linear relationships were observed between these variables for a similar $D_s\ D_0^{-1}$ range in peat and bark substrates [18,48,49]. For instance, Allaire et al. [18] established a relative gas diffusivity threshold of 0.015 $m^2\ s^{-1}\ m^{-2}$ s to promote the growth of *Prunus* × *cistena* in 5 L (21 cm high × 18 cm diameter) containers. Meanwhile, Boudreault et al. [20] suggested, based on modeled O_2 profiles in root plugs (310 cm^3 cavities; 12 cm high × 6.5 cm diameter with ~5 cm^2 hole at bottom for drainage and air exchange) and observed $D_s\ D_0^{-1}$–root growth relationship for the same tree seedling species (see Figure 5 in [20]), that nursery substrates should have a relative gas diffusivity greater than ~0.004 $m^2\ s^{-1}\ m^{-2}$ s when matric potential is maintained within the 0 to −5 kPa range. This apparent discrepancy between relative gas diffusivity minimal values may be due to the fact that the seedlings from forest nurseries were grown in cavities of reduced height and radius compared to the pots commonly used in horticultural nurseries (1 L and more). Unlike horticultural pots, the cavities of containers used by forest nurseries had a greater surface area available for gas exchange relative to volume. This surface is made up of the upper part (top) of the root plug, but also of the contour of the cavity, where gas exchange capacity is accentuated by root water uptake during substrate drying (see Boudreault et al. [20]). Moreover, the bottom of the root plug is directly exposed to air, as opposed to the 5 L containers in Allaire et al. [18], which had five small drainage holes at the bottom of the container not directly exposed to air. Hence, although the $D_s\ D_0^{-1}$ values measured here were typically below the 0.015 $m^2\ s^{-1}\ m^{-2}$ s threshold, they were higher than the 0.004 $m^2\ s^{-1}\ m^{-2}$ threshold obtained by Boudreault et al. [20]), and therefore the white spruce seedling growth was unlikely limited by aeration.

The level of peat humification, reduction in particle size and compaction of the substrate at potting can increase bulk density, which also generally increases during the growing season [9,11]. A significant positive correlation was observed between bulk density and root dry mass (Table 4; Figure 6). Low bulk density could affect plant water supply, resulting in lower total plant mass [20]. Unsaturated hydraulic conductivity is also low in peat substrates under dry conditions [46,50,51]. On the other hand, large particle size combined with low bulk density can reduce unsaturated hydraulic conductivity, which could limit water transfer for substrates under dry irrigation management, as suggested by our data and previous work for the same species [20].

Total porosity values for sampled substrates were similar to those reported in the literature for blond and brown fibric peat [11,16,52–54]. Moreover, substrates from the eight forest nurseries showed desorption curves similar to those observed in other studies [10,54]. Overall, the available and easily available water content varied little among substrates (Table 2). This is not surprising, as the volumetric water content at −5 and −10 kPa were similar, except for the substrate from nursery 1, which had a generally higher water content (Figure 4). Bernier and Gonzalez [9] observed a positive correlation between root dry mass and easily available water. However, this relationship was not observed in the present study, which may be due to irrigation management and environmental conditions between nurseries.

Given the importance of bulk density when packing growing media in seedling trays at the nursery operation scale, the shape of the curve in Figure 6 is of practical interest and should be analyzed cautiously. First, the results suggest a plateau around 0.10 to 0.11 g cm^{-3}, with a possible decrease beyond 0.11 g cm^{-3}. Higher values may possibly lead to growth decreases, as reported by Boudreault et al. [20] for their second-year (2009) data, when seedlings were exposed to outdoor conditions. Further, the conclusion applies to the first year only, as with time the substrate will significantly evolve and become more compacted. Seedling responses will also be altered in the second year, when plants will be brought into outdoor conditions for their second-year growth, which may expose the substrate to prolonged wet conditions [20]. Boudreault et al. [20] consistently reported results similar to those above for the first year, but observed important aeration limitations when substrates were more compacted in the second year.

Fertilization applied during the growing season influenced seedling growth and development, as demonstrated by correlations between substrate fertility (N_{total} and P concentrations) and shoot dry mass, height and root-collar diameter (Table 4). The amount of mineral nutrients needed in each forest nursery during the growing season (Table 1) is calculated based on growth standards to be achieved (height, root and shoot dry masses, etc.) in order to optimize the concentration of nutrients in the substrate and in shoot and root tissues [24]. Hence, the different fertilization regimes used by nursery growers may have masked some of the relationships between the physicochemical properties of the substrates and the growth variables of white spruce seedlings. The concentrations of mineral nutrients in shoots were, however, within recommended ranges for the growth and physiological processes (net photosynthesis, etc.) of white spruce seedlings in all nurseries (for instance, foliar nitrogen was 2.4% to 3%; Table 3) [55–57]. Before the delivery of seedlings to reforestation sites and payment to forest nurseries in Quebec, seedlings are evaluated and must meet 27 morphophysiological quality standards, including leaf nitrogen concentration [3]. To this end, nurseries must respect the optimal thresholds of the main cultural practices (peat substrate, irrigation, fertilization, short-day treatment, etc.) to achieve these 27 quality standards and reduce seedling losses and production costs [4,58,59]. Another factor to be considered when examining relations between the physical properties of substrates and conifer seedling biomass is root growth throughout the growing season. As roots develop, they occupy the pore space of the growing medium, consume oxygen and affect gas diffusion in the rhizosphere, which may also impact root and shoot growth [60–62].

5. Conclusions

This characterization of the physicochemical properties of substrates from eight forest nurseries highlighted a range of quality in local peat. While substrate chemical properties were similar, their physical properties varied significantly. The low intrinsic aeration of substrates did not limit the root and shoot growth of white spruce seedlings (1+0). Conversely, seedling growth was lower in substrates with high air-filled porosity. Our findings suggest that the air-filled porosity at container capacity of forest nursery substrates should be between 0.03 and 0.10 $cm^3\ cm^{-3}$ to improve the root and shoot growth of white spruce seedlings (1+0). Moreover, a substrate bulk density between 0.07 and 0.115 $g\ cm^{-3}$ had a positive effect on root biomass. This is likely due to irrigation management resulting in drier substrate conditions during this first year of growth under a polyethylene tunnel, where controlled irrigation maintains relatively low water content and adequate air-filled porosity. However, a few forest nursery substrates had low air-filled porosity values at container capacity and low pore effectiveness coefficients, which could hinder root growth and development during the second growing season when seedlings are cultivated outdoors and thus exposed to prevailing weather conditions (i.e., rainfall). This hypothesis should be tested by further study of their second year of growth.

Author Contributions: Conceptualization, S.P., J.C., M.S.L. and S.B.; methodology, S.P., J.C., S.B. and M.S.L.; software, S.B.; validation, S.B., S.P., J.C. and M.S.L.; formal analysis, S.B.; investigation, S.B., S.P., J.C. and M.S.L.; resources, S.P.; data curation, S.B.; writing—original draft preparation, S.B., S.P., M.S.L. and J.C.; writing—review and editing, S.B., S.P., M.S.L. and J.C.; visualization, S.B.; supervision, S.P., J.C. and M.S.L.; project administration, S.P.; funding acquisition, S.P. All authors have read and agreed to the published version of the manuscript.

Funding: Funding for this project was provided by the Fonds de recherche du Québec—Nature et technologies (Partnerships Research Program on Forest Management and Environment III: grant n° 2008-FT-124361), awarded to S. Pepin.

Data Availability Statement: Not applicable.

Acknowledgments: The authors would like to thank the forest nurseries that provided us with potting substrates and tree seedlings, as well as Jean-François Bernier, Anne-Isabelle Bonifassi, Marie-Ève Giroux and Andrée-Ann Prince for their involvement in this project. We would also like to thank Carole Boily (Laval University), Bertrand Fecteau and Luc Godin (pépinière Pampev Inc.) and the staff of the Organic and Inorganic Chemistry Laboratory of the Forest Research Branch of the Quebec Ministry of Natural Resources and Forests for their participation in this study.

Conflicts of Interest: The authors declare no conflict of interest.

References

1. Lefebvre, S. La Production de Plants Forestiers: Enjeux et Perspectives. In Proceedings of the Conférence Midi du Ministère des Ressources Naturelle et des Forêts, Québec, QC, Canada, 16 March 2022.
2. Myelinated Retinal Nerve Fibers (MRNF). *Guide de Terrain. Inventaire de Qualification des Plants Résineux Cultivés en Récipients*; Gouvernement du Québec: Québec, QC, Canada, 2019; p. 114.
3. Myelinated Retinal Nerve Fibers (MRNF). *Monographie du Secteur des Plants Forestiers au Québec*; Gouvernement du Québec: Québec, QC, Canada, 2016; p. 37.
4. Lamhamedi, M.S.; Renaud, M.; Desjardins, P.; Veilleux, L. Root growth, plug cohesion, mineral nutrition, and carbohydrate content of (1+0) *Picea mariana* seedlings in response to a short-day treatment. *Tree Plant. Notes* **2013**, *56*, 35–46.
5. Girard, D.; Gagnon, J. Wireless network of electronic scales to monitor the substrate volumetric water content for managing irrigation of containerized seedlings produced in forest nurseries. *Tree Plant. Notes* **2016**, *59*, 9–19.
6. Lamhamedi, M.S.; Lambany, G.; Margolis, H.A.; Renaud, M.; Veilleux, L.; Bernier, P.Y. Growth, physiology and leachate losses in *Picea glauca* seedlings (1+0) grown in air-slit containers under different irrigation regimes. *Can. J. For. Res.* **2001**, *31*, 1968–1980. [CrossRef]
7. Lamhamedi, M.S.; Labbé, L.; Margolis, H.A.; Stowe, D.C.; Blais, L.; Renaud, M. Spatial variability of substrate water content and growth of white spruce seedlings. *Soil Sci. Soc. Am. J.* **2006**, *70*, 108–120. [CrossRef]
8. Stowe, D.C.; Lamhamedi, M.S.; Carles, S.; Fecteau, B.; Margolis, H.A.; Renaud, M.; Bernier, P.Y. Managing irrigation to reduce nutrient leaching in containerized white spruce seedling production. *New For.* **2010**, *40*, 185–204. [CrossRef]
9. Bernier, P.Y.; Gonzalez, A. Effects of the physical properties of Sphagnum peat on the nursery growth of containerized *Picea mariana* and *Picea glauca* seedlings. *Scan. J. For. Res.* **1995**, *10*, 176–183. [CrossRef]
10. Bernier, P.Y.; Stewart, J.D.; Gonzalez, A. Effects of the physical properties of Sphagnum peat on water stress in containerized *Picea mariana* seedlings under simulated field conditions. *Scan. J. For. Res.* **1995**, *10*, 184–189. [CrossRef]
11. Heiskanen, J. Favourable water and aeration conditions for growth media used in containerized tree seedling production: A review. *Scan. J. For. Res.* **1993**, *8*, 337–358. [CrossRef]
12. McMillin, J.D.; Wagner, M.R. Effects of water stress on biomass partitioning of ponderosa pine seedlings during primary root growth and shoot growth periods. *For. Sci.* **1995**, *41*, 594–610.
13. Heiskanen, J. Effect of peat-based two-component growth media on the growth of containerized Scots pine seedlings. *Suo* **1994**, *45*, 17–29.
14. Bunt, A.C. *Media and Mixes for Container-Grown Plants: A Manual on the Preparation and Use of Growing Media for Pot Plants*; Springer: Dordrecht, The Netherlands, 1988; p. 309. [CrossRef]
15. Landis, T.D.; Tinus, R.W.; Barnett, J.P. *The Container Tree Nursery Manual. Vol. 2: Containers and Growing Media*; Handbook Agriculture No 674; USDA Forest Service: Washington, DC, USA, 1989; 87p.
16. Caron, J.; Rivière, L.M. Quality of Peat Substrates for Plants Grown in Containers. In *Organic Soils and Peat Materials for Sustainable Agriculture*; Parent, L.-E., Ilnicki, P., Eds.; CRC Press: Boca Raton, FL, USA, 2003; pp. 67–93. [CrossRef]
17. Hammond, R.F. The Origin, Formation and Distribution of Peatland Resources. In *Peat in Horticulture*; Robinson, D.W., Lamb, J.G.D., Eds.; Academic Press: London, UK, 1975; pp. 1–22.
18. Allaire, S.; Caron, J.; Duchesne, I.; Parent, L.E.; Rioux, J.A. Air-filled porosity, gas relative diffusivity and tortuosity: Indices of Prunus x cistena growth in peat substrate. *J. Am. Soc. Hort. Sci.* **1996**, *121*, 236–242. [CrossRef]
19. Folk, R.S.; Timmer, V.R.; Scarratt, J.B. Evaluating peat as a growing medium for jack pine seedlings. 1. Conventional quality indices. *Can. J. For. Res.* **1992**, *22*, 945–949. [CrossRef]
20. Boudreault, S.; Pepin, S.; Caron, J.; Lamhamedi, M.S.; Paiement, I. Substrate aeration properties and growth of containerized white spruce: A case study. *Vadose Zone J.* **2014**, *13*, 1–15. [CrossRef]
21. Nkongolo, N.V.; Caron, J. Bark particle sizes and the modification of the physical properties of peat substrates. *Can. J. Soil Sci.* **1999**, *79*, 111–116. [CrossRef]
22. Caron, J.; Nkongolo, N.V. Assessing gas diffusion coefficients in growing media from in situ water flow and storage measurements. *Vadose Zone J.* **2004**, *3*, 300–311. [CrossRef]
23. Lamhamedi, M.S.; Veilleux, L.; Renaud, M. Élaboration des seuils de tolérance au gel des plants d'épinette blanche (1+0) en pépinière forestière selon les régions écologiques du Québec. Gouvernement du Québec, ministère des Forêts, de la Faune et des Parcs. *Mémoire De Rech. For.* **2005**, *147*, 50.
24. Langlois, C.G.; Gagnon, J. A Global Approach to Mineral Nutrition Based on the Growth Needs of Seedlings Produced in Forest Tree Nurseries. In *Plant Nutrition—From Genetic Engineering to Field Practice*; Barrow, N.J., Ed.; Kluwer Academic Publishers: Perth, Australia, 1993; pp. 303–306.
25. Girard, D.; Gagnon, J.; Langlois, C.-G. PLANTEC: Un logiciel pour gérer la fertilisation des plants dans les pépinières forestières. Gouvernement du Québec, ministère des Ressources naturelles, Direction de la recherche forestière. *Note Rech. For.* **2001**, *111*, 1–8.

26. Lambany, G.; Robidas, L.; Ballester, P. Measurement of soil water content in a peat-vermiculite medium using time domain reflectometry (TDR): A laboratory and field evaluation. *Tree Plant. Notes* **1997**, *47*, 88–93.
27. Lamhamedi, M.S.; Renaud, M.; Desjardins, P.; Veilleux, L. Mise à l'échelle opérationnelle du traitement hâtif de jours courts sur la morpho-physiologie et l'insuffisance racinaire des plants d'épinette noire (1+0) produits en tunnel. Gouvernement du Québec, ministère des Forêts, de la Faune et des Parcs. *Mémoire Rech. For.* **2009**, *154*, 28.
28. Caron, J.; Parent, L.E.; Elrick, D.E.; Naasz, R. Physical Properties of Organic Soils and Growing Media: Water and Air Storage and Flow Dynamics. In *Soil Sampling and Methods of Analysis*; Carter, M.R., Gregorich, E.G., Eds.; CRC Press: Boca Raton, FL, USA, 2008; pp. 885–912.
29. Kemper, W.D.; Roseneau, R.C. Aggregate Stability and Size Distribution. In *Methods of Soil Analysis*; Klute, A., Ed.; American Society of Agronomy-Soil Science Society of America: Madison, WI, USA, 1986; pp. 425–442. [CrossRef]
30. Paquet, J.M.; Caron, J.; Banton, O. In-situ determination of the water desorption characteristics of peat substrates. *Can. J. Soil Sci.* **1993**, *73*, 329–339. [CrossRef]
31. Nemati, M.R.; Caron, J.; Banton, O.; Tardif, P. Determining air entry value in peat substrates. *Soil Sci. Soc. Am. J.* **2002**, *66*, 367–373. [CrossRef]
32. Caron, J.; Rivière, L.M.; Guillemain, G. Gas diffusion and air-filled porosity: Effect of some oversize fragments in growing media. *Can. J. Soil Sci.* **2005**, *85*, 57–65. [CrossRef]
33. De Boodt, M.; Verdonck, O. The physical properties of the substrate in horticulture. *Acta Hort.* **1972**, *26*, 37–44. [CrossRef]
34. Milk, R.R.; Fonteno, W.C.; Larson, R.A. Hydrology of horticultural substrates: I. Mathematical models for moisture characteristics of horticultural container media. *J. Am. Soc. Hort. Sci.* **1989**, *114*, 48–52. [CrossRef]
35. King, J.A.; Smith, K.A. Gaseous diffusion through peat. *J. Soil Sci.* **1987**, *38*, 173–177. [CrossRef]
36. Carter, M.R.; Gregorich, E.G. *Soil Sampling and Methods of Analysis*, 2nd ed.; Canadian Society of Soil Science: Pinawa, MB, Canada; CRC Press: Boca Raton, FL, USA, 2008; p. 1224. [CrossRef]
37. Walinga, I.; Van Der Lee, J.J.; Houba, V.J.G.; Van Vark, W.; Novosamsky, I. *Plant Analysis Manual*; Kluwer Academic Publishers: Dordrecht, The Netherlands, 1995; pp. 1–247.
38. Timmer, V.R.; Parton, W.J. Optimum nutrient levels in a container growing medium determined by saturated aqueous extract. Comm. *Soil Sci. Plant Anal.* **1984**, *15*, 607–618. [CrossRef]
39. Ministère des Forêts, de la Faune et des Parcs. *Spécifications Techniques: Sphaigne de Tourbe*; Gouvernement du Québec: Québec, QC, Canada, 2017; p. 9.
40. Carlson, L.W. *Conseils pour la Culture en Récipient des Semis de Conifères dans les Provinces des Prairies*; Rapport d'Information NOR-X-214; Environnement Canada, Service Canadien des Forêts, Centre de Recherche Forestière du Nord: Edmonton, AB, Canada, 1983; p. 72.
41. Heiskanen, J. Irrigation regime affects water and aeration conditions in peat growth medium and the growth of containerized Scots pine seedlings. *New For.* **1995**, *9*, 181–195. [CrossRef]
42. Heiskanen, J. Water status of Sphagnum peat and a peat-perlite mixture in containers subjected to irrigation regimes. *Hort Sci.* **1995**, *30*, 281–284. [CrossRef]
43. Wesseling, J.; Wijk, W.R. Land Drainage in Relation to Soil and Crops. In *Drainage of Agricultural Lands*; Luthin, J.N., Ed.; American Society of Agronomy: Madison, WI, USA, 1957; pp. 461–504. [CrossRef]
44. Vomocil, J.A.; Flocker, W.J. Effect of soil compaction on storage and movement of soil air and water. *Trans. Am. Soc. Agr. Eng.* **1961**, *4*, 242–246. [CrossRef]
45. Johnson, P. *Horticultural and Agricultural Uses of Sawdust and Soil Amendments*; Soil and Plant Laboratory: Anaheim, CA, USA, 1968; pp. 1–46.
46. Caron, J.; Xu, H.L.; Bernier, P.Y.; Duchesne, I.; Tardif, P. Water availability in three artificial substrates during *Prunus* × *cistena* growth: Variable threshold values. *J. Am. Soc. Hortic. Sci.* **1998**, *123*, 931–936. [CrossRef]
47. Caron, J.; Morel, P.; Rivière, L.-M.; Guillemain, G. Identifying appropriate methodology to diagnose aeration limitations with large peat and bark particles in growing media. *Can. J. Soil Sci.* **2010**, *90*, 481–494. [CrossRef]
48. Nkongolo, N.V.; Caron, J. Pore space organization and plant response in peat substrates: I. Prunus × cistena and Spiraea japonica. *Sci. Res. Essays* **2006**, *1*, 77–86.
49. Nkongolo, N.V.; Caron, J. Pore space organization and plant response in peat substrates: II. Dendrathemum morifolium Ramat. *Sci. Res. Essays* **2006**, *1*, 93–102.
50. Stowe, D.C.; Lamhamedi, M.S.; Margolis, H. Water relations, cuticular transpiration, and bud characteristics of air-slit containerized *Picea glauca* seedlings in response to controlled irrigation regime. *Can. J. For. Res.* **2001**, *31*, 1922–1929. [CrossRef]
51. da Silva, F.F.; Wallach, R.; Chen, Y. Hydraulic properties of Sphagnum peat moss and tuff (scoria) and their potential effect on water availability. *Plant Soil* **1993**, *154*, 119–126. [CrossRef]
52. Heiskanen, J. Hydrological properties of container media based on sphagnum peat and their potential implications for availability of water to seedlings after outplanting. *Scan. J. For. Res.* **1999**, *14*, 78–85. [CrossRef]
53. Puustjarvi, V.; Robertson, R.A. Physical and Chemical Properties. In *Peat in Horticulture*; Robinson, D.W., Lamb, J.G.D., Eds.; Academic Press: London, UK, 1975; pp. 23–38.
54. Heiskanen, J. Mechanical container filling alters texture and water retention of peat growth media. *Hort Technol.* **1994**, *4*, 286–288. [CrossRef]

55. Heiskanen, J. Physical properties of two-component growth media based on Sphagnum peat and their implications for plant-available water and aeration. *Plant Soil* **1995**, *172*, 45–54. [CrossRef]
56. Brix, H. Effects of nitrogen fertilizer source and application rates on foliar nitrogen concentration, photosynthesis, and growth of Douglas-fir. *Can. J. For. Res.* **1981**, *11*, 775–780. [CrossRef]
57. Gagnon, J.; Lamhamedi, M.S. Les Concentrations Foliaires en Azote Recommandées au Québec pour les Essences Résineuses Produites en Récipients Sont-Elles Adéquates? In Proceedings of the Production de Plants Forestiers au Québec: La Culture de L'innovation, Colloque de Transfert de Connaissances et de Savoir-Faire, Carrefour Forêt Innovations, Québec, QC, Canada, 4–6 October 2011; pp. 47–51.
58. Lamhamedi, M.S.; Desjardins, P.; Renaud, M.; Veilleux, L. Techniques culturales de production pour améliorer la qualité morpho-physiologique des plants forestiers et la rentabilité des pépinières forestières au Québec. In *Journée de Formation et de Transfert de Connaissances, D'expertises et de Savoir-Faire Auprès des Pépinières Forestières du Québec Membres de la Fédération Québécoise des Coopératives Forestières*; Gouvernement du Québec, Ministère des Ressources Naturelles, Direction de la Recherche Forestière: Québec, QC, Canada, 2013; pp. 1–130.
59. Lamhamedi, M.S.; Renaud, M.; Desjardins, P.; Veilleux, L. L'utilisation des toiles claires peut-elle augmenter la croissance des racines des plants d'épinette blanche (1+0) en pépinière forestière? In *Production de Plants Forestiers au Québec: La Culture de L'innovation, Proceedings of the Colloque de Transfert de Connaissances et de Savoir-Faire, Carrefour Forêt Innovations, Québec, QC, Canada, 4–6 October 2011*; Colas, F., Lamhamedi, M.S., Eds.; Gouvernement du Québec: Québec, QC, Canada, 2011; pp. 87–95.
60. Ben-Noah, I.; Friedman, S.P. Review and evaluation of root respiration and of natural and agricultural processes of soil aeration. *Vadose Zone J.* **2018**, *17*, 1–47. [CrossRef]
61. Fields, J.S.; Owen, J.S., Jr.; Stewart, R.D.; Heitman, J.L.; Caron, J. Modeling water fluxes through containerized soilless substrates using HYDRUS. *Vadose Zone J.* **2020**, *19*, e20031. [CrossRef]
62. Landl, M.; Schnepf, A.; Uteau, D.; Peth, S.; Athmann, M.; Kautz, T.; Perkons, U.; Vereecken, H.; Vanderborght, J. Modeling the impact of biopores on root growth and root water uptake. *Vadose Zone J.* **2019**, *18*, 1–20. [CrossRef]

forests

MDPI

Article

The Use of Compost Tea in a Containerized Urban Tree Nursery Shows No Evident Benefits to Tree Growth or Mycorrhizal Colonization

Dan Du [1,*], Stephen J. Livesley [1], Stefan K. Arndt [1], Camille Truong [2] and Rebecca E. Miller [1,2]

[1] School of Agriculture, Food and Ecosystem Sciences, The University of Melbourne, Richmond, VIC 3121, Australia; sjlive@unimelb.edu.au (S.J.L.)

[2] Royal Botanic Gardens Victoria, Melbourne, VIC 3004, Australia

* Correspondence: rebeccadd@gmail.com

Abstract: Compost tea is a liquid organic amendment that has been reported to benefit plant growth and performance through positive effects on microbial communities and plant nutrition. However, few studies have demonstrated this for containerized plants produced in tree nurseries. Five common urban tree species (*Acer negundo, Corymbia maculata, Ficus platypoda, Hymenosporum flavum, Jacaranda mimosifolia*) were grown in a containerized experiment to investigate the effects of compost tea application on tree growth and root mycorrhizal colonization over six months. The microbial composition of compost tea was also determined with 16S (bacteria) and ITS1 (fungi) metabarcoding. No significant positive effect of compost tea on plant growth or root mycorrhizal colonization was observed. Roots of all tree species were colonized by one type of mycorrhizal fungi, ectomycorrhizae (ECM), or vesicular–arbuscular mycorrhizae (VAM). However, no relationship between the mycorrhizal colonization percentage and plant growth was detected. Thus, there was no evidence that a once-off application of compost tea had benefits for mycorrhizal colonization and growth of containerized trees in a nursery setting. Further research is needed to investigate whether any benefit from compost tea is evident once containerized trees are planted into urban landscapes where growth conditions may be more challenging.

Keywords: nutrient; fertilizer; mycorrhizae; ectomycorrhizae; endomycorrhizae; vesicular; arbuscular

Citation: Du, D.; Livesley, S.J.; Arndt, S.K.; Truong, C.; Miller, R.E. The Use of Compost Tea in a Containerized Urban Tree Nursery Shows No Evident Benefits to Tree Growth or Mycorrhizal Colonization. *Forests* **2023**, *14*, 1195. https://doi.org/10.3390/f14061195

Academic Editors: Mohammed S. Lamhamedi, Damase P. Khasa, Steeve Pépin and Cate Macinnis-Ng

Received: 6 April 2023
Revised: 18 May 2023
Accepted: 2 June 2023
Published: 9 June 2023

1. Introduction

Compost tea is an organic soil amendment that is gathering attention due to its reported benefits and positive effects on plants due to its direct nutrient value [1–4] and the promotion of microbial communities in soil or artificial substrates [2,3,5–9]. These improved microbial communities can further promote plant performance by improving soil properties [10], recycling nutrients [11,12], defending against pests and pathogens [1,8,13], and promoting mutualistic associations with plants [2].

Root symbiosis with mycorrhizal fungi is one type of microbial association that has been observed in 75%–90% of terrestrial plants studied so far [14]. Mycorrhizae can be categorized into four types based on their morphology and the fungal and plant groups involved: ectomycorrhizae (ECM) and vesicular–arbuscular mycorrhizae (VAM), which are found in a wide diversity of plant taxa, and ericoid and orchid mycorrhizae, which are limited to members of Ericaceae and Orchidaceae, respectively [14,15]. Mycorrhizal associations provide multiple benefits to plants, including enhanced uptake of water and nutrients [10,16], and protection against both abiotic and biotic stresses, such as heavy metals [17–19], salinity [20,21], and soil pathogens [22,23]. They can also improve soil structure by contributing to the formation of long-lasting soil aggregates [10], and even increase plant root longevity [22].

All these benefits from mycorrhizal fungi are of particular importance to trees that are planted and grown in potentially harsh urban soil conditions, which may include water stress, nutrient deficiency, and compaction [24]. Furthermore, trees grown in urban environments often have lower levels of mycorrhizal colonization and are associated with fewer fungal species compared to trees grown in natural forest landscapes [25–27]. It is unknown if compost tea application during the standard tree nursery production process promotes both tree growth rates and root mycorrhizal associations. If compost tea is facilitating mycorrhizal associations, then those trees could be better equipped when they are planted into harsh urban soil environments.

Studies investigating effects of compost tea on plants adopt a single application [28–30] or multiple regular applications [2,4,9,31], in experiments conducted over weeks to months. To date, most studies working with compost tea and/or mycorrhizal fungi have focused on agricultural crops [2,4,8,28–30,32] and very few have investigated similar applications in the horticultural industry [13], including for advanced tree production in a nursery setting [6,33]. Further, few studies have analysed the microbial composition of compost tea [2,6] and whether it contains microorganisms to facilitate potential positive (e.g., mycorrhizae) or negative (e.g., pathogens) effects on tree growth both in the short term (during nursery production) and long term (once planted in the landscape). Working in collaboration with an advanced tree nursery located near Melbourne (VIC, Australia), this study aimed to investigate the microbial composition of their compost tea and the effects of a once-off application of such compost tea to promote mycorrhizal association with tree roots and tree growth, in a standard containerized nursery production system following standard nursery operation practice. Specifically, we aimed to test the following hypotheses:

- Compost tea application will have a positive effect on containerized tree growth rates;
- Compost tea application will have a positive effect on root mycorrhizal colonization percentages;
- Percentages of root mycorrhizal colonization will positively correlate with rates of tree growth.

2. Materials and Methods

2.1. Compost Tea Formula

The compost was prepared according to the nursery's standard practice with ingredients from both green wastes produced within the nursery premises and commercially purchased products (Table 1). Root ball material was added specifically as a potential source of mycorrhizal inoculum. A large (5 kg) compost tea bag was made from a grain bag used in the beer brewing industry. The compost tea bag was placed into a 200 L barrel containing 175 L of water, 1.5 L of fish hydrolysate, 1.5 L of cold-pressed kelp, and 1.5 L of molasses. A water pump was used to create a vortex and to provide aeration at the same time, for approximately 24 h. The compost tea was then added to containerized trees on the same day.

2.2. Focal Species

We selected five tree species belonging to five different plant families, including native Australian and exotic species [34], as well as deciduous and evergreen species (Table 2). The type of their mycorrhizal associations (VAM and/or ECM) was defined according to Soudzilovskaia et al. [35]. The functioning of mycorrhizae in these five species and the main constraints in artificial substrates and urban soils are currently unclear and rarely reported.

2.3. Experimental Design and Maintenance

For each tree species, 10 trees of similar size were selected for the experiment (n = 50). All the trees used in this study were between 1 and 3 years old when the experiment started. They were randomly assigned into either the control (n = 5) or compost tea treatment (n = 5) groups. The substrate used for this experiment was made by the nursery with composted pine bark, wood fibre, humate, and sand (Table S1) with air-filled porosity between 18%–21%. The compost tea was added to the treatment group in early spring on 8 September 2020. Based on the plant size, *Acer negundo* was kept in 27 L pots, and the other four species were kept in 14 L pots. The compost tea was applied to the treatment group based on growing substrate volume in each pot, approximately 20 mL per L of growing substrate. The growth of all 50 trees, treated and control, was closely monitored and the trees were maintained according to nursery standard practices for six months. All 50 containerized trees were fertilized every three months with 2 g of inorganic fertilizer (N:P:K = 9:4:12) per L of substrate applied as top dressing to each pot. Manual weeding was conducted during the experiment when needed. Structural pruning was used to maintain a dominant leader. Daily irrigation was provided to each container via a drip line.

Table 1. Ingredients for compost used for this study. The core temperature of the compost pile was closely monitored to prevent overheating. Once the temperature reached 60–75 °C, the pile was manually turned. After it was turned four times, the pile was left to mature for two months. The compost was used to make compost tea after maturation.

Ingredient	Amount (by Volume)	Additional Information
Leaf trimmings	10%–15%	Green waste on site
Weeds	20%–25%	Green waste on site
Grass clippings	5%–10%	Green waste on site
Dynamic lifter	5%	Commercial product
Horse manure	5%	-
Seamungus	5%	Commercial product
Rock dust	2%	-
Wood chips	25%–30%	-
Root balls	10%–15%	From discarded mixed tree species on site
Mature compost	5%	From an old compost pile made previously in the same way

Table 2. Tree species included in the experiment, including information on their mycorrhizal types, i.e., vesicular–arbuscular mycorrhizae (VAM) and/or ectomycorrhizae (ECM).

Family	Species	Type	Origin	Mycorrhizal Type
Sapindaceae	*Acer negundo* L.	Deciduous	Exotic	VAM
Myrtaceae	*Corymbia maculata* (Hook.) K.D. Hill & L.A.S. Johnson	Evergreen	Native	Dual (VAM/ECM)
Moraceae	*Ficus platypoda* (Miq.) A. Cunn. ex Miq.	Evergreen	Native	VAM
Pittosporaceae	*Hymenosporum flavum* F. Muell.	Evergreen	Native	VAM
Bignoniaceae	*Jacaranda mimosifolia* D. Don	Deciduous	Exotic	VAM

2.4. Plant Growth Measurements

To investigate any effect of compost tea treatment on plant growth, tree growth was measured as stem height and stem diameter for six months after the compost tea application (from 8 September 2020 in spring to 19 March 2021 in autumn). Studies working with compost tea and/or mycorrhizal associations vary greatly in experiment duration, ranging from weeks to months [2,6,21]. Furthermore, studies on mycorrhizae indicate that colonization after exposure to inoculum occurs within 2–6 weeks [36]. Six months would therefore be long enough for mycorrhizal associations to have developed and plant growth response to be evident. Stem height above substrate surface in the container was measured using a 5 m measuring tape, accurate to 1 cm. The maximum stem diameter at 30 cm above the substrate surface in the container was measured using a digital Vernier calliper, accurate to 0.01 mm. Initial measurements of tree size (Table S2) were taken on 14 September 2020, six days after compost tea treatment. Measurements were made on eight occasions during the six-month experiment, with final measurements taken on 19 March 2021. The relative growth rates (RGRs) (in mm mm^{-1} day^{-1}) were calculated according to Equations (1) and (2). All measurements were transformed into mm, and the time period (T) was in days; H_0 and D_0 indicate the initial stem height and diameter, respectively, and H_t and D_t are the final stem height and diameter.

$$\text{RGR of Height} = \frac{\log(H_t) - \log(H_0)}{T} \quad (1)$$

$$\text{RGR of Diameter} = \frac{\log(D_t) - \log(D_0)}{T} \quad (2)$$

2.5. Root Sample Collection and Processing

Root samples were collected and processed four months after compost tea application, from 18–28 January 2021. A steel corer with inner diameter of 4 cm was used to collect one core containing substrate and fine roots from each container, except for *C. maculata*, which required two cores per container. The corer was cleaned with water between sampling of each container. The cores were placed into Ziplock bags and stored in a chilled cool box for transport to the laboratory. In the laboratory, core samples were soaked in water for approximately 30 min and fine roots were separated from surrounding substrate particles. Each core sample was sufficient to obtain >2 g fresh weight of fine roots. The cleaned root samples were preserved in 70% ethanol solution in water (v/v) within 24 h of collection, prior to mycorrhizal colonization assessment.

2.6. Assessment of Mycorrhizal Colonization

To quantify root mycorrhizal colonization, the clean root samples were processed using a modified ink and vinegar method developed by Vierheilig et al. [37]. The roots kept in 70% ethanol water solution (v/v) were rinsed with distilled water, then cleared by soaking in a 10% KOH solution at room temperature. The clearing times differed among species, from 4 days to 3 weeks. Cleared roots were rinsed in acidified distilled water (containing a few drops of white vinegar) and dyed with a 5% ink–vinegar solution overnight at room temperature. Excess stain was removed by washing roots in diluted white vinegar (5%). Roots were stored in 50% glycerol solution in water (v/v) for microscopic observation. Mycorrhizal colonization was assessed following the gridline intersection method [15,38] (Figure S1). VAM and ECM fungal colonization percentages were assessed separately on a random subsample of stained roots, based on key features, including vesicles, arbuscules, and nonseptate hyphae as diagnostic structures for VAM, and fungal mantle and Hartig net for ECM [15]. When a piece of root crossed a gridline, the presence or absence of mycorrhizae at that intersection was noted based on the identifying features observed. At least 100 intersections per subsample were counted to calculate the percentage of mycorrhizal colonization (Equation (3)).

$$\text{Mycorrhizal colonization percentage} = \frac{\text{Intersection count with mycorrhizae presence}}{\text{Total intersection count}} \times 100\% \quad (3)$$

2.7. Microbial Community Profiling of Compost Tea

To verify the presence of mycorrhizal fungi in the compost tea, and potential for pathogen introduction, we characterized the bacterial and fungal communities present in the compost tea using DNA meta-barcoding. A sample was frozen at −80 °C and sent to the Australian Genome Research Facility (AGRF) for DNA extraction and amplicon sequencing using Illumina MiSeq 2 × 300 bp. Briefly, DNA was extracted using the DNeasy PowerLyzer Power Soil Kit (Qiagen), and the V3–V4 region of bacterial 16S rDNA was amplified by PCR with primers 341F/806R; for fungi, the ITS1 region of nuclear rDNA was amplified with primers ITS1F/ITS2 [39]. Amplicons were generated using the Platinum SuperFi II mastermix (Life Technologies, Australia) with the following conditions: Initial denaturation at 98 °C for 30 s, followed by 30 cycles of 98 °C for 30 s, 60 °C for 10 s, and 72 °C for 30 s, with a final elongation step at 72 °C for 5 min. Because our data did not include control samples to account for potential cross-contamination during library preparation and multiplexing, we used stringent quality filters and only considered taxa with relative abundance of >0.5% of the total read count in the sample. Raw reads were quality-filtered in Trimmomatic [40] (SLIDINGWINDOW:4:30, MINLEN:100), followed by primers/adaptors trimming, denoising, and OTU clustering at 97% similarity using AMPtk [41]. Taxonomy assignment of OTUs was performed using a "hybrid" approach in AMPtk against the SILVA 138.1 (16S) and UNITE v.9.0 (ITS) reference databases, and fungal functional guilds were assigned according to FungalTraits [42].

2.8. Data Analysis

Data analyses were performed using Rstudio (R version 4.2.1 and packages including tidyr, dplyr, ggpubr, and performance) [43]. The main three numeric variables, RGRs of stem height and diameter, and percentage of mycorrhizal colonization, were checked for normal distribution using the Shapiro–Wilk test, and homogeneity of variance was also checked using the performance package of R. In order to satisfy the general linear model (GLM) assumptions, the mycorrhizal colonization percentage was logistic-transformed using the formula $\log(X/(1-X))$. Multifactor GLMs were then used to investigate (1) the effects of treatment and species on RGRs of stem height and diameter and (2) the effects of treatment and species on mycorrhizal colonization percentage. Linear regression models were used to investigate the relationships between mycorrhizal colonization percentage and plant growth rates.

3. Results

3.1. Plant Growth Response to Compost Tea Treatment

There was no significant positive effect of compost tea treatment on either the RGR of stem height or stem diameter (Figure 1). The exception was *A. negundo*, where compost tea treatment had a significant ($p = 0.03$) negative effect on the RGR of stem diameter (Figure 1B). There were significant interspecific differences in RGR of stem height ($p < 0.001$) but not in RGR of stem diameter ($p = 0.676$).

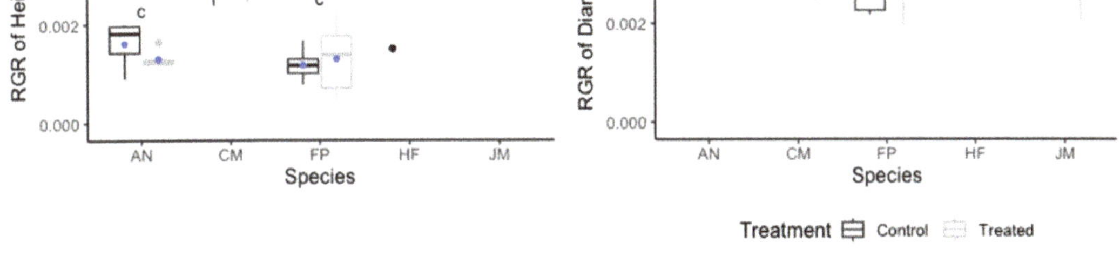

Figure 1. Relative growth rates (RGRs) of (**A**) stem height and (**B**) stem diameter at 30 cm above the substrate level for five tree species for six months after compost tea treatment (treated) or no treatment (control). Tree species are *Acer negundo* (AN), *Corymbia maculata* (CM), *Ficus platypoda* (FP), *Hymenosporum flavum* (HF), and *Jacaranda mimosifolia* (JM). For each box plot ($n = 5$ trees), the top and bottom boundaries are the third and first quartiles, the middle line is the median, and the blue dot is the mean. The light and dark grey dots are outliers. Results of 2-factor GLMs are shown; different letters indicate significant differences at $p < 0.05$ between species determined using Tukey's HSD post hoc test.

3.2. Root Mycorrhizal Colonization Response to Compost Tea Treatment

The roots of all 50 trees were colonized with mycorrhizal fungi regardless of treatment. The compost tea treatment had no overall significant positive effect on mycorrhizal colonization percentage (Figure 2). The exception was *C. maculata*, where the compost tea treatment had a significant negative effect upon the mycorrhizal colonization percentage ($p = 0.022$). The extent of colonization ranged from 2% to 78% across all individual trees and varied between pots of the same species. Mycorrhizal colonization percentage differed significantly among species ($p < 0.01$), but there was no significant interaction between treatment and species. Additionally, no significant relationship was detected between mycorrhizal colonization percentage and plant growth for any species (Figure S2), with RGR of stem height ($p = 0.213$) or stem diameter ($p = 0.399$).

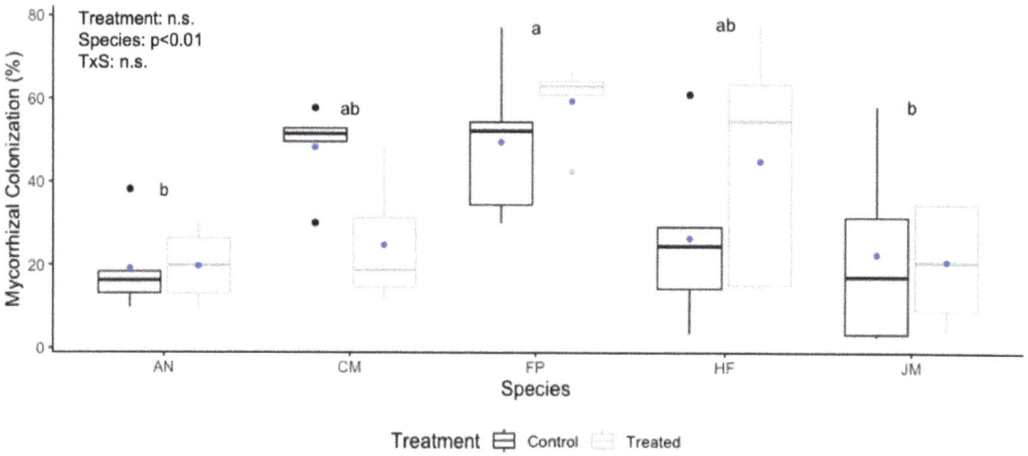

Figure 2. Mycorrhizal colonization percentage (%) of five tree species four months after compost tea treatment (treated) or no treatment (control). Tree species are *Acer negundo* (AN), *Corymbia maculata* (CM), *Ficus platypoda* (FP), *Hymenosporum flavum* (HF), and *Jacaranda mimosifolia* (JM). For each box plot ($n = 5$ trees), the top and bottom boundaries are the third and first quartiles, the middle line is the median, and the blue dot is the mean. The light and dark grey dots are outliers. Results of a 2-factor GLM are shown; different letters indicate significant differences at $p < 0.05$ between species determined using Tukey's HSD post hoc test.

3.3. Mycorrhizal Colonization Types

Acer negundo, *F. platypoda*, *H. flavum*, and *J. mimosifolia* formed VAM associations, with visible arbuscules, vesicles, and associated internal and external nonseptate hyphae (Figure S3). *Corymbia maculata* formed ECM associations, with visible mantles and dense hyphae, both internal and external (Figure S3); no evidence of dual VAM/ECM associations was observed for *C. maculata* roots.

3.4. The Bacterial and Fungal Composition of the Compost Tea

A total of 20,284 fungal reads (ITS1) passed quality filtering and clustered into 19 taxa (Table 3). Two taxa could only be identified at family level, in addition to one unknown fungus. The compost tea sample was mostly composed of saprotroph species (yeast, mould, soft rot, and filamentous fungi commonly found in soils), with some endophytic species and putative plant and animal pathogens. No mycorrhizal VAM or ECM fungal taxa were detected in the compost tea. The number of bacterial reads (16S) that passed quality filtering was 35,273. They clustered into 17 species that were mostly decomposers and potential pathogens (Table 4).

Table 3. Fungal taxa (read relative abundance > 0.5%) detected in the compost tea sample using ITS1 meta-barcoding, with their primary/secondary functions and lifeforms according to FungalTraits [42].

ID	Primary Lifestyle	Secondary Lifestyle	Relative Abund. (%)	Additional Information
Barnettozyma californica	Saprotroph (various substrates)		25.07	Yeast, sugar-rich substrates
Mortierella indohii	Soil saprotroph	Root-associated	14.53	Filamentous mycelium
Penicillium sp.	Saprotroph (various substrates)	Foliar endophyte	11.82	Mould, some species are toxin-producing, animal parasites or mycoparasites
Chaetomium piluliferum	Litter saprotroph	Foliar endophyte	9.44	Soft rot
Trichothecium roseum	Plant pathogen	Litter saprotroph	8.60	Filamentous mycelium
Thermomyces sp.	Soil saprotroph	Litter saprotroph	5.64	Mould, potential plant pathogenicity
Unknown Trichosporonaceae	–	–	4.16	–
Rhodotorula sp.	Saprotroph (various substrates)	Foliar endophyte	3.89	Yeast
Candida sp.	Saprotroph (various substrates)		3.62	Yeast, sugar-rich substrates
Byssochlamys sp.	Saprotroph (various substrates)		3.13	Mould, some species are food spoilage agents
Mortierella reticulata	Soil saprotroph	Root-associated	2.52	Filamentous mycelium
Unknown Fungi	–	–	2.22	–
Fusarium oxysporum	Plant pathogen	Litter saprotroph	1.24	Soft rot
Mortierella sp.	Soil saprotroph	Root-associated	1.00	Filamentous mycelium
Unknown Nectriaceae	–	–	0.76	–
Aspergillus fumigatus	Saprotroph (various substrates)	Foliar endophyte	0.67	Mould
Cutaneotrichosporon sp.	Animal parasite	Animal decomposer	0.58	Yeast
Wickerhamomyces anomalus	Litter saprotroph	Saprotroph	0.57	Yeast, sugar-rich substrates
Acremonium sp.	Saprotroph (various substrates)	Foliar endophyte	0.56	Soft rot, potential plant pathogenicity

Table 4. Bacterial taxa (read relative abundance > 0.5%) detected in the compost tea sample using 16S meta-barcoding, with their function, relative abundance (%), and additional information for each species listed based on cited reports from the literature.

ID	Function	Relative Abund. (%)	Additional Information
Weissella sp.	Probiotic/pathogen	25.60	Prolific in environment, probiotic or pathogenic [44]
Acinetobacter sp.	Pathogen/degrader	22.27	Can be pathogenic to humans [45]
Acinetobacter guillouiae	Unknown	11.05	An environmental species [46]
Pseudomonas sp.	Pathogen/degrader	10.30	Commonly exist in soil and can be plant pathogens [11]
Unknown Enterobacteriaceae	Unknown	4.15	
Pseudomonas veronii	Degrader	3.52	A bioremediation of contaminated soils [47]
Acinetobacter sp.	Pathogen/degrader	3.50	Can be pathogenic to human [45]
Enterobacter sp.	Probiotic	3.36	Might be nitrogen fixing bacteria [48]
Acetinobacter sp.	Pathogen/degrader	3.32	Can be pathogenic to human [45]
Pseudomonas sp.	Pathogen/degrader	3.22	Commonly exist in soil and can be plant pathogens [11]
Hafnia sp.	Pathogen	2.14	Opportunistic pathogen
Rahnella aquatilis	Pathogen	1.67	Opportunistic pathogen [49]
Serratia sp.	Pathogen	1.67	Opportunistic pathogen
Bacillus sp.	Unclear	1.47	Commonly exist in soil and might fix nitrogen [11]
Lactococcus sp.	Lactic acid bacteria	1.04	They are generally safe and produce lactic acid [50]
Arthrobacter sp.	Opportunistic pathogen	0.87	Commonly exists in soil [11]
Acetinobacter sp.	Pathogen/degrader	0.86	Can be pathogenic to human [45]

4. Discussion

4.1. Plant Growth Response to Compost Tea

Our first hypothesis was that the compost tea would positively impact containerized plant growth. However, we did not measure a significant positive effect of compost tea treatment on growth in any of the five species (Figure 1). There are several possible explanations which relate to both the properties of the compost tea and the nursery production practices during this study. While compost tea can provide a direct source of nutrients to stimulate plant growth [1,4], it is unlikely that the single application of compost tea in this study would have stimulated growth due to any direct increase of nutrient supply. It has been shown that compost teas can contain low levels of macro- and micronutrients required for plant growth [1]. Consistent with that, Edenborn et al. [30] similarly reported no effect of a single compost tea application on eggplant growth, whereas a high frequency of application (weekly) of compost tea was able to promote pak choi plant growth in a containerized experiment [4]. In this study, we assessed how standard nursery practices affected plant growth which included a single application of compost tea as well as the application of inorganic fertilizer every three months. As such, it is unlikely that any small macro- or micronutritional benefit from the compost tea application would have been detected beyond the larger inorganic fertilizer benefit provided to both control and treated groups. Similarly, it is likely that any positive nutrient contribution from any enhanced microbial mineralization processes would have also been masked by the larger inorganic fertilizer benefits. In fact, the addition of inorganic nutrients, such as nitrogen, can impede soil microbial decomposition processes [51]. Studies have reported positive [2,4,6,8], negative [31,52], and no effects [30,33] from compost tea application upon plant growth of different plant species grown in various media. The relative growth rate of height varied greatly across the five studied tree species, and future studies should consider higher replication as this could test the significance of any effect of compost tea on tree stem height growth in a variety of species.

4.2. Mycorrhizal Colonization Response to Compost Tea Treatment

Our second hypothesis that compost tea would positively affect mycorrhizal colonization percentages was based on the observed positive effects of compost tea on soil microbiology [2,5,8]. This hypothesis assumes that, firstly, there may be some mycorrhizal fungi propagules (spores and/or hyphae) in the compost tea derived from composted root material and the environment, and, secondly, that other microbes, such as mycorrhizal "helper" bacteria, would support enhanced formation of mycorrhizal associations [53]. However, we did not observe a significant positive effect of compost tea treatment on mycorrhizal colonization percentage in any species, regardless of their type (VAM or ECM) of mycorrhizal association (Figure 2). Similar to our results, Ou-Zine et al. [32] reported no effect of compost tea on arbuscular mycorrhizal colonization in maize. In our study, trees from the control group showed similar and even higher percentage of mycorrhizal colonization compared with the treated group, which indicates that the standard nursery practices in place already allow for mycorrhizal colonization of these plants, even with the application of inorganic fertilizers, which can alter and inhibit mycorrhizal colonization [54,55]. Whether the significantly lower ECM colonization of *C. maculata* roots of plants treated with compost tea reflects the negative effect of compost tea on ectomycorrhizal colonization, through enhanced nutrient input, or antagonistic interactions among microbes [9] requires further investigation; however, highly variable mycorrhizal colonization of young eucalypt roots has been observed, even in field soils [56].

Importantly, we did not detect any mycorrhizal fungus in the compost tea (Table 3); therefore, it could not be a direct source of mycorrhizal fungal inoculum. It is to be noted that ITS1 and the primer set used may not be the most suitable to detect VAM that are usually studied using specific primers targeting the 18S or 28S regions, but ECM fungi are routinely detected using ITS1 [39]. Secondly, no known mycorrhizae helper bacteria were found in the compost tea. Additionally, it has to be recognized that DNA sequencing identifies both living and dead genetic materials and therefore can indicate the presence but not necessarily functional activity of microbes in a sample.

Although studies have shown positive effects of compost teas in promoting microbial communities in various growing media [2,5,6,8,9], the outcomes of compost tea application can be very different, even contradictory, perhaps due to the lack of unified standards for compost tea [3]. Few of these studies have reported the specific fungal and/or microbial composition of the compost teas applied [2,6], and little is known about the interactions among microbes from compost tea and the ones existing in the growing media. Both positive synergistic [53] and antagonistic interactions [9] among the microbial communities have been observed. Although the meta-barcoding analyses revealed that the compost tea in this study contained beneficial bacteria and fungi that could be working as decomposers (Tables 3 and 4), it was not within the scope of this study to determine whether these probiotic microbes were established in the substrate after the one-time application. It may be that any nutritional and growth benefit from compost tea application is detectable once trees have been planted into the landscape, and nursery practices including fertilization cease.

4.3. *Mycorrhizal Colonization and Plant Growth in a Container Production System*

Our third hypothesis, that there would be a significant correlation between mycorrhizal colonization percentage and plant growth rate, was based on the well-known benefits of mycorrhizae for plant growth in both containerized and natural settings [10,16,25,57]. However, there was no evidence to support this hypothesis over the six months of this study. Perhaps most importantly, any plant growth response to mycorrhizal colonization might be impeded by existing nutrient availability in the growing media and by the inorganic fertilizer additions every three months. Nutrient levels, especially plant available phosphorus (P) concentrations in substrates, can significantly impact plant growth promotion by mycorrhizae [58–60]. Specifically, when the P level is sufficient, the contribution by mycorrhizal fungi gathering P for plants is insignificant, and growth responses to mycorrhizae are consequently more frequently reported under lower nutrients levels. For example, Wu and Xia [16] reported a significant positive effect of mycorrhizal colonization on plant growth and nutrient concentration in plant tissues in containerized citrus, with no fertilizer addition, while Sylvia et al. [61] reported no effect of mycorrhizal colonization on plant growth in 26 tree species when fertilizer was added, and Corkidi et al. [62] reported the same result in corn. Positive plant physiological and growth responses to mycorrhizal colonization are also more often reported under water deficit than well-watered conditions [63]. Given the regular fertilization and irrigation provided in this experiment, it is reasonable to conclude that beneficial mycorrhizal fungi activities were not realised, as no significant effect from mycorrhizal colonization on plant growth was identified.

In addition, the absence of any significant effect of mycorrhizal colonization percentage on plant growth rates may arise as different tree species can vary in their level of mycorrhizal dependency or responsiveness [64,65]. Little is known about mycorrhizal dependency and responsiveness of these and many other landscape tree species [61].

Even though our third hypothesis of a positive relationship between mycorrhizal colonization percentages and tree growth rates was rejected, there was also no evidence of a negative impact associated with mycorrhizal colonization. Unlike independent soil microbes, mycorrhizal fungi rely on plants to provide carbohydrates, but this usually comes at low cost to the plants that often release large amounts of excess carbon as exudates into the rhizosphere [66].

4.4. *Other Effects of Compost Tea and Future Research*

Despite the lack of a positive tree growth response or root mycorrhizal colonization percentage following compost tea application, there are several important findings for nursery container production systems. Firstly, all 50 trees in the experiment were colonized by mycorrhizal fungi, regardless of species, mycorrhizal types, or treatment. The good levels of mycorrhizal colonization suggest that VAM and ECM inoculum is readily entering the nursery production system. According to the experiment setup, sources of mycorrhizal inoculum entering the tree pots include the substrate itself, the irrigation water, or simply airborne deposition from the surrounding vegetation and environment. Secondly, although trees were raised in an artificial medium in containers, they showed similar percentages of root mycorrhizal colonization to those reported for trees grown in natural soil [27,54].

While no benefits of compost tea application were detected in terms of plant growth, there may be other benefits for plants from root mycorrhizal colonization, such as defence against soil pathogens [22,23], but this was not part of our study. Studies on a range of agricultural species indicate that compost tea can assist with disease control [7,8,13] and pathogen suppression [5,29]. Although the composition of the compost tea used in this study included potential plant pathogens (Tables 3 and 4), we did not observe a negative effect of compost tea application on plant growth (Figure 1). The potential for improved disease resistance in tree production systems and further studies of the potential promotion of symbiotic associations [2] are worthwhile. Future research could include following the performance of trees from treated and nontreated groups that are planted in different urban landscapes (i.e., with degraded soil, lack of resources, and less diverse microbial communities) to investigate the impact of compost tea at the time of planting a containerized nursery tree into the harsh urban soil landscape. The nutritional needs of the trees under such conditions and the role of microbial communities, especially mycorrhizal fungi, in stress mitigation and resource acquisition could also provide valuable information for urban forest managers [67]. Future research could also investigate the interaction between mycorrhizal fungi naturally occurring in the soil with the microbial communities of compost tea, from the perspective of plant benefits, given that we detected several potential pathogens (Tables 3 and 4) and that antagonistic effects between the two have been determined in certain conditions [9]. The compost tea used for this study did not have any mycorrhizal fungal inoculum (Table 3), despite the addition of tree root material as one compost ingredient. Future research could explore other methods of effective mycorrhizal inoculation and their impact on tree nursery production and seedling performance [68,69]. Particular care should also be taken regarding introducing exotic microbial species into the environment when inoculating plants with artificial or food waste sources. Future research could also investigate the mycorrhizal dependency and responsiveness of tree species in urban environments. Given that trees grown in urban settings generally have lower mycorrhizal colonization percentages and fewer colonizing fungal species compared with their counterparts in natural forests [27], such knowledge would provide directions for advanced tree production and scientific guidance for improving the urban environment for tree establishment.

Supplementary Materials: The following supporting information can be downloaded at: https://www.mdpi.com/article/10.3390/f14061195/s1, Table S1: Ingredients for substrates; Table S2: Initial measurement of experiment trees; Figure S1: Method for mycorrhizal colonization percentage measurement; Figure S2: Scatterplots of mycorrhizal colonization percentage and plant growth; Figure S3: Images of roots colonized by mycorrhizal fungi in this study.

Author Contributions: Conceptualization, R.E.M. and D.D.; methodology, R.E.M. and D.D.; formal analysis, D.D. and C.T.; investigation, D.D.; resources, R.E.M. and S.J.L.; data curation, D.D.; writing—original draft preparation, D.D.; writing—review and editing, R.E.M., S.J.L., S.K.A., C.T. and D.D.; visualization, D.D.; supervision, R.E.M., S.J.L. and S.K.A.; funding acquisition, R.E.M. and D.D. All authors have read and agreed to the published version of the manuscript.

Funding: This research received no external funding.

Data Availability Statement: Not applicable.

Acknowledgments: The authors wish to thank Speciality Trees Nursery for the kind support, including all plants, media, the compost tea, and management of the experimental plants on their site. We also thank Hamish Mitchell, Ryan Bell, Sascha Andrusiak, Rowan Berry, Lisa Wittick, Vicky Waymouth, and Genevieve Hehir for their technical support. We thank Bihan Guo, Paul Cheung, Claire Kenefick, and Lavinia Chu for their volunteering help during the experiment. This research received support from the Madeleine Selwyn-Smith Memorial Award and Betty Elliott Horticulture Scholarship.

Conflicts of Interest: The authors declare no conflict of interest. The funders had no role in the design of the study; in the collection, analyses, or interpretation of data; in the writing of the manuscript; or in the decision to publish the results.

References

1. Quilty, J.R.; Cattle, S.R. Use and Understanding of Organic Amendments in Australian Agriculture: A Review. *Soil Res.* **2011**, *49*, 1–26. [CrossRef]
2. Kim, M.J.; Shim, C.K.; Kim, Y.K.; Hong, S.J.; Park, J.H.; Han, E.J.; Kim, J.H.; Kim, S.C. Effect of Aerated Compost Tea on the Growth Promotion of Lettuce, Soybean, and Sweet Corn in Organic Cultivation. *Plant Pathol. J.* **2015**, *31*, 259–268. [CrossRef]
3. Islam, M.K.; Yaseen, T.; Traversa, A.; Ben Kheder, M.; Brunetti, G.; Cocozza, C. Effects of the Main Extraction Parameters on Chemical and Microbial Characteristics of Compost Tea. *Waste Manag.* **2016**, *52*, 62–68. [CrossRef]
4. Pant, A.P.; Radovich, T.J.K.; Hue, N.V.; Paull, R.E. Biochemical Properties of Compost Tea Associated with Compost Quality and Effects on Pak Choi Growth. *Sci. Hortic.* **2012**, *148*, 138–146. [CrossRef]
5. Shaban, H.; Fazeli-Nasab, B.; Alahyari, H.; Alizadeh, G.; Shahpesandi, S. An Overview of the Benefits of Compost Tea on Plant and Soil Structure. *Adv. Biores.* **2015**, *6*, 154–158.
6. Otero, M.; Salcedo, I.; Txarterina, K.; González-Murua, C.; Duñabeitia, M.K. Quality Assessment of *Pinus radiata* Production under Sustainable Nursery Management Based on Compost Tea. *J. Plant Nutr. Soil Sci.* **2019**, *182*, 356–366. [CrossRef]
7. Stewart-Wade, S.M. Efficacy of Organic Amendments Used in Containerized Plant Production: Part 1—Compost-Based Amendments. *Sci. Hortic.* **2020**, *266*, 108856. [CrossRef]
8. Larkin, R.P. Relative Effects of Biological Amendments and Crop Rotations on Soil Microbial Communities and Soilborne Diseases of Potato. *Soil Biol. Biochem.* **2008**, *40*, 1341–1351. [CrossRef]
9. Molineux, C.J.; Connop, S.P.; Gange, A.C. Manipulating Soil Microbial Communities in Extensive Green Roof Substrates. *Sci. Total Environ.* **2014**, *493*, 632–638. [CrossRef]
10. van der Heijden, M.G.A.; Martin, F.M.; Selosse, M.-A.; Sanders, I.R. Mycorrhizal Ecology and Evolution: The Past, the Present, and the Future. *New Phytol* **2015**, *205*, 1406–1423. [CrossRef]
11. Killham, K. *Soil Ecology*; Cambridge University Press: Cambridge, MA, USA; New York, NY, USA, 1994; ISBN 978-0-521-43517-8.
12. Paul, E.A. (Ed.) *Soil Microbiology, Ecology, and Biochemistry*, 3rd ed.; Academic Press: Amsterdam, The Netherlands; Boston, MA, USA, 2007; ISBN 978-0-12-546807-7.
13. Litterick, A.M.; Harrier, L.; Wallace, P.; Watson, C.A.; Wood, M. The Role of Uncomposted Materials, Composts, Manures, and Compost Extracts in Reducing Pest and Disease Incidence and Severity in Sustainable Temperate Agricultural and Horticultural Crop Production—A Review. *Crit. Rev. Plant Sci.* **2004**, *23*, 453–479. [CrossRef]
14. Brundrett, M.C.; Tedersoo, L. Evolutionary History of Mycorrhizal Symbioses and Global Host Plant Diversity. *New Phytol.* **2018**, *220*, 1108–1115. [CrossRef]
15. Brundrett, M.; Bougher, N.; Dell, B.; Grove, T.; Malajczuk, N. *Working with Mycorrhizas in Forestry and Agriculture*; ACIAR Monograph Series; Australian Centre for International Agricultural Research: Canberra, Australia, 1996; ISBN 978-1-86320-181-0.
16. Wu, Q.-S.; Xia, R.-X. Arbuscular Mycorrhizal Fungi Influence Growth, Osmotic Adjustment and Photosynthesis of Citrus under Well-Watered and Water Stress Conditions. *J. Plant Physiol.* **2006**, *163*, 417–425. [CrossRef]
17. Emamverdian, A.; Ding, Y.; Mokhberdoran, F.; Xie, Y. Heavy Metal Stress and Some Mechanisms of Plant Defense Response. *Sci. World J.* **2015**, *2015*, 756120. [CrossRef]
18. Leyval, C.; Turnau, K.; Haselwandter, K. Effect of Heavy Metal Pollution on Mycorrhizal Colonization and Function: Physiological, Ecological and Applied Aspects. *Mycorrhiza* **1997**, *7*, 139–153. [CrossRef]
19. Schützendübel, A.; Polie, A. Plant Responses to Abiotic Stresses: Heavy Metal-Induced Oxidative Stress and Protection by Mycorrhization. *J. Exp. Bot.* **2002**, *53*, 1351–1365. [CrossRef]
20. Giri, B.; Kapoor, R.; Mukerji, K.G. Improved Tolerance of *Acacia nilotica* to Salt Stress by Arbuscular Mycorrhiza, *Glomus fasciculatum* May Be Partly Related to Elevated K/Na Ratios in Root and Shoot Tissues. *Microb. Ecol.* **2007**, *54*, 753–760. [CrossRef]
21. Hajiboland, R.; Aliasgharzadeh, N.; Laiegh, S.F.; Poschenrieder, C. Colonization with Arbuscular Mycorrhizal Fungi Improves Salinity Tolerance of Tomato (*Solanum lycopersicum* L.) Plants. *Plant Soil* **2010**, *331*, 313–327. [CrossRef]
22. Eissenstat, D.M.; Wells, C.E.; Yanai, R.D.; Whitbeck, J.L. Building Roots in a Changing Environment: Implications for Root Longevity: REVIEW Building Absorptive Roots and Root Life Span. *New Phytol.* **2000**, *147*, 33–42. [CrossRef]
23. Jung, S.C.; Martinez-Medina, A.; Lopez-Raez, J.A.; Pozo, M.J. Mycorrhiza-Induced Resistance and Priming of Plant Defenses. *J. Chem. Ecol.* **2012**, *38*, 651–664. [CrossRef]

24. Smith, K.; May, P. The Challenge of Urban Soils: Evaluation and Implications for Tree Growth. *Landsc. Aust.* **1998**, *20*, 38–42.
25. Stabler, L.B.; Martin, C.A.; Stutz, J.C. Effect of Urban Expansion on Arbuscular Mycorrhizal Fungal Mediation of Landscape Tree Growth. *J. Arboric.* **2001**, *27*, 11. [CrossRef]
26. Wiseman, P.E.; Wells, C. Soil Inoculum Potential and Arbuscular Mycorrhizal Colonization of *Acer rubrum* in Forested and Developed Landscapes. *J. Arboric.* **2005**, *31*, 7.
27. Bainard, L.D.; Klironomos, J.N.; Gordon, A.M. The Mycorrhizal Status and Colonization of 26 Tree Species Growing in Urban and Rural Environments. *Mycorrhiza* **2011**, *21*, 91–96. [CrossRef]
28. Savarese, C.; Cozzolino, V.; Verrillo, M.; Vinci, G.; De Martino, A.; Scopa, A.; Piccolo, A. Combination of Humic Biostimulants with a Microbial Inoculum Improves Lettuce Productivity, Nutrient Uptake, and Primary and Secondary Metabolism. *Plant Soil* **2022**, *481*, 285–314. [CrossRef]
29. On, A.; Wong, F.; Ko, Q.; Tweddell, R.J.; Antoun, H.; Avis, T.J. Antifungal Effects of Compost Tea Microorganisms on Tomato Pathogens. *Biol. Control* **2015**, *80*, 63–69. [CrossRef]
30. Edenborn, S.L.; Johnson, L.M.K.; Edenborn, H.M.; Albarran-Jack, M.R.; Demetrion, L.D. Amendment of a Hardwood Biochar with Compost Tea: Effects on Plant Growth, Insect Damage and the Functional Diversity of Soil Microbial Communities. *Biol. Agric. Hortic.* **2018**, *34*, 88–106. [CrossRef]
31. Radin, A.M.; Warman, P.R. Effect of Municipal Solid Waste Compost and Compost Tea as Fertility Amendments on Growth and Tissue Element Concentration in Container-Grown Tomato. *Commun. Soil Sci. Plant Anal.* **2011**, *42*, 1349–1362. [CrossRef]
32. Ou-Zine, M.; El Hilali, R.; Haggoud, A.; Achbani, E.H.; Bouamri, R. Effects and Relationships of Compost Dose and Organic Additives on Compost Tea Properties, Efficacy Against Fusarium Oxysporum and Potential Effect on Endomycorrhization and Growth of *Zea Mays* L. *Waste Biomass Valor* **2022**, *13*, 4431–4445. [CrossRef]
33. Sax, M.S.; Scharenbroch, B.C. Assessing Alternative Organic Amendments as Horticultural Substrates for Growing Trees in Containers. *J. Environ. Hortic.* **2017**, *35*, 66–78. [CrossRef]
34. Atlas of Living Australia. Available online: https://www.ala.org.au (accessed on 13 May 2023).
35. Soudzilovskaia, N.A.; Vaessen, S.; Barcelo, M.; He, J.; Rahimlou, S.; Abarenkov, K.; Brundrett, M.C.; Gomes, S.I.F.; Merckx, V.; Tedersoo, L. FungalRoot: Global Online Database of Plant Mycorrhizal Associations. *New Phytol.* **2020**, *227*, 955–966. [CrossRef] [PubMed]
36. Lovelock, C.E.; Miller, R. Heterogeneity in Inoculum Potential and Effectiveness of Arbuscular Mycorrhizal Fungi. *Ecology* **2002**, *83*, 823–832. [CrossRef]
37. Vierheilig, H.; Coughlan, A.P.; Wyss, U.; Piché, Y. Ink and Vinegar, a Simple Staining Technique for Arbuscular-Mycorrhizal Fungi. *Appl. Environ. Microbiol.* **1998**, *64*, 5004–5007. [CrossRef] [PubMed]
38. Giovannetti, M.; Mosse, B. An Evaluation of Techniques for Measuring Vesicular Arbuscular Mycorrhizal Infection in Roots. *New Phytol.* **1980**, *84*, 489–500. [CrossRef]
39. Tedersoo, L.; Bahram, M.; Zinger, L.; Nilsson, R.H.; Kennedy, P.G.; Yang, T.; Anslan, S.; Mikryukov, V. Best Practices in Metabarcoding of Fungi: From Experimental Design to Results. *Mol. Ecol.* **2022**, *31*, 2769–2795. [CrossRef]
40. Bolger, A.M.; Lohse, M.; Usadel, B. Trimmomatic: A Flexible Trimmer for Illumina Sequence Data. *Bioinformatics* **2014**, *30*, 2114–2120. [CrossRef] [PubMed]
41. Palmer, J.M.; Jusino, M.A.; Banik, M.T.; Lindner, D.L. Non-Biological Synthetic Spike-in Controls and the AMPtk Software Pipeline Improve Mycobiome Data. *PeerJ* **2018**, *6*, e4925. [CrossRef] [PubMed]
42. Põlme, S.; Abarenkov, K.; Henrik Nilsson, R.; Lindahl, B.D.; Clemmensen, K.E.; Kauserud, H.; Nguyen, N.; Kjøller, R.; Bates, S.T.; Baldrian, P.; et al. FungalTraits: A User-Friendly Traits Database of Fungi and Fungus-like Stramenopiles. *Fungal Divers.* **2020**, *105*, 1–16. [CrossRef]
43. R Development Core Team. *R: A Language and Environment for Statistical Computing*; R Foundation for Statistical Computing: Vienna, Austria, 2020.
44. Fusco, V.; Quero, G.M.; Cho, G.-S.; Kabisch, J.; Meske, D.; Neve, H.; Bockelmann, W.; Franz, C.M.A.P. The Genus *Weissella*: Taxonomy, Ecology and Biotechnological Potential. *Front. Microbiol.* **2015**, *6*, 155. [CrossRef]
45. Finley, R.L.; Collignon, P.; Larsson, D.G.J.; McEwen, S.A.; Li, X.-Z.; Gaze, W.H.; Reid-Smith, R.; Timinouni, M.; Graham, D.W.; Topp, E. The Scourge of Antibiotic Resistance: The Important Role of the Environment. *Clin. Infect. Dis.* **2013**, *57*, 704–710. [CrossRef]
46. Yoon, E.-J.; Goussard, S.; Touchon, M.; Krizova, L.; Cerqueira, G.; Murphy, C.; Lambert, T.; Grillot-Courvalin, C.; Nemec, A.; Courvalin, P. Origin in *Acinetobacter guillouiae* and Dissemination of the Aminoglycoside-Modifying Enzyme Aph(3′)-VI. *mBio* **2014**, *5*, e01972-14. [CrossRef]
47. de Lima-Morales, D.; Chaves-Moreno, D.; Wos-Oxley, M.L.; Jáuregui, R.; Vilchez-Vargas, R.; Pieper, D.H. Degradation of Benzene by *Pseudomonas veronii* 1YdBTEX2 and 1YB2 Is Catalyzed by Enzymes Encoded in Distinct Catabolism Gene Clusters. *Appl. Environ. Microbiol.* **2016**, *82*, 167–173. [CrossRef]
48. Hayat, R.; Ali, S.; Amara, U.; Khalid, R.; Ahmed, I. Soil Beneficial Bacteria and Their Role in Plant Growth Promotion: A Review. *Ann. Microbiol.* **2010**, *60*, 579–598. [CrossRef]
49. Domann, E.; Hong, G.; Imirzalioglu, C.; Turschner, S.; Kuhle, J.; Watzel, C.; Hain, T.; Hossain, H.; Chakraborty, T. Culture-Independent Identification of Pathogenic Bacteria and Polymicrobial Infections in the Genitourinary Tract of Renal Transplant Recipients. *J. Clin. Microbiol.* **2003**, *41*, 5500–5510. [CrossRef]

50. Salminen, S. Demonstration of Safety of Probiotics—A Review. *Int. J. Food Microbiol.* **1998**, *44*, 93–106. [CrossRef]
51. Janssens, I.A.; Dieleman, W.; Luyssaert, S.; Subke, J.-A.; Reichstein, M.; Ceulemans, R.; Ciais, P.; Dolman, A.J.; Grace, J.; Matteucci, G.; et al. Reduction of Forest Soil Respiration in Response to Nitrogen Deposition. *Nat. Geosci.* **2010**, *3*, 315–322. [CrossRef]
52. Carballo, T.; Gil, M.V.; Calvo, L.F.; Morán, A. The Influence of Aeration System, Temperature and Compost Origin on the Phytotoxicity of Compost Tea. *Compos. Sci. Util.* **2009**, *17*, 127–139. [CrossRef]
53. Mestre, M.C.; Pastorino, M.J.; Aparicio, A.G.; Fontenla, S.B. Natives Helping Foreigners?: The Effect of Inoculation of Poplar with Patagonian Beneficial Microorganisms. *J. Soil Sci. Plant Nutr.* **2017**, *17*, 1028–1039. [CrossRef]
54. Zaouchi, Y.; Bahri, N.B.; Rezgui, S.; Bettaieb, T. Effects of Arbuscular Mycorrhizal Inoculation and Fertilization on Mycorrhizal Statute of *Jacaranda mimosifolia* D.Don Cultivated in Nurseries. *Comptes Rendus Biol.* **2013**, *336*, 493–499. [CrossRef]
55. Tomao, A.; Antonio Bonet, J.; Castaño, C.; de-Miguel, S. How Does Forest Management Affect Fungal Diversity and Community Composition? Current Knowledge and Future Perspectives for the Conservation of Forest Fungi. *For. Ecol. Manag.* **2020**, *457*, 117678. [CrossRef]
56. Adams, F.; Reddell, P.; Webb, M.J.; Shipton, W.A. Arbuscular Mycorrhizas and Ectomycorrhizas on *Eucalyptus grandis* (Myrtaceae) Trees and Seedlings in Native Forests of Tropical North-Eastern Australia. *Aust. J. Bot.* **2006**, *54*, 271. [CrossRef]
57. Estaún, V.; Camprubí, A.; Calvet, C.; Pinochet, J. Nursery and Field Response of Olive Trees Inoculated with Two Arbuscular Mycorrhizal Fungi, Glomus Intraradices and Glomus Mosseae. *J. Am. Soc. Hortic. Sci.* **2003**, *128*, 767–775. [CrossRef]
58. Morrison, S.J.; Nicholl, P.A.; Hicklenton, P.R. VA Mycorrhizal Inoculation of Landscape Trees and Shrubs Growing under High Fertility Conditions. *J. Environ. Hortic.* **1993**, *11*, 64–71. [CrossRef]
59. Graham, J.H.; Eissenstat, D.M. Field Evidence for the Carbon Cost of Citrus Mycorrhizas. *New Phytol.* **1998**, *140*, 103–110. [CrossRef]
60. Albornoz, F.E.; Dixon, K.W.; Lambers, H. Revisiting Mycorrhizal Dogmas: Are Mycorrhizas Really Functioning as They Are Widely Believed to Do? *Soil Ecol. Lett.* **2021**, *3*, 73–82. [CrossRef]
61. Sylvia, D.; Alagely, A.; Kent, D.; Mecklenburg, R. Mycorrhizae of Landscape Trees Produced in Raised Beds and Containers. *J. Arboric.* **1998**, *24*, 308–315. [CrossRef]
62. Corkidi, L.; Allen, E.B.; Merhaut, D.; Allen, M.F.; Downer, J.; Bohn, J.; Evans, M. Assessing the Infectivity of Commercial Mycorrhizal Inoculants in Plant Nursery Conditions. *J. Environ. Hortic.* **2004**, *22*, 149–154. [CrossRef]
63. Fini, A.; Frangi, P.; Amoroso, G.; Piatti, R.; Faoro, M.; Bellasio, C.; Ferrini, F. Effect of Controlled Inoculation with Specific Mycorrhizal Fungi from the Urban Environment on Growth and Physiology of Containerized Shade Tree Species Growing under Different Water Regimes. *Mycorrhiza* **2011**, *21*, 703–719. [CrossRef]
64. Wiseman, P.E.; Wells, C.E. Arbuscular Mycorrhizal Inoculation Affects Root Development of *Acer* and *Magnolia* Species. *J. Environ. Hortic.* **2009**, *27*, 70–79. [CrossRef]
65. Janos, D.P. Plant Responsiveness to Mycorrhizas Differs from Dependence upon Mycorrhizas. *Mycorrhiza* **2007**, *17*, 75–91. [CrossRef]
66. Prescott, C.E. Sinks for Plant Surplus Carbon Explain Several Ecological Phenomena. *Plant Soil* **2022**, *476*, 689–698. [CrossRef]
67. Hoeksema, J.D.; Chaudhary, V.B.; Gehring, C.A.; Johnson, N.C.; Karst, J.; Koide, R.T.; Pringle, A.; Zabinski, C.; Bever, J.D.; Moore, J.C.; et al. A Meta-Analysis of Context-Dependency in Plant Response to Inoculation with Mycorrhizal Fungi. *Ecol. Lett.* **2010**, *13*, 394–407. [CrossRef] [PubMed]
68. Rodriguez-Ramos, J.C.; Cale, J.A.; Cahill, J.F.; Erbilgin, N.; Karst, J. Soil Transfers from Intact to Disturbed Boreal Forests Neither Alter Ectomycorrhizal Fungal Communities nor Improve Pine Seedling Performance. *J. Appl. Ecol.* **2022**, *59*, 2430–2439. [CrossRef]
69. Wasyliw, J.; Fellrath, E.G.; Pec, G.J.; Cale, J.A.; Franklin, J.; Thomasson, C.; Erbilgin, N.; Karst, J. Soil Inoculation of Lodgepole Pine Seedlings Alters Root-associated Fungal Communities but Does Not Improve Seedling Performance in Beetle-killed Pine Stands. *Restor. Ecol.* **2023**, *31*, e13663. [CrossRef]

forests

MDPI

Article

Plant Growth and Microbiota Structural Effects of Rhizobacteria Inoculation on Mahogany (*Swietenia macrophylla* King [Meliaceae]) under Nursery Conditions

Flor I. Trujillo-Elisea [1,2], Natalia Y. Labrín-Sotomayor [1], Patricia A. Becerra-Lucio [1], Angel A. Becerra-Lucio [1], Jorge E. Martínez-Heredia [3,†], Ana T. Chávez-Bárcenas [2,*] and Yuri J. Peña-Ramírez [1,*]

1 El Colegio de la Frontera Sur (ECOSUR) unidad Campeche, Av. Rancho. Poligono 2ª, Colonia Ciudad Industrial Lerma, Campeche 24500, Mexico
2 Facultad de Agrobiología "Presidente Juárez", Universidad Michoacana de San Nicolás de Hidalgo, Uruapan 60170, Mexico
3 Tecnológico Nacional de México Campus China, Av. 11 Entre 22 y 28, China, Campeche 24520, Mexico
* Correspondence: tztziqui.chavez@umich.mx (A.T.C.-B.); ypena@ecosur.mx (Y.J.P.-R.); Tel.: +52-452-5250971 (A.T.C.-B.); +52-981-1273720 (ext. 2306) (Y.J.P.-R.)
† In memoriam.

Abstract: *Swietenia macrophylla* is a tropical timber species of ecological and economic importance. However, its slow vegetative growth and root development in nurseries strongly limit its production. This study evaluated the effect of 10 rhizobacteria strains during the early stages of production of *S. macrophylla*. Superficially disinfected seeds were inoculated with rhizobacteria under commercial nursery conditions. Inoculation was complemented by initial fertilization without growth regulators, fungicides, or bactericides. The results indicate that the rhizobacteria strains induce different responses in plants. Significant differences in plant biomass and root architecture were found. Treatments inoculated with *Bacillus* sp., *Bacillus polyfermenticus*, and *Bacillus siamensis* strains; showed an increase of up to 41% (dry weight). Plants increased root biomass by 30% when inoculated with *S. siamensis*. All inoculated strains were identified as members of the genus *Bacillus* spp., and their presence three months after inoculation was assessed by 16S rRNA gene-based amplicon massive sequencing. We found that *Bacillus* sp. genus was only present in inoculated treatments, suggesting that inoculated bacteria could establish themselves successfully as part of the microbiota. These results support the advantages of using PGPRs in commercial tropical tree production.

Keywords: plant-microorganism interaction; root development; tropical forest trees

Citation: Trujillo-Elisea, F.I.; Labrín-Sotomayor, N.Y.; Becerra-Lucio, P.A.; Becerra-Lucio, A.A.; Martínez-Heredia, J.E.; Chávez-Bárcenas, A.T.; Peña-Ramírez, Y.J. Plant Growth and Microbiota Structural Effects of Rhizobacteria Inoculation on Mahogany (*Swietenia macrophylla* King [Meliaceae]) under Nursery Conditions. *Forests* **2022**, *13*, 1742. https://doi.org/10.3390/f13101742

Academic Editors: Mohammed S. Lamhamedi, Damase P. Khasa and Steeve Pépin

Received: 9 September 2022
Accepted: 16 October 2022
Published: 21 October 2022

Publisher's Note: MDPI stays neutral with regard to jurisdictional claims in published maps and institutional affiliations.

1. Introduction

Swietenia macrophylla King. (Meliaceae), commonly known as Mahogany is a neotropical forest species participating in critical ecological roles as a habitat for animal species, soil conservation, and carbon capture dynamics [1]. This species also provides economic benefits for its logging, reaching a value of up to five times more than coniferous woods [2]. For these reasons, it has become a priority species grown in nurseries for restoration and commercial production purposes. However, it has been reported that early-stage *S. macrophylla* plants show slow growth and poor root system development [3,4], affecting the plant's height increase and overall quality, compromising survival during later stages in nurseries and plantations.

Several studies have identified edaphic bacteria, commonly named plant growth-promoting rhizobacteria, are capable of colonizing and significantly inducing the development of roots in plants. Described mechanisms include growth regulators production, such as auxins, cytokinins and gibberellins [5–7]; modulating ethylene levels [8]; increasing the availability of nutrients through the biological fixation of atmospheric nitrogen [9];

and mineral solubilization [10]. These properties have been primarily demonstrated in experimental and agricultural species rather than species of forestry interest.

Recent data show that *Bacillus subtillis* inoculation, in addition to fertilization in nurseries, promotes a significant increase in the biomass of *Fraxinus americana* L. [11]. This response was similar to the results of applying *Hydrogenophaga pseudoflava* to *Picea glauca* Moench. *x engelmannii*, which promoted increases of 25% in the dry matter weight of the roots and the number of shoots, as well as a 30% increase in the number of branches generated during the first year after the transplant to the field with respect to untreated plants [12]. In tropical forest species, rhizobacteria inoculation has been evaluated under uncontrolled nursery or field conditions in *Schizolobium parahyba* [13], *Shorea selanica* [14], *Tectona grandis*, *Eucalyptus* spp. [15], and *Cedrela fissilis* [16], another Meliaceae family member closely related to *S. macrophylla*. These works have demonstrated a positive impact on seedling performance [14,16,17], phytosanitary improvement [15], and wood yield [13], contributing to a growing acceptance by foresters as a component of their technological package for nursery and field establishment [18,19]. The mid-to-long-term establishment and benefits of inoculated rhizobacteria under commercial conditions are often assumed with certainty. However, bacterial prevalence depends on factors like nutrients, soil properties, plant exudates, and competition [20]. Recent work focused on evaluating rhizobacterial survival under commercial conditions demonstrated by qPCR the rapid decay within 22 to 34 days of inoculated strains, becoming undetectable by day 41 [21]. The evaluation of the microbiota architecture on inoculated plants under commercial conditions has been recently explored on annual crops [22], however forest models, where long-term plant-microbe interactions are essential, remain unexplored.

In the present work, ten non-commercial rhizobacteria strains were selected to evaluate the plant-growth-promoting effect on *S. macrophylla*, particularly exploring their impact on plant root structure under commercial nursery phase. At the end of the trial, 16S massive sequencing was used to evaluate the capabilities of inoculated rhizobacteria to get established as a long-term microbiota member in the rhizosphere and their possible effect on the establishment of rhizospheric bacterial microbiota architecture.

2. Materials and Methods

2.1. Bacterial Strains Identification

Ten rhizospheric bacterial strains (identified as 25, 29, 35, 38, 42, 46, 47, 49, 50 and 52) isolated from *Persea americana* Mill. in the "Laboratorio Planta Ambiente" from "Facultad de Agrobiología Presidente Juarez" described previously [23,24], were selected for evaluation. Molecular identification of strains was performed as follows. Genomic DNA was extracted using the modified enzymatic technique reported by [25]. The 16S rRNA gene region was amplified using the universal primers 27F (5-AGRGTTTGATYMTGGCTCAG-3′) and 338R (5′-TGCWGCCWCCCGTAGGWGT-3′) [26]. Polymerase chain reaction (PCR) amplification was performed in a final volume of 20 μL containing 1 μL of dimethyl sulfoxide, 10 μM each primer, 250 ng of DNA and 10 μL (2×) of GoTaq®Green Master Mix (Promega, Madison, WI, USA). Thermal cycling was performed as follows: an initial denaturation at 94 °C for 5 min followed by 30 denaturation cycles at 94 °C for 30 s, annealing at 58 °C for 30 s, and extension at 72 °C for 1.5 min, with a final extension at 74 °C for 5 min. Amplicons were sequenced using the Sanger method by Macrogen (Rockville, MD, USA). Sequences were analyzed with the search program Basic Local Alignment Search Tool (BLAST) to determine their identity, and sequences were deposited in the GenBank database (Supplementary Table S1).

2.2. Rhizobacteria Functional Validation Bioassay

The methodology proposed previously [23] was followed to produce the bacterial cultures as an inoculum source in the Mahogany bioassay. Briefly, the bacterial strains were grown in a potato-sucrose medium at 28 °C, with constant agitation at 160 rpm to reach a OD_{600} nm = 0.6, equivalent to 10^8 CFU mL^{-1}, and used immediately. Inoculation time varied to synchronize all cultures to reach desired OD. *Arabidopsis thaliana* L. ecotype Columbia (Col-0) seeds were superficially disinfected. Subsequently, six seeds were sown in line per plate, using 10% Murashigue and Skoog (MS) medium (Caisson Labs, UT, USA). Each bacterial strain was inoculated by streaking at a distance of 5 cm from the seeds in a parallel opposing line. The trial was established under a completely randomized experimental design, with three replicates per treatment, including a control treatment. Finally, the plates were transferred to a growth chamber (ECOSHEL Mod. C800D, Mc Allen, TX, USA); placed vertically at an angle of 70 degrees, and randomly ordered. Primary root length and number of secondary roots were measured 10 and 20 days after sowing (DAS) by plate photograph and software analysis using Rhizo2 [27]. Plants were grown under 16:8 h light:dark photoperiod, light intensity of 200 µmol m^{-2} s^{-1}; 25 °C, and 50% relative humidity.

2.3. Functional Evaluation of Rhizobacteria on *S. macrophylla* under Nursery Conditions

After collection from mature, healthy trees, *S. macrophylla* seeds were superficially disinfected with sequential washes of 10% (v) of commercial sodium hypochlorite solution and sterile water following procedures reported by [28]. Surface sterilization was corroborated by culturing a sample of seeds in Petri dishes containing semisolid potato-dextrose agar (PDA) medium and incubating for 48 h at 37 °C. Subsequently, seeds were submerged in corresponding liquid bacterial inoculum supplemented with carboxymethylcellulose (CMC) (DEIMAN, Mexico) in a 6 mg·mL^{-1} ratio and with constant agitation at 190 rpm for 30 min. Seeds were sown in a substrate based on expanded perlite, exfoliated vermiculite (Agrolita, Mexico), and peat moss (Premier, Quebec, Canada) at a ratio of 3:3:4 (v/v), with 3.7 g·L^{-1} Multicote™ fertilizer (Haifa Iberia S.L., Madrid, Spain). The mixture was arranged in trays with 150-mL cavities and supplemented with the corresponding bacterial inoculum at a 1:10 (v/v) ratio at the sowing time. The nursery experiment was performed in a commercial nursery with 50% shade. Fifteen days later, a second inoculation was conducted in which 10 mL of the bacterial suspension was placed [1:7 ratio (v/v)] directly on the substrate that covered the seeds, as reported previously [29]. Plant irrigation started 24 h later with crude well water (approximately 50 m^3/Ha every 48 h). For nursery trials, a completely randomized design was used, with 15 plants per replicate and 3 replicates per treatment. The duration of the experiment was three months. The whole experiment was repeated a year later (furtherly referred to as the first and second experiments). Measurements of stem height and diameter were performed each month. The plant biomass in dry weight and the number of dead plants were determined at the end of each experiment.

2.4. Bacterial Structure

Bacterial diversity and relative abundance were determined at the end of the second experiment as follows: A composite sample of the roots of ten plants from each treatment was collected and milled using liquid nitrogen and mortar. Subsequently, 0.25 g of each sample was used to extract total DNA using a PowerSoil® DNA isolation kit (MoBio Laboratories, Carlsbad, CA, USA), following the manufacturer's instructions. For 16S rRNA gene sequencing of the samples, 300 to 400 ng of total DNA was sent to an external sequencing service (RTLGenomics) using an Illumina MiSeq 2000 platform and 27F/338R set of primers. Fastq files were processed using QIIME 2 2020.6 [30] following the Moving Pictures tutorial as a pipeline. The sequences were grouped into operational taxonomic units (OTUs), and classifications were assigned using the GreenGenes 13-8 dataset release as a reference [31].

2.5. Statistical Analyses

The software Infostat 2017 (National University of Cordoba, Argentina) was used to analyze the data by means of the probabilistic Q-Q plot, as well as the homogeneity of variances by means of the Levene test, later the analysis of variance was performed (ANOVA) and Tukey's means test used to evaluate differences between treatments inoculated with PGPR bacteria.

3. Results

3.1. Taxonomic Assignation and Validation of In Vitro Growth-Promoting Effect of Rhizobacteria

Molecular identification of strains resulted in their assignation to *Bacillus* genus of bacteria (Firmicutes). Species-level was reached only (identity > 99%) for *Bacillus subtillis* (strain-25); *B. polifermenticus* (strain-38); *B. pumilis* (strain-49); and *B. siamensis* (strain-52), whereas the other six strains were identified only to genus level (*Bacillus* spp.). The plant growth-promoting functional properties of strains were firstly validated under controlled conditions in vitro on a model plant as *S. macrophylla* seedlings resulted very difficult to evaluate under these conditions. The root structure of *A. thaliana* seedlings growing in Petri dishes inoculated with each strain showed significant differences ($p \leq 0.001$ and $F \geq 7.61$). Plantlets in inoculated plates developed shorter primary roots, reducing their length to 35% compared to plants growing in control plates. Secondary root production was also affected by rhizobacteria presence, increasing up to 367% the number of these structures in relation to plants developing in non-inoculated plates 20 days after sowing (Figures 1a and 2).

3.2. Effect of Rhizobacteria Strains Inoculation on S. macrophylla

The plant-promoting effect of rhizobacteria strains was evaluated in two separate experiments. *S. macrophylla* seeds embedded with rhizobacteria were evaluated under nursery conditions. After three months of evaluation, plants showed no altered phenotype (Figure 1b). Inoculated plants reach the same germination rate compared to control non-inoculated individuals (~75%). Other parameters such as stem height and root-collar diameter showed subtle, but not statistically significant differences with respect to control plants during the experiments (Figure 1e). In contrast, dry weight evaluated three months after sowing, consistently showed significant increases ($p \leq 0.05$ and $F \geq 2.18$) in leaves and roots dry weight when compared to non-inoculated plants (Figures 1f and 3a). Notably, those plants inoculated with *B. siamensis* showed an increase of 75% in leaves and 41% in roots dry weight. Stem dry weight presented subtler differences (Figure 3a).

3.3. Bacterial Communities in the Rhizosphere of Seedling S. macrophylla in the Nursery

Bacterial microbiota from the rhizospheric environment from healthy plants was analyzed by the end of the second experiment. Microbiota analysis yields an average of 24,856 bacterial sequences per sample, with lengths of $\geq$ 280 nucleotides, yielding a total of 298,237 quality sequences taxonomically grouped into 14 phyla, 33 classes, 85 orders, 149 families, 238 genera, and 285 species. In all treatments and control samples, the most abundant phyla were Proteobacteria (37.6 to 64.3%), followed by Cyanobacteria (6.1 to 32.3%) Actinobacteria (9.8 to 17.6%), Bacteroidetes (0.5 to 2.2%), Gemmatimonadetes (0.2% to 1.3%) and Firmicutes (0.03 to 1.6%) (Figure 4a). No significant shifts were found at this taxonomic level; however, at the Class level, the Firmicutes represented by members of *Pullulanibacillus* spp. (27%), *Alicyclobacillus* spp. (0.8 to 88.2%) and *Bacillus* spp. (27.2 to 95.8%) genera were identified in all cases, except in the case of *Bacillus* spp. members that were detected in all treatments except for the control treatment (Figure 4b).

Figure 1. General aspects of key procedures and results. Photographic evidence of (**a**) *A, thaliana* in vitro assay showing the different phenotypes found in seedlings without rhizobacteria inoculates (**left**), where straight roots with scarce secondary roots are formed in comparison to seedlings growing in the presence of one of the strains used in this work (**right**) *B. siamensis* (strain-52). (**b**) Surface-disinfected *S. macrophylla* seeds (see Section 2). The white bar in the bottom-left of the picture corresponds to 1 cm. (**c**) In-nursery re-inoculation of rhizobacteria strains. Notice that at this point, seeds are already covered by the substrate. (**d**) Nursery once established the bioassay within a commercial production process. (**e**) *S. macrophylla* plants two months after germination growing in the nursery. (**f**) An aspect of plants three months after germination. Scale in the left is expressed in cm. (**g**) Representative dried roots of *S. macrophylla* plants corresponding to non-inoculated plants (**left**) and plants inoculated with *B. siamensis* (strain-52) (**right**). The white bar in the bottom-left of the picture corresponds to one cm.

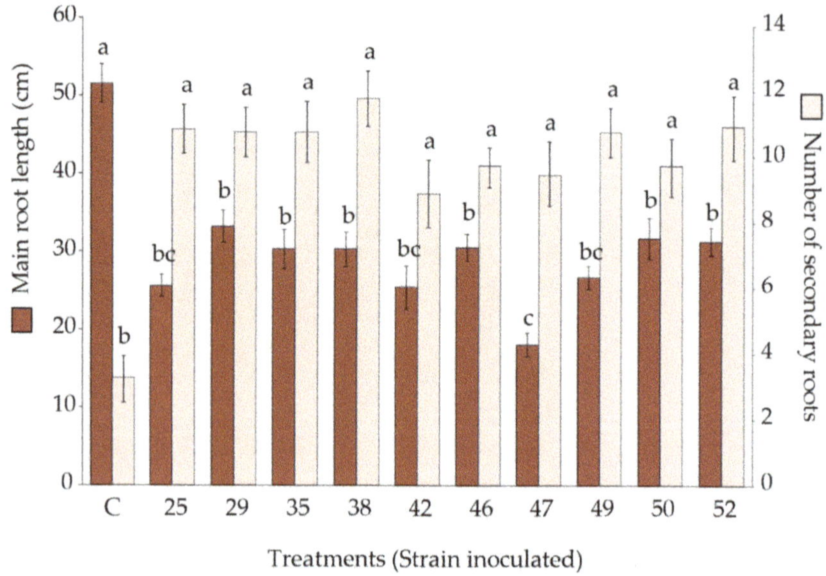

Figure 2. Effect of rhizobacteria strains on the root system of *A. thaliana*. Main root length (dark bars) and the number of secondary roots (light bars) were evaluated in vitro 20 days after germination in the presence of different bacteria strains. The number in each treatment corresponds to the number of each strain. Error bars represent the standard error. Letters over error bars indicate significant differences between the means (n = 30) of the treatments using Tukey's post hoc test ($p \leq 0.05$). Letter C refers to identical treatment of seeds and plants except for absence of bacterial inoculation.

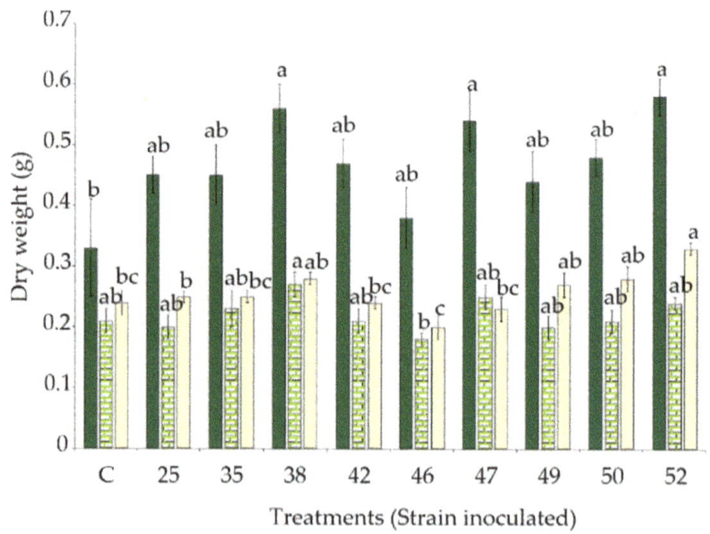

Figure 3. Effect of rhizobacteria strains on biomass and survival over nursery *S. macrophylla*. Total dry weight of leaves (dark bars), stems (patterned-filled bars), and roots (light bars) was determined three months after germination and inoculation with different rhizobacteria strains. C letter refers to control treatment (non-inoculated). Letters over error bars indicate significant differences between the means (n = 30) of the treatments using Tukey's post hoc test ($p \leq 0.05$).

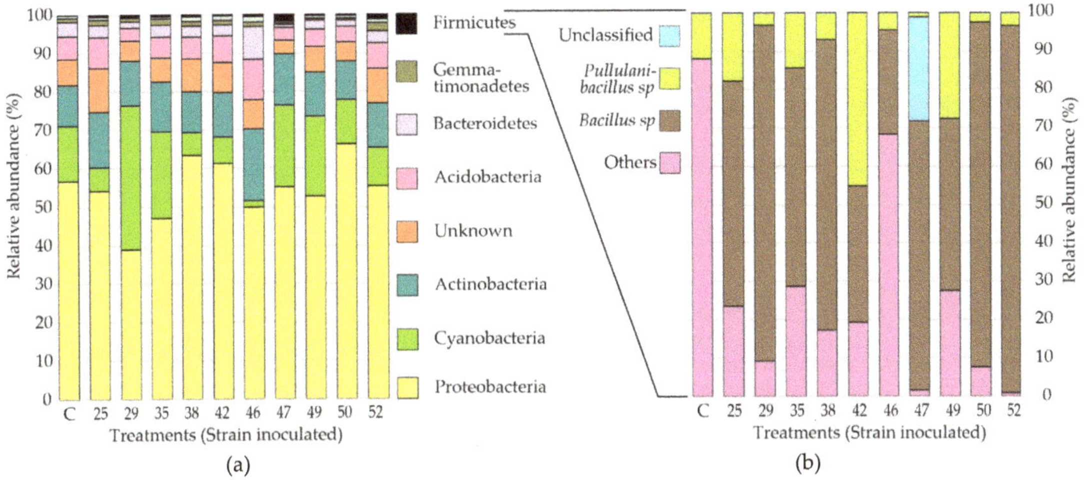

Figure 4. Composition of the bacterial community of the rhizosphere of *S. macrophylla*. 16S amplicon assessment of the microbiota present in the rhizosphere of plants three months after germination and inoculation with different rhizobacteria strains. Relative abundance (> 1%) and diversity of the primary bacterial strains in the rhizosphere at (**a**) Phylum or (**b**) Genus level graph showing only phylotypes belonging to the Firmicutes Phylum. The number in each treatment corresponds to the number of each strain. C letter refers to control treatment (non-inoculated).

4. Discussion

The use of rhizobacteria to improve the performance of forest trees during the commercial nursery phase is widely accepted [18]. This work evaluated the effect on *S. macrophylla* of ten rhizobacteria strains under nursery conditions. The results of this study demonstrate the root structure altering effect of evaluated rhizobacteria on *S macrophylla* plants. The phenotypic impact reported here suggests that these strains have a physiological effect on the root performance of seedlings under nursery conditions. Results derived from in vitro validation using *A. thaliana*, also demonstrated clear effects of rhizobacteria inoculation on root architecture. Similar morphological changes in response to rhizobacteria inoculation have been reported in a wide number of tropical tree species as *Cedrela fissilis* [16], *Acacia auriculiformis* [19], *Cecropia pachystachya, Heliocarpus popayanensis Trema micrantha, Cabralea canjerana, Cariniana estrellensis, Trichilia elegans* [32], *Eucalyptus nitens* [33], *Jatropha curcas* [34] and others. Reported changes in root architecture are associated with the modulation of cell division and the differentiation of the apical meristem and lateral root primordia sites [35]. This phenomenon is primarily due to auxin-dependent mechanisms [36]. Shifts in the concentration of plant growth regulators can be promoted by the activity of several rhizobacteria species through the production of molecules such as indole-3-acetic acid [29,37], N-acyl-homoserine lactones [38], cyclodipeptides [39], and volatile compounds [6,40]. In this regard, although many studies have performed in vitro analyses of rhizobacteria using model plants, few studies have evaluated the effect of these strains on forest species under commercial conditions.

The inoculation of rhizobacteria in forest species has demonstrated several benefits, like increases in height, stem diameter, and total biomass in *Pinus pinea*, and *Quercus ilex* seedlings where the inoculation of *Pseudomonas latus* and *Chryseobacterium balustinum* increased the biomass of shoot dry weight by 25% and 35.5% respectively four months after inoculation [41]. In *Eucalyptus nitens*, treatments inoculated with strains belonging to the genus *Bacillus* spp. promoted increases close to 140% in the aerial biomass and 130% in the root biomass [29], whereas in *Fraxinus americana*, *B. subtilis* added with fertilizer increased root biomass by 19.6 %, 22.9 % in shoot biomass, and 19.4% in leaf biomass [11].

The results here reported relative to the increasing effect of root and leaves dry weight on *S. macrophylla* plants inoculated with rhizobacteria strains are consistent with previous reports, clearly showing that these microorganisms changed root system structure. Even root confinement to tray cavities may impact overall root architecture; the observed effect of inoculated rhizobacteria on root dry weight could indicate an increase of secondary root number (Figures 1g and 3a). The effect of evaluated rhizobacteria at different levels suggests the presence of various molecular mechanisms. The mode of action of the most promising strains, such as *B. polifermenticus* (strain-38) or *B. siamensis* (strain-52), should be explored in the future. The positive effect on *S. macrophylla* plants by the *Bacillus* spp. strains evaluated in this work was observed even when the strains were initially isolated from *Persea Americana* (Lauraceae); a phylogenetically distant species; this observation supports the idea that plant-growth promotion by rhizobacteria is a peculiarity of each strain, that can be expressed in taxonomically-different plant hosts. This kind of non-specific plant-microbe association suggests the existence of adaptative plasticity, previously demonstrated by the functional capabilities of commercial strains isolated from annual crops on tropical tree species [42].

The ability of inoculated bacteria to establish in the rhizosphere and promote physiological responses in the long-term depends on their fitness within the native microbiota structure [43]; for this reason, the elucidation composition of the bacterial community after the inoculation of rhizobacteria becomes relevant, and new methods are being developed to monitor population dynamics [21]. The 16S-based metagenomic analysis performed to *S. macrophylla* rhizosphere showed a very low relative abundance of sequences associated with the genus *Bacillus* spp. ranging below 1% of total reads in all inoculated treatments in contrast to non-inoculated plants, where sequences belonging to this taxon were absent. Assuming that *Bacillus* sp. sequences belong to inoculated strains, this may suggest that regardless of the high amount of inoculated CFU, in the long-term, only a low proportion remains, probably displaced by environmental-borne microbial populations more capable of occupying ecological niches in the rhizosphere. However, even the low relative abundance of *Bacillus* sp. found on inoculated treatments appears to be enough to induce the observed phenotypes on inoculated plants, which supports the idea that rhizobacteria establish a tight relation with the hosting plants in which the quality, and not the microbial population size is relevant for plant fitness. Moreover, inoculated rhizobacteria may induce subtle microbiota shifts or indirect guild shifts of more abundant representatives causing observed phenotypes. As has been previously demonstrated, several rhizobacteria species are capable to produce secondary metabolites capable to modulate both microbiota structure and plant-microbe interactions [44].

The high diversity of the microbiota found is interesting, since the substrate used is from a non-edaphic origin and is of completely foreign to tropical region, hence any substantial contribution of bacteria that could establish an interaction with *S. macrophylla* plants was unexpected to be substrate-borne. The significant diversity of the communities observed may be probably originated from the environment through water irrigation, the air, as well as the possible contribution from seed endophytic bacteria [45]. However, it should be noted that the relative abundance of certain specific groups found in the microbiota of pristine habitats, where *S. macrophylla* is distributed, differs significantly from that present in the microbiota analyzed in the commercial substrate at the end of the nursery experiment. For example, the relative abundance of the phylum Firmicutes exhibited a relative abundance of between 3 and 7.6% in tropical forest samples [46], but only 0.03 to 1.6% in samples from the present study, whereas the phylum Proteobacteria exhibited an abundance of close to 30% in samples from other studies in the same area [46–48]. These results demonstrate that the genus *Bacillus* spp., -to which many rhizobacteria belong-used in this study was not detected in the control treatment. This result suggests that the inoculated bacteria may have been established in the communities of these treatments and were directly or indirectly responsible for the phenotypic effect observed for the plants in the different treatments. In addition, an analysis of the data shows that the communities present in treatments show differences in diversity and abundance at distinct taxonomic levels, both between treatments and with respect to the control, suggesting that the effect of bacterial strains is intrinsic and that the impact on the phenotype of plants, may be due to synergistic contributions that can cause subtle changes in the abundance of the inoculated bacteria. Similar observations were reported recently for *Zea mays* rhizospheric communities, where massive 16S sequencing reveals that rhizobacterial inoculation has a subtle impact in the native microbiota [22] A deeper examination of the microbiota (including fungi, protozoans, microalgae, lichens, and viruses) and time courses monitoring in nursery developing plants [21] could reveal complex networks of microscopic consortia contributing to the successful early development of *S. macrophylla* under commercial conditions in nursery and plantations.

5. Conclusions

Evaluated rhizobacteria strains were able to promote growth of *S. macrophylla* plants under forest nursery conditions. The activity of these strains was demonstrated by the increase of aerial and root biomass. Serendipitous evidence suggests that some of the evaluated strains may also provide a health-promoting effect. Finally, the analysis of microbiota structure of *S. macrophylla* rhizosphere evaluation demonstrated that three months after sowing, a complex bacterial community had been established, including members of *Bacillus* sp. genus present only in those treatments where rhizobacteria strains were inoculated.

Supplementary Materials: The following supporting information can be downloaded at: https://www.mdpi.com/article/10.3390/f13101742/s1, Supplementary Table S1: Taxonomic assignation.

Author Contributions: Conceptualization, A.T.C.-B. and Y.J.P.-R.; investigation, F.I.T.-E., P.A.B.-L., A.A.B.-L. and J.E.M.-H.; data curation and statistical analysis N.Y.L.-S.; writing—original draft preparation, F.I.T.-E. and Y.J.P.-R.; funding acquisition, Y.J.P.-R., A.T.C.-B., F.I.T.-E. and N.Y.L.-S. All authors have read and agreed to the published version of the manuscript.

Funding: This research was funded by Consejo Estatal de Investigación Científica y Desarrollo Tecnológico de Campeche (COESICYDET), Research grant YPR 2015-2018. El Colegio de la Frontera Sur, Research grant 5103711808. Consejo Nacional de Ciencia y Tecnología. F.I.T.E. Scholarship grant 624355.

Data Availability Statement: The samples analyzed may be available upon request after a share transfer agreement. The datasets generated during the current study are available from the corresponding author upon reasonable request.

Acknowledgments: The authors thank Alberto Can Ek for his training and support in producing mahogany plants together with the donation of supplies to carry out the bioassay. To Edwin Pérez-Sanchez for technical assistance granted in the process of training in tropical forestry and selection of individual suppliers of the seeds used in the mahogany bioassay. To Luisa Alhucema López-Ochoa for providing the *Arabidopsis thaliana* (ecotype Col-0) seeds used in the in vitro bioassay, to José López-Bucio for his contributions establishing the *A. thaliana* and Rhizobacteria interaction bioassay, to L. A. Morales-Baltazar for molecular identification of IP strains, to Issac A. Salmeron-Santiago for his help in genomic DNA extraction and the molecular identification of bacterial strains and Hugo Ruíz-González, Monserrat Rodríguez-López, Gabriel Chan-Coba, Elsy Cahuich-Morales, Esteban Hernández-Medina and Tzeltzin Sánchez-Magaña for support in the establishment of the mahogany bioassay and sampling analysis.

Conflicts of Interest: The authors declare no conflict of interest.

References

1. Silva Guzmán, J.A.; Ramírez Arango, A.M.; Fuentes Talavera, F.J.; Rodriíguez Anda, R.; Turrado Saucedo, J.; Georg Richter, H. Diagnóstico de La Industria de Transformación Primaria de Las Maderas Tropicales de México. *Rev. Mex. Ciencias For.* **2018**, *6*, 202–221. [CrossRef]
2. SEMARNAT. *Anuario Estadístico de La Producción Forestal 2017*, 1st ed.; Secretaría de Medio Ambiente y Recursos Naturales: Guadalajara, México, 2018; 284p.

3. Gilman, E.F.; Harchick, C. Root System Morphology Influences Lateral Stability of Swietenia Mahagoni. *Arboric. Urban For.* **2014**, *40*, 27–35. [CrossRef]
4. Mexal, J.G.; Cuevas Rangel, R.A.; Negreros-Castillo, P.; Paraguirre Lezama, C. Nursery Production Practices Affect Survival and Growth of Tropical Hardwoods in Quintana Roo, Mexico. *For. Ecol. Manag.* **2002**, *168*, 125–133. [CrossRef]
5. García de Salamone, I.E.; Hynes, R.K.; Nelson, L.M. Cytokinin Production by Plant Growth Promoting Rhizobacteria and Selected Mutants. *Can. J. Microbiol.* **2001**, *47*, 404–411. [CrossRef]
6. López-Bucio, J.; Campos-Cuevas, J.C.; Hernández-Calderón, E.; Velásquez-Becerra, C.; Farías-Rodríguez, R.; Macías-Rodríguez, L.I.; Valencia-Cantero, E. Bacillus Megaterium Rhizobacteria Promote Growth and Alter Root-System Architecture Through an Auxin- and Ethylene-Independent Signaling Mechanism in Arabidopsis Thaliana. *Mol. Plant-Microbe Interact.* **2007**, *20*, 207–217. [CrossRef] [PubMed]
7. Kang, S.-M.; Khan, A.L.; You, Y.-H.; Kim, J.-G.; Kamran, M.; Lee, I.-J. Gibberellin Production by Newly Isolated Strain Leifsonia Soli SE134 and Its Potential to Promote Plant Growth. *J. Microbiol. Biotechnol.* **2014**, *24*, 106–112. [CrossRef]
8. Penrose, D.M.; Glick, B.R. Methods for Isolating and Characterizing ACC Deaminase-Containing Plant Growth-Promoting Rhizobacteria. *Physiol. Plant.* **2003**, *118*, 10–15. [CrossRef]
9. Zhang, L.; Hurek, T.; Reinhold-Hurek, B. A NifH-Based Oligonucleotide Microarray for Functional Diagnostics of Nitrogen-Fixing Microorganisms. *Microb. Ecol.* **2007**, *53*, 456–470. [CrossRef]
10. Khan, A.A.; Jilani, G.; Akhtar, M.S.; Saqlan, S.M.; Rasheed, M. Phosphorus Solubilizing Bacteria: Occurrence, Mechanisms and Their Role in Crop Production. *J. Agric. Biol. Sci.* **2009**, *1*, 48–58. [CrossRef]
11. Liu, F.; Xing, S.; Ma, H.; Du, Z.; Ma, B. Plant Growth-Promoting Rhizobacteria Affect the Growth and Nutrient Uptake of Fraxinus Americana Container Seedlings. *Appl. Microbiol. Biotechnol.* **2013**, *97*, 4617–4625. [CrossRef]
12. Chanway, C.P.; Holl, F.B. First Year Field Performance of Spruce Seedlings Inoculated with Plant Growth Promoting Rhizobacteria. *Can. J. Microbiol.* **1993**, *39*, 1084–1088. [CrossRef]
13. Cely, M.V.T.; Siviero, M.A.; Emiliano, J.; Spago, F.R.; Freitas, V.F.; Barazetti, A.R.; Goya, E.T.; Lamberti, G.D.S.; dos Santos, I.M.O.; De Oliveira, A.G.; et al. Inoculation of Schizolobium Parahyba with Mycorrhizal Fungi and Plant Growth-Promoting Rhizobacteria Increases Wood Yield under Field Conditions. *Front. Plant Sci.* **2016**, *7*, 1–13. [CrossRef] [PubMed]
14. Sitepu, I.R.; Aryanto; Ogita, N.; Osaki, M.; Santoso, E.; Tahara, S.; Hashidoko, Y. Screening of Rhizobacteria from Dipterocarp Seedlings and Saplings for the Promotion of Early Growth of Shorea Selanica Seedlings. *Tropics* **2007**, *16*, 245–252. [CrossRef]
15. Mafia, R.G.; Alfenas, A.C.; Maffia, L.A.; Ferreira, E.M.; Binoti, D.H.B.; Mafia, G.M.V. Plant Growth Promoting Rhizobacteria as Agents in the Biocontrol of Eucalyptus Mini-Cutting Rot. *Trop. Plant Pathol.* **2009**, *34*, 10–17. [CrossRef]
16. Trápani, A.I.; Ulla, E.L.; Amigo, J.A.; Jimenez, P. Response of Cedrela Fissilis Vell. to Inoculation with Plant Growth-Promoting Rhizobacteria in Nurseries. *Revista Agronómica del Noroeste Argentina* **2021**, *41*, 93–98.
17. Alexandre, F.S.; Della Flora, L.V.; Henrique, I.G.; da Silva, D.C.; Mercedes, A.P.; Cardoso Silva, A.; Silva de Oliveira, A.; Bondespacho da Silva, M.P.; Formelh Ronning, B.P.; Dreher, D.R.; et al. Arbuscular Mycorrhizal Fungi (Rhizophagus Clarus) and Rhizobacteria (Bacillus Subtilis) Can Improve the Clonal Propagation and Development of Teak for Commercial Plantings. *Front. Plant Sci.* **2021**, *12*, 1–12. [CrossRef]
18. Asif, M.; Mughal, A.H.; Bisma, R.; Mehdi, Z.; Saima, S.; Ajaz, M.; Malik, M.A.; Masood, A.; Sidique, S. Application of Different Strains of Biofertilizers for Raising Quality Forest Nursery. *Int. J. Curr. Microbiol. Appl. Sci.* **2018**, *7*, 3680–3686. [CrossRef]
19. Muthukumar, T.; Udaiyan, K. Coinoculation of Bioinoculants Improve Acacia Auriculiformis Seedling Growth and Quality in a Tropical Alfisol Soil. *J. For. Res.* **2018**, *29*, 663–673. [CrossRef]
20. Tailor, A.J.; Joshi, B.H. Harnessing Plant Growth Promoting Rhizobacteria Beyond Nature: A Review. *J. Plant Nutr.* **2014**, *37*, 1534–1571. [CrossRef]
21. Gao, N.; Zhang, H.; Xiong, R.; Fang, L.; Shen, W.; Senoo, K. Different Strategies for Colonization and Prevalence after Inoculation with Plant Growth-Promoting Rhizobacteria Revealed by a Monitoring Method. *Soil Sci. Plant Nutr.* **2022**, *68*, 442–453. [CrossRef]
22. Vasconcellos, R.L.F.; Romagnoli, E.M.; Taketani, R.G.; Santos, S.N.; Zucchi, T.D.; Melo, I.S. Impact of Inoculation with Pseudomonas Aestus CMAA 1215T on the Non-Target Resident Bacterial Community in a Saline Rhizosphere Soil. *Curr. Microbiol.* **2021**, *78*, 218–228. [CrossRef] [PubMed]
23. Trujillo-Elisea, F.I. Capacidad de Solubilización de Fosfato de Aluminio Por Bacterias Edáficas Promotoras de Crecimiento Vegetal. Bachelor's Thesis, Universidad Michoacana de San Nicolás de Hidalgo, Uruapan, Mexico, 2015; 96p.
24. Becerra-Lucio, A.A. Capacidad Solubilizadora de Fósforo En Bacterias Edáficas Promotoras Del Crecimiento Vegetal Aisladas En Uruapan, Michoacán. Bachelor's Thesis, Universidad Michoanana de San Nicolás de Hidalgo, Uruapan, Mexico, 2015; 86p.
25. Cutting, S.; Vander Horn, P. Genetic Analysis. In *Molecular Biological Methods for Bacillus (Modern Microbiological Methods)*; Hardwood, C.R., Cutting, S.M., Eds.; Wiley: Chichester, UK, 1990; p. 581. ISBN 0471923931.
26. Weisburg, W.G.; Barns, S.M.; Pelletier, D.A.; Lane, D.J. 16S Ribosomal DNA Amplification for Phylogenetic Study. *J. Bacteriol.* **1991**, *173*, 697–703. [CrossRef]
27. Shahzad, Z.; Kellermeier, F.; Armstrong, E.M.; Rogers, S.; Lobet, G.; Amtmann, A.; Hills, A. EZ-Root-VIS: A Software Pipeline for the Rapid Analysis and Visual Reconstruction of Root System Architecture. *Plant Physiol.* **2018**, 217. [CrossRef]

28. Martínez, C.R.C.; Gutiérrez, E.J.C.; Arrazate, C.H.A. (Eds.) *Protocolo de Multiplicación y Conservación in Vitro de Cuatro Especies Forestales Tropicales de Semillas Recalcitrantes CASO: Caoba (Swietenia Macrophylla)*, 1st ed. 2019. Available online: https://www.gob.mx/cms/uploads/attachment/file/126297/Protocolo_de_multiplicacion_y_conservacion_in_vitro_de_cuatro_especies_forestales_tropicales_de_semillas_recalcitrantes.pdf (accessed on 1 January 2019).
29. Angulo, V.C.; Sanfuentes, E.A.; Rodríguez, F.; Sossa, K.E. Caracterización de Rizobacterias Promotoras de Crecimiento En Plántulas de Eucalyptus Nitens. *Rev. Argent. Microbiol.* **2014**, *46*, 338–347. [CrossRef]
30. Caporaso, J.G.; Kuczynski, J.; Stombaugh, J.; Bittinger, K.; Bushman, F.D.; Costello, E.K.; Fierer, N.; Peña, A.G.; Goodrich, J.K.; Gordon, J.I.; et al. QIIME Allows Analysis of High-Throughput Community Sequencing Data. *Nat. Methods* **2010**, *7*, 335–336. [CrossRef] [PubMed]
31. Bokulich, N.A.; Kaehler, B.D.; Rideout, J.R.; Dillon, M.; Bolyen, E.; Knight, R.; Huttley, G.A.; Gregory Caporaso, J. Optimizing Taxonomic Classification of Marker-Gene Amplicon Sequences with QIIME 2's Q2-Feature-Classifier Plugin. *Microbiome* **2018**, *6*, 90. [CrossRef]
32. de Souza, N.L.; Rocha, S.S.; Narezzi, N.T.; Tiepo, A.N.; de Oliveira, A.L.M.; Oliveira, H.C.; Bianchini, E.; Pimenta, J.A.; Stolf-Moreira, R. Differential Impacts of Plant Growth-Promoting Bacteria (PGPB) on Seeds of Neotropical Tree Species with Contrasting Tolerance to Shade. *Trees* **2020**, *34*, 121–132. [CrossRef]
33. González, P.; Sossa, K.; Rodríguez, F.; Sanfuentes, E. Rhizobacteria Strains as Promoters of Rooting in Hybrids of Eucalyptus Nitens × Eucalyptus Globulus. *Chil. J. Agric. Res.* **2018**, *78*, 3–12. [CrossRef]
34. Jha, K.S.; Saraf, M. Effect of Plant Growth Promoting Rhizobacteria on Seed Germination Behaviour and Seedling Vigor of Jatropha Curcas. *Int. J. Biotechnol. Biosci.* **2011**, *1*, 161–169.
35. Verbon, E.H.; Liberman, L.M. Beneficial Microbes Affect Endogenous Mechanisms Controlling Root Development. *Trends Plant Sci.* **2016**, *21*, 218–229. [CrossRef]
36. Benková, E.; Bielach, A. Lateral Root Organogenesis—From Cell to Organ. *Curr. Opin. Plant Biol.* **2010**, *13*, 677–683. [CrossRef] [PubMed]
37. Dobbelaere, S.; Croonenborghs, A.; Thys, A.; Vande Broek, A.; Vanderleyden, J. Phytostimulatory Effect of Azospirillum Brasilense Wild Type and Mutant Strains Altered in IAA Production on Wheat. *Plant Soil* **1999**, *6*, 155–164. [CrossRef]
38. Ortíz-Castro, R.; Martínez-Trujillo, M.; López-Bucio, J. N -Acyl-L-Homoserine Lactones: A Class of Bacterial Quorum-Sensing Signals Alter Post-Embryonic Root Development in Arabidopsis Thaliana. *Plant. Cell Environ.* **2008**, *31*, 1497–1509. [CrossRef]
39. Ortiz-Castro, R.; López-Bucio, J.S.; López-Bucio, J. Physiological and Molecular Mechanisms of Bacterial Phytostimulation. In *Advances in PGPR Research*; Sing, H.B., Sarma, B.K., Keswani, C., Eds.; CABI: Wallingford, UK, 2017; pp. 16–28.
40. Pérez-Flores, P.; Valencia-Cantero, E.; Altamirano-Hernández, J.; Pelagio-Flores, R.; López-Bucio, J.; García-Juárez, P.; Macías-Rodríguez, L. Bacillus Methylotrophicus M4-96 Isolated from Maize (Zea Mays) Rhizoplane Increases Growth and Auxin Content in Arabidopsis Thaliana via Emission of Volatiles. *Protoplasma* **2017**, *254*, 2201–2213. [CrossRef] [PubMed]
41. García, J.A.L.; Domenech, J.; Santamaría, C.; Camacho, M.; Daza, A.; Mañero, F.J.G. Growth of Forest Plants (Pine and Holm-Oak) Inoculated with Rhizobacteria: Relationship with Microbial Community Structure and Biological Activity of Its Rhizosphere. *Environ. Exp. Bot.* **2004**, *52*, 239–251. [CrossRef]
42. dos Santos, R.F.; da Cruz, S.P.; Botelho, G.R.; Flores, A.V. Inoculation of Pinus Taeda Seedlings with Plant Growth-Promoting Rhizobacteria. *Floresta e Ambient.* **2018**, *25*, 1–7. [CrossRef]
43. Bulgarelli, D.; Schlaeppi, K.; Spaepen, S.; van Themaat, E.V.L.; Schulze-Lefert, P. Structure and Functions of the Bacterial Microbiota of Plants. *Annu. Rev. Plant Biol.* **2013**, *64*, 807–838. [CrossRef]
44. Patel, J.S.; Yadav, S.K.; Bajpai, R.; Teli, B.; Rashid, M. PGPR Secondary Metabolites: An Active Syrup for Improvement of Plant Health. In *Molecular Aspects of Plant Beneficial Microbes in Agriculture*; Sharma, V., Salwan, R., Al-Ani, L.K.T., Eds.; Academic Press: Cambridge, MA, USA, 2020; pp. 195–208. ISBN 978-0-12-818469-1.
45. Vivas, M.; Kemler, M.; Slippers, B. Maternal Effects on Tree Phenotypes: Considering the Microbiome. *Trends Plant Sci.* **2015**, *20*, 541–544. [CrossRef]
46. Becerra-Lucio, A.A.; Labrín-Sotomayor, N.Y.; Becerra-Lucio, P.A.; Trujillo-Elisea, F.I.; Chávez-Bárcenas, A.T.; Machkour-M'Rabet, S.; Peña-Ramírez, Y.J. Diversity and Interactomics of Bacterial Communities Associated with Dominant Trees during Tropical Forest Recovery. *Curr. Microbiol.* **2021**. [CrossRef]
47. Estrada-Medina, H.; Canto-Canché, B.B.; De los Santos-Briones, C.; O'Connor-Sánchez, A. Yucatán in Black and Red: Linking Edaphic Analysis and Pyrosequencing-Based Assessment of Bacterial and Fungal Community Structures in the Two Main Kinds of Soil of Yucatán State. *Microbiol. Res.* **2016**, *188–189*, 23–33. [CrossRef]
48. Pajares, S.; Campo, J.; Bohannan, B.J.M.; Etchevers, J.D. Environmental Controls on Soil Microbial Communities in a Seasonally Dry Tropical Forest. *Appl. Environ. Microbiol.* **2018**, *84*, e00342-18. [CrossRef]

forests

MDPI

Article

Nutrient Availability Has a Greater Influence than Pot Host on Seedling Development of Hemiparasitic Hawaiian Sandalwood (*Santalum paniculatum* Hook. and Arn.)

Tawn M. Speetjens and Douglass F. Jacobs *

Department of Forestry and Natural Resources, Purdue University, West Lafayette, IN 47907, USA
* Correspondence: djacobs@purdue.edu

Abstract: Sandalwood (*Santalum* spp.) has been overharvested throughout its range, including the Hawaiian Islands, where 6 of the 19 species *Santalum* spp. are endemic. As hemiparasitic plant species, Hawaiian sandalwoods require a host plant for optimal forest establishment, yet the importance of a host during seedling development is unclear. Furthermore, understanding interactions between pot hosts and nutrient availability on sandalwood seedling development during nursery culture will help to promote the production of high-quality sandalwood seedlings for restoration and commercial purposes. We evaluated the effects of controlled-release fertilizer (CRF), chelated Fe treatments, and two pot host species (*Acacia koa* and *Dodonaea viscosa)* on the seedling development of Hawaiian sandalwood (*Santalum paniculatum*). Increased nutrient availability (CRF) led to increased dry mass, root collar diameter, shoot height, chlorophyll index, and nutrient status values, confirming that the hemiparasitic *S. paniculatum* can be successfully grown in early stages of cultivation by providing adequate mineral fertilizers. There was a significant interaction between the nutrient availability and chelated iron treatments associated with increased height, root collar diameter, dry mass, chlorophyll index, Fe concentration, and Fe content when chelated Fe was applied (vs. not) in a nutrient-limiting environment. The pot host treatment did not affect any growth metrics, but it did affect the total count of haustoria, with *A. koa*-hosted seedlings developing 60.3% more haustoria than *D. viscosa*-hosted and control seedlings. Our results demonstrate that high-quality *S. paniculatum* seedlings can be grown in containers by providing adequate nutrition and that *S. paniculatum* in a nutrient-limiting growing environment may benefit from chelated iron fertilizers.

Keywords: *Santalum paniculatum*; hemiparasite; fertilizer; pot host; nursery culture; controlled-release fertilizer; chelated iron fertilizer; *Acacia koa*; *Dodonaea viscosa*

Citation: Speetjens, T.M.; Jacobs, D.F. Nutrient Availability Has a Greater Influence than Pot Host on Seedling Development of Hemiparasitic Hawaiian Sandalwood (*Santalum paniculatum* Hook. and Arn.). *Forests* **2023**, *14*, 458. https://doi.org/10.3390/f14030458

Academic Editors: Mohammed S. Lamhamedi, Damase P. Khasa and Steeve Pépin

Received: 18 January 2023
Revised: 14 February 2023
Accepted: 16 February 2023
Published: 23 February 2023

1. Introduction

Members of the *Santalum* L. genus are root hemiparasitic woody shrubs or trees that produce aromatic oil-rich heartwood and are collectively referred to as sandalwoods [1]. Indian sandalwood (*Santalum album* L.) has been coveted for its high-quality heartwood and has been traded and used for 2500 years in China and 3200 years in India [2]. Indian sandalwood was historically burned as incense in Hindu and Buddhist ceremonies, carved into deities' sculptures, and used as fuel for funeral pyres for revered religious figures such as Mahatma Gandhi [1]. Sandalwood species grow naturally in India, Australia, Indonesia, and numerous islands throughout the Pacific Ocean [3,4]. The high value of the heartwood has led to overharvesting and the exploitation of the genus throughout its range, including the Hawaiian Islands, where six of the 19 sandalwood species are endemic [5,6].

In Hawaii, sandalwood is known locally as 'iliahi and was used for medicines and to add scent to plant fiber cloths [7]. A Hawaiian sandalwood trade occurred from 1790 to 1840 and ultimately extirpated Hawaiian sandalwood species from an estimated 90% of their natural range [7,8]. All known Hawaiian sandalwood species are still extant, although they occur in much smaller populations, and one variety, *Santalum freycinetianum* var.

lanaiense Rock, is endangered [6,9]. Historic distribution models show that sandalwood populations covered a broad elevational gradient (0–2550 m) and grew in most Hawaii forest types before being overharvested [10,11]. After the collapse of the sandalwood trade, forests that once hosted sandalwood populations were converted to agricultural use, predominantly for sugar cane and cattle grazing [7]. The introduction of invasive species associated with the conversion of native forests to an agricultural landscape contributed to the displacement and eventual extinction of unknown numbers of native plant and animal species in Hawaii, ultimately leading to the deterioration of and the disruption of ecosystem services [12–16]. Except for the coastal species, *Santalum ellipticum* Gaudich., remnant populations of Hawaiian sandalwood primarily grow in the high-elevation regions on six of the main Hawaiian Islands [5,17]. Remnant populations persist across a patchwork of public and private lands, with varying levels of disturbance and management intensity [18–20].

Many areas that once hosted Hawaiian sandalwood forests are actively targeted for restoration planting, and there is also a burgeoning interest in practicing commercial sandalwood forestry in Hawaii [18,21]. The historical uses for sandalwood persist in the modern day, while new uses are evolving in the essential oil and pharmaceutical industry [1,22,23]. The global demand for sandalwood has steadily outpaced the supply, causing prices to soar and creating an opportunity for savvy sandalwood foresters [24]. The Hawaiian Islands are uniquely poised to host a burgeoning sandalwood industry that could restore ecological, economic, and cultural value to a landscape that has historically been overexploited for its natural resources. The Hawaii Island endemic, *Santalum paniculatum* Hook. and Arn., produces commercial grade heartwood oil and is of particular commercial interest [25].

Commercial production and restoration planting will require reliable propagation protocols to produce high-quality seedlings. The seedling quality is closely linked to survival and performance at outplanting and can be influenced by nursery culture practices, including but not limited to container selection, fertilization, irrigation, and competition reduction [26]. Therefore, understanding how different cultural practices affect specific aspects of seedling quality is imperative to designing effective and efficient propagation protocols. Such protocols are established for other sandalwood species outside of Hawaii, such as *S. album*. The literature has shown that co-planting *S. album* seedlings with a pot host enhances growth during the nursery grow-out phase and after field planting [27,28]. Additionally, providing essential nutrients improves the growth of *S. album* seedlings [29–31]. However, Hawaiian sandalwood propagators cannot adopt many of the established methods because of differences in sandalwood species physiology, available host species, and growing environments [5,27,28,32–36]. Seed germination protocols developed for *S. album* have been successfully adapted for Hawaiian species, but protocols for growing seedlings to planting maturity require further development [37]. The nursery propagation of sandalwood is complicated by its hemiparasitic nature and the resulting uncertainty concerning best practices for integrating hosts and fertilizers into nursery culture practices to produce the highest quality seedlings. Understanding how nutrient availability and co-planted hosts should be utilized to produce high-quality *S. paniculatum* seedlings is the first of many steps in developing comprehensive propagation protocols.

As root hemiparasites, members of the *Santalum* genus are capable of autotrophic carbon gain through photosynthesis and heterotrophic carbon gain through parasitic attachments to the root xylem tissues of neighboring host plants [38,39]. In addition to carbohydrates, sandalwoods can acquire water, mineral nutrients, amino acids, and other organic xylem solutes from host plants [40,41]. Although sandalwood can grow without a host, attachment to a host increases its growth rates and capacity to grow competitively [32]. Nitrogen-fixing plants were initially thought to be superior hosts, although growth trials with *S. album* have since shown varying levels of benefit from different N-fixing species, with some providing less benefit than some non-N-fixing host species [28,42,43]. Sandalwood trees use a specialized organ called haustorium to attach to the host roots, penetrate the root epidermis, and gain access to the host root xylem, where they extract resources [31,40,43]. The host-derived acquisition of resources is reliant on haustoria connections, and there is potential for the host to negatively affect the sandalwood through competition if they do not become attached [28,40].

Controlled-release fertilizer (CRF) has been used to produce high-quality seedlings with Hawaiian sandalwood species [37,44,45]. These are water-soluble fertilizers encased in a semi-permeable polymer coat that allows mineral ions to leach through at a controlled rate [46]. The regulated release of nutrients may lead to reduced leaching and improved fertilizer use efficiency compared to conventional non-encapsulated fertilizers [47,48]. Controlled-release fertilizers have been used effectively for agricultural crops and forest tree species, although the effect of CRF on Hawaiian sandalwood has not been quantified [46,49–51]. In addition to CRF, propagators of Hawaiian sandalwood have found that chelated iron treatments increase the growth and correct commonly occurring iron deficiencies of Hawaiian sandalwood seedlings [37,44,45]. Chelated iron fertilizers are composed of a Fe^{2+} ion bound to a synthetic chelating agent such as ethylenediamine di-(o-hydroxyphenyl acetic) acid (EDDHA) [52]. Chelated forms of iron remain water-soluble and available for plants in calcareous soils, whereas unchelated forms of iron typically do not [53]. Chelated iron is provided to the plant in a form that is readily absorbed by the plant roots or leaf tissue and rapidly corrects iron deficiencies in numerous crops and Hawaiian sandalwood [45,54,55]. Pot hosts and fertilizers have been used to successfully increase the seedling quality of *S. album,* although further investigation is required to determine their effects on Hawaiian sandalwood seedlings [3,23,28,31,56]. We conducted an experiment to quantify the independent and interacting effects of CRF, chelated iron, and pot host species on the seedling quality of Hawaiian sandalwood (*S. paniculatum*).

Our experiment utilized *S. paniculatum,* which has the most extensive potential range and the largest remnant population of the Hawaiian sandalwoods [10,17,57]. *S. paniculatum* is endemic to Hawaii Island, the largest of the main Hawaiian Islands and it is the only Hawaiian species that is commercially cultivated or harvested. The oil extracted from *S. paniculatum* heartwood is high in santalols, increasing the value of the oil to be comparable to Indian sandalwood oil [25]. It is a broadleaf evergreen tree that forms a single bole trunk and can reach heights of 13 m to 20 m when mature [11]. It grows in moderately wet to dry forests, although remnant populations primarily grow in high-elevation mesic and dry forests on the west side of Hawaii Island [5].

Specifically, we hypothesize that (i) nutrients available from CRF will improve the quality of *S. paniculatum* seedlings through increased nutrient acquisition, decreasing the chance that nutrients will be a limiting factor to growth; (ii) the interaction between the CRF and chelated iron will have a significant effect on seedling quality because the chelated Fe will ensure adequate Fe concentration; and (iii) the host treatment will improve the seedling quality by providing a host to parasitize and extract resources to support growth. Specifically, the nitrogen-fixing *Acacia koa* A. Gray, host will provide the greatest benefit to sandalwood seedling quality.

2. Materials and Methods

2.1. Experimental Site

The experiment was conducted at the Hāloa ʻĀina Reforestation Project (HARP) native plant nursery (N 19°32′16.550″, W 155°48′22.101″, elevation: 1462 m) in Kealakekua on Hawaii Island. The HARP manages degraded sandalwood forests on 1164 hectares of privately owned land and 435 hectares of lease land in the montane regions of Kealakekua. The experiment occurred in a 7 m × 8 m shade house with a 50% shade film roof and walls of 70% shade cloth. The top was 4 m tall on the east end and tapered down to 3 m on the west. We recorded the temperature and humidity at a weather station installed 30 m from the experiment site. The average annual temperature and relative humidity of the growing area during the experiment (9 August 2020–5 August 2021) were 18.3 °C and 70% RH, respectively, and the minimum and maximum temperatures were 3.3 °C and 27.7 °C, respectively.

2.2. Experimental Design and Treatments

The experiment was a randomized complete block design with a full factorial combination of CRF (two levels; applied, control), chelated iron fertilizer (two levels; applied, control), and host species (three levels; *A. koa*, *Dodonaea viscosa* Jacq., control) as the predicting factors. We defined twelve unique treatments by combining the three predicting factors. The experiment had four blocks containing one representation of each of the twelve treatments. Each treatment was represented by a tray of 32 seedlings subsamples that received the same combination of CRF, chelated iron, and host species. We randomly selected twelve of the 32 seedlings from each treatment for destructive sampling at the end of the grow-out period. We grew 1536 (32 * 12 * 4) seedlings and randomly selected 576 (12 * 12 * 4) for destructive sampling. The treatment trays were randomly distributed within each block and were rearranged within the block every three months to maintain independence between treatments and limit edge effects.

S. paniculatum seeds were germinated by first soaking them for 24 h in a solution of 400 ppm gibberellic acid (GA3) (ProGibb T and O®, Valent BioSciences, Libertyville, IL, USA) and distilled water. After the GA3 solution was rinsed from seeds, they were sown into a bed of vermiculite one inch from the surface and watered top-down twice per week. We transplanted the *S. paniculatum* seedlings from the vermiculite bed into grow-out containers at the 4–6 true leaf stage on 6–7 August 2020. The mean height and root collar diameter (RCD) of the *S. paniculatum* seedlings at the time of transplantation was 8.6 cm ($\pm$0.23) and 3.01 mm ($\pm$0.07), respectively. The seedlings were grown in a 1540 mL container (MT49BT; 10 cm $\times$ 10 cm $\times$ 23 cm depth) (Stuewe and Sons Inc., Tangent, OR, USA) in media composed of equal parts of PRO-MIX MP MYCORRHIZAE ORGANIK® potting soil (Premier Tech Horticulture®, Quakertown, PA, USA), perlite, and fine black cinder. The host seedlings were transplanted into designated growing containers with *S. paniculatum* seedlings on 16 December 2020. The mean shoot height and RCD of the hosts when transplanted was 10.3 cm ($\pm$3.1) and 2.1 mm ($\pm$0.18) for *A. koa* (n = 25) and 8.8 cm ($\pm$1.6) and 0.9 mm ($\pm$0.01) for *D. viscosa* (n = 25), respectively. The hosts were co-planted when the *S. paniculatum* seedlings were four months old, which provided a competitive advantage to the sandalwood. We replaced host seedlings that died until 20 January 2021, and excluded subsamples with dead hosts in the random sampling. The host seedlings were not pruned.

Controlled-release fertilizer was applied to designated treatments by incorporating Osmocote plus® (5–6-month release, at 21.1 °C)(Scotts Co.® Marysville, OH, USA) into the soil at medium bag rates (6.2 kg/m^3), which was then reapplied at six months by mixing 9.2 g per container into the surface of the soil. Osmocote plus® is composed of 15% total nitrogen, 9% available phosphate, 12% soluble potash, 1.3% Mg, 6.0% S, 0.02% B, 0.05% Cu, 0.46 % Fe (0.01% chelated), 0.06% Mn, 0.02% Mo, and 0.05% Zn by weight. The fertilizer and soil were mixed thoroughly with a shovel until the fertilizer was considered evenly distributed via visual inspection. Once the grow-out container trays were filled with growing medium, the treatments were randomly assigned to trays. Chelated iron was applied to the designated treatments one month after we transplanted *S. paniculatum* seedlings to growth containers. The chelated iron was applied by dusting 4.8 g of fertilome® EDDHA chelated iron (6% water soluble iron) (Voluntary Purchasing Groups®, Bonham, TX, USA) onto the soil and watering it through until all powder was dissolved from the surface. The chelated iron was first applied on 14 September 2020, then reapplied with the same method on 28 December 2021 and 12 April 2021.

Each container received approximately 265 mL of tap water applied top-down with a hand watering wand once per week for the first six months, then twice per week for the remaining duration of the experiment. The same volume (265 mL) of a dilute solution (374 ppm) of Miracle-gro® water-soluble fertilizer (18% N, 18% P, 21% K, 0.5% Mg, 0.02% B, 0.05% Cu, 0.1% Fe (0.1% chelated Fe), 0.05% Mn, 0.0005% Mo, 0.05% Zn) (Scotts Co.® Marysville, OH, USA) was applied to every plant in place of irrigation water one day every other week to ensure the control treatment plants did not die and grew large enough to provide leaf samples for the nutrient analysis.

2.3. Plant Material

S. paniculatum seeds were collected from a site 37.6 km northeast of the experiment site (N 19°48′22.6″, W 155°36′12.8″, elevation of 2025 m) in an ecoregion similar to the experiment site. *Acacia koa* A. Gray, a nitrogen-fixing tree, and *D. viscosa*, a non-nitrogen-fixing tree shrub, were chosen as the species for the host treatment. They are commonly associated with *S. paniculatum* in the surrounding forest and are frequently used in restoration planting practices in Hawaii [58–60]. *A. koa* is endemic to Hawaii, and while *D. viscosa* is a native Hawaiian species, it is widespread and also associated with *S. album* in India [11,61]. All host seedlings are grown from seeds collected within 2 km of the experiment site. The *A. koa* seeds were scarified by soaking them in 95 °C water for 24 h as they cooled, then sown into a tray of vermiculite 7 cm deep. The *D. viscosa* seeds were soaked in distilled water for 24 h and sown with the same method. All seedlings were inoculated with rhizobium collected from potted *A. koa* seedlings from the Hāloa ʻĀina nursery. A 124 mL slurry of *A. koa* nitrogen-fixing nodules was diluted into 190 L of water and applied to all pots with a watering wand.

2.4. Measurements

Destructive sampling occurred by block starting on 24 June 2021 and lasting until 5 August 2021, which spread the age at sampling from 322 to 364 days. Thirty-four seedlings were girdled by rats or slugs and were not included in the sampling and analysis. Chlorophyll index measurements were collected using an atLEAF+® chlorophyll meter. The atLEAF+® chlorophyll meter measures the greenness of a leaf sample to produce a chlorophyll index value similar to a SPAD® meter [62]. The chlorophyll index generated by the atLEAF+ and SPAD® meters correlates with the chlorophyll concentration of the measured leaf [62]. We used the mean of five readings from each seedling for the analysis. Chlorophyll readings were taken from the newest pair of fully mature leaves on the terminal shoot, at two-thirds the distance from the leaf base to the apex [63]. The leaves used for chlorophyll readings were collected, dried, and ground for the tissue nutrient analysis for the concentrations of N (Total), P, K, S, Mg, Ca, Na, B, Zn, Mn, Fe, Cu, and Cl [64]. Sample leaves for the nutrient analysis were collected from each of the twelve subsamples to form a composite sample representing each treatment in a block. We calculated the foliar N and Fe contents for each seedling by multiplying the treatment's nutrient concentration by the seedling's dry mass.

The growing medium was washed from seedlings using a water bath. The removed soil was filtered to catch severed root fragments to be incorporated into the root dry mass (g) measurement. We measured the shoot length (cm) from the root collar to the last node on the terminal shoot using a meter stick, and we measured the RCD (mm) using dial calipers. The shoot and root were severed at the root collar and dried in a convection drying oven at 70 °C for 72 h to achieve a constant mass [65]. The dry mass (g) of the shoots, roots, and leaves were measured using a Mettler Toledo® AB104-S analytical balance. We calculated the shoot/root ratio for each seedling by dividing the shoot dry mass by the root dry mass. We visually counted the number of haustoria on the roots of each sandalwood seedling and the number of haustoria attached to the host for paired seedlings.

2.5. Statistical Analysis

The response variables (RCD, height, dry mass, shoot: root, chlorophyll index, haustoria count, foliar N concentration, foliar N content, foliar Fe concentration, and foliar Fe content) were analyzed separately using a linear mixed effects model with the block and treatment ID as a random factor. The treatment ID was added as a random factor to eliminate pseudo-replication in the model from subsampling within a treatment replicate. The samples for foliar nutrient concentrations (N and Fe) were composites of the treatment, meaning only the block was included as a random factor in the analysis. We used a three-way analysis of variance (ANOVA), and a type III sum of squares was calculated on each model to examine comparisons defined in the a priori hypotheses. We evaluated the residual values of each model to ensure the assumptions of normality and homogeneity of variance were met before the analysis. The response variables that did not meet the assumptions of the linear model were log-transformed to satisfy the model assumptions. A comparison of the estimated marginal means comparison (emmeans) with Tukey's *p*-value adjustment was used to determine significant differences ($\alpha = 0.05$) among treatments when detected with the ANOVA ($p < 0.05$). All statistical analyses were performed with R software (R Version 3.2.4, the R Foundation for Statistical Computing Platform).

3. Results

3.1. Plant Morphology

The mean dry mass of *S. paniculatum* seedlings was significantly affected by the CRF ($F_{1,33.17} = 439.58$, $p < 0.001$) and chelated iron ($F_{1,3.19} = 50.47$, $p < 0.001$), but the host treatment had no effect (Table 1). However, there was a significant interaction between the CRF and chelated iron treatments ($F_{1,33.20} = 34.46$, $p < 0.001$) (Table 1), which was associated with an increase in mean dry mass when we applied chelated iron without CRF, but not when applied with CRF (Figures 1 and 2). The mean dry mass of the seedlings that received chelated iron only was 3.09 g (±0.26) compared to 1.51 g (±0.26) for the seedlings that did not receive CRF or chelated iron (Figure 1).

Table 1. The *p* values resulting from a three-way ANOVA performed on mixed effect models fitted for each response variable. CRF = controlled-release fertilizer; Fe = chelated iron fertilizer; S/R = shoot/root ratio; Chl. = chlorophyll index.

	Dry Mass	Collar Diameter	Shoot Height	S/R	Chl.	N Cont.	N Conc.	Fe Cont.	Fe Conc.	Total Haustoria
CRF	**<0.001**	**<0.001**	**<0.001**	0.442	**<0.001**	**<0.001**	**<0.001**	**<0.001**	**<0.001**	0.211
Fe	**<0.001**	**<0.001**	**<0.001**	0.400	**<0.001**	**<0.001**	**<0.001**	**<0.001**	**<0.001**	0.832
Host	0.145	0.291	0.640	0.249	**0.004**	0.198	0.876	0.435	**0.014**	**0.015**
CRF × Fe	**<0.001**	**0.032**	**0.002**	0.951	**<0.001**	**0.016**	**<0.001**	**<0.001**	0.703	0.319
CRF × Host	0.656	0.845	0.452	0.533	0.227	0.178	0.268	0.166	0.090	0.663
Fe × Host	0.214	0.389	0.658	0.708	0.120	0.170	0.575	0.634	0.528	0.905
CRF × Fe × Host	0.701	0.532	0.884	0.506	0.138	0.156	0.270	0.856	0.736	0.592

Bolded values represent a significant effect on the response variable.

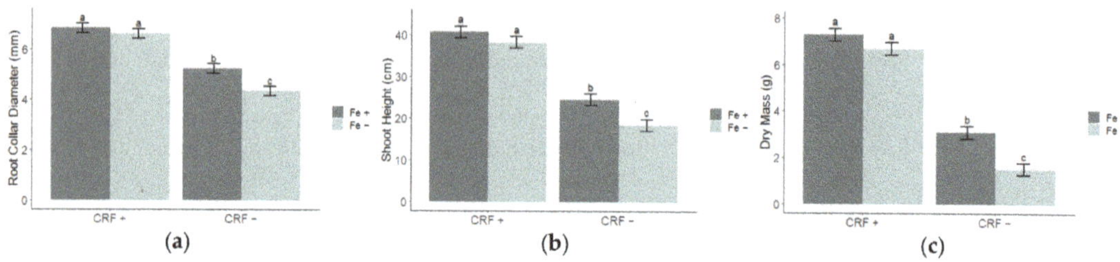

Figure 1. The mean (±SE) (**a**) dry mass, (**b**) root collar diameter, and (**c**) shoot height values of *S. paniculatum* seedlings at the end of the experiment. Different letters indicate significant differences among treatments ($\alpha = 0.05$). CRF improved growth in all treatments, although chelated iron only improved growth when applied without CRF.

Figure 2. Pictures of representative seedlings from each of the twelve unique treatments show that seedlings that received CRF were significantly larger than seedlings that did not receive CRF. The *D. viscosa* hosts were much smaller than the *A. koa* hosts at the end of the experiment. Fe+ = chelated Fe applied; Fe− = control; CRF+ = CRF applied; CRF− = control.

The mean root collar diameter was significantly affected by CRF ($F_{1,33.72} = 165.64$, $p < 0.001$) and chelated iron ($F_{1,33.73} = 13.62$, $p < 0.001$) but was not affected by the host treatment (Table 1). There was a significant interaction between CRF and chelated iron ($F_{1,33.20} = 5.0$, $p = 0.032$) (Table 1) that was associated with an increase in RCD when we applied chelated iron without CRF but not when applied with CRF (Figures 1 and 2). The mean root collar diameter of treatments that only received chelated iron was 5.25 mm (±0.18) compared to a mean of 4.34 mm (±0.18) in treatments that did not receive chelated iron or CRF (Figure 1).

There was a significant effect of CRF ($F_{1,32.97} = 285.79$, $p < 0.001$) and chelated iron ($F_{1,32.98} = 24.41$, $p < 0.001$) on the seedling dry mass, but the effect of the host treatment was not significant (Table 1). The interaction between CRF and chelated Fe had a significant effect ($F_{1,32.99} = 11.14$, $p = 0.002$) (Table 1) associated with an increase in dry mass when we applied chelated iron without CRF but not when applied with CRF (Figures 1 and 2). The treatment plants that only received chelated iron had a significantly greater mean shoot height of 24.05 cm (±1.38) compared to 18.4 cm (±1.38) from those that did not receive CRF or chelated iron (Figure 1). The mean shoot root ratio was not significantly affected by CRF ($F_{1,33.05} = 1.00$, $p = 0.324$), chelated iron ($F_{1,33.05} = 1.74$, $p = 0.196$), or the host ($F_{1,33.05} = 1.57$, $p = 0.223$).

3.2. Chlorophyll Index

The mean chlorophyll index was significantly affected by the CRF ($F_{1,21.69} = 1263.8$, $p < 0.001$), chelated iron ($F_{1,21.71} = 157.52$, $p < 0.001$), and host ($F_{1,21.81} = 157.52$, $p = 0.004$) treatments (Table 1). The mean chlorophyll index values of *A. koa* and control (49.4 ± 0.7) plants were greater than for the treatment plants that received *D. viscosa* (47.0 ± 0.7) (Figure 3). The interaction between the CRF and chelated iron had a significant effect ($F_{1,21.69} = 59.60$, $p < 0.001$) (Table 1) on the chlorophyll index, which was associated with an increase in the mean chlorophyll index when we applied chelated iron without CRF, but not when CRF was applied (Figure 3).

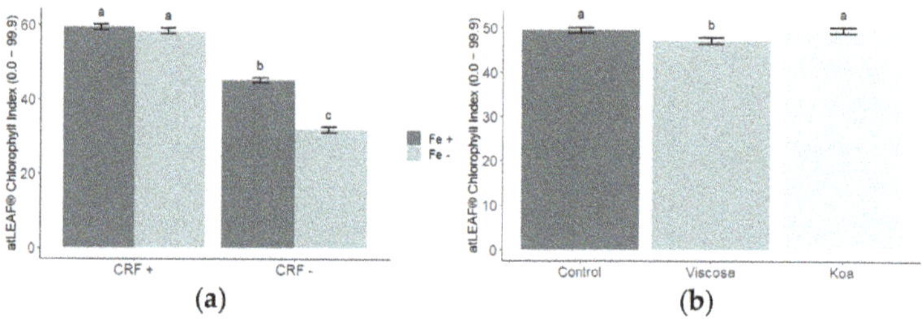

Figure 3. Mean (± SE) chlorophyll index values. (**a**) The interaction between CRF and chelated iron significantly affected the chlorophyll index, causing chelated iron to have an effect only when applied without CRF. (**b**) Seedlings paired with *D. viscosa* had a lower chlorophyll index value than *A. koa*-paired and control seedlings. Different letters indicate significant differences among treatments ($\alpha = 0.05$). Fe+ = chelated Fe applied; Fe− = control; CRF+ = CRF applied; CRF− = control.

3.3. Foliar Nitrogen and Iron

The mean nitrogen concentration was significantly affected by the CRF ($F_{1,33} = 315.38$, $p < 0.001$) and the chelated iron ($F_{1,33} = 10.26$, $p = 0.003$), but it was not affected by the host treatment (Table 1). The interaction between the CRF and chelated iron had a significant effect ($F_{1,33} = 9.77$, $p < 0.001$) (Table 1) associated with an increase in the mean N concentration when we applied chelated iron without CRF but not when applied with CRF (Figure 4). The mean N concentration of the seedlings that received only chelated Fe was 1.70% (±0.16) compared to 2.44% (±0.16) for the treatments that did not receive CRF or chelated iron (Figure 4).

The mean nitrogen content of the sandalwood seedlings was significantly affected by the CRF ($F_{1,33} = 315.38$, $p < 0.001$) and chelated iron ($F_{1,33} = 10.26$, $p = 0.003$) but was not affected by the host treatment (Table 1). The interaction between the CRF and chelated iron had a significant effect on the mean N content ($F_{1,32.73} = 6.48$, $p < 0.001$) (Table 1) associated with an increase in mean nitrogen content when we applied chelated iron without CRF but not when applied with CRF (Figure 4). The mean N content of seedlings that only received chelated Fe was 21.9 mg (±5.8) compared to 15.1 mg (±5.8) for seedlings that did not receive CRF or chelated iron (Figure 4).

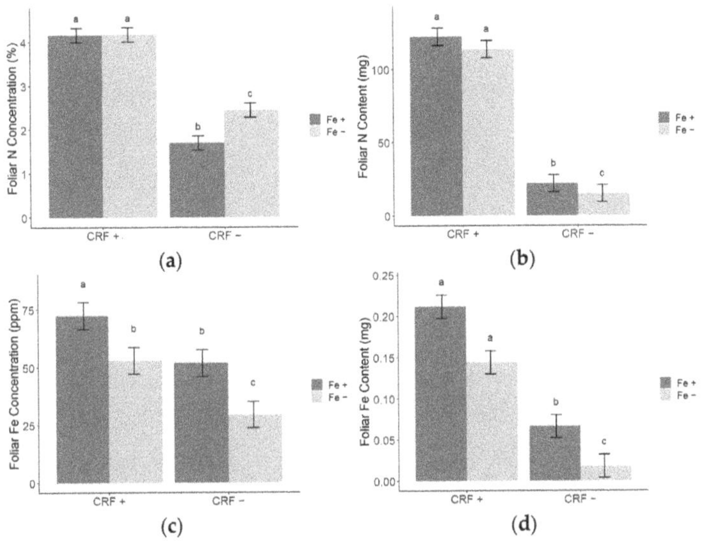

Figure 4. The mean (±SE) foliar nitrogen (N) (concentration (%) and content (mg)) and foliar iron (Fe) (concentration (ppm) and content (mg)) values of *S. paniculatum* seedlings. (**a**) The N concentration (%), (**b**) N content (mg), and (**d**) Fe content was increased in all plants that received CRF. The chelated iron had a significant effect when applied without CRF, decreasing the N concentration, but increasing the N content and Fe content. (**c**) The foliar Fe concentration was the same in treatments that only received CRF and treatments that only received chelated iron. Different letters indicate significant differences among treatments ($\alpha = 0.05$). Fe+ = chelated Fe applied; Fe− = control; CRF+ = CRF applied; CRF− = control.

The iron concentration (ppm) was significantly affected by CRF ($F_{1,33} = 31.80$, $p < 0.001$), chelated iron ($F_{1,33} = 28.74$, $p < 0.001$), and the host ($F_{2,33} = 4.80$, $p = 0.015$) treatment (Table 1). The mean Fe concentration of treatments that received CRF was 62.7 ppm (±5.1) compared to 40.7 ppm (±5.1) for those that did not. The mean Fe concentration of treatments that received chelated iron was 62.2 (±5.1) compared to 41.2% (±5.1) for those that did not. The mean Fe concentration of seedlings that received the *A. koa* host treatment (60.2 (±5.46)) was significantly greater than the mean for *D. viscosa* (46.6 (±5.35) and control (48.4 (±5.46)) (Figure 4). None of the interactions between predictors significantly affected the mean Fe concentration (Table 1)

3.4. Haustoria Abundance

The host treatment significantly affected the mean total haustoria present on a sandalwood seedling ($F_{2,32.39} = 4.83$, $p = 0.015$) (Table 1). The mean number of haustoria on *A. koa*-hosted sandalwood seedlings (19.5 (±2.53)) was greater compared to the mean number of haustoria found on the *D. viscosa*-hosted (12.5 (±2.53)) and control (11.0 (±2.53)) seedlings. There was no difference between the *D. viscosa*-hosted and control treatment levels. The count of the attached haustorium revealed that 54.7% (±5.8) of the sandalwood seedlings paired with *A. koa* were attached and 14.4% (±3.8) of the seedlings paired with *D. viscosa* were attached.

4. Discussion

Our results confirm that CRF, chelated Fe, and pot host species significantly affect several aspects of *S. paniculatum* seedling quality, including the height, collar, dry mass, chlorophyll concentration, and nutrient status. Although each predictor affected the seedling quality differently, the nutrient availability provided by CRF had the most significant effect. In support of our hypothesis (i), we found that nutrients provided by CRF significantly increased in shoot height, RCD, dry mass, chlorophyll index, and foliar nutrient (N and Fe) concentrations and contents relative to the non-fertilized control. Increased macronutrient availability resulting from applying CRF has been shown to improve growth in other Hawaiian forest species [51,66]. Similar to studies with other sandalwood species, our findings demonstrate that nutrient availability is a limiting factor for growth and that simply providing nutrients to the hemiparasitic sandalwood can produce high-quality seedlings [29,31]. In further support of our findings, a study that controlled nutrient levels through fertigation demonstrated that providing macronutrients (N, P, K, S, Ca) increases the shoot height and dry mass in nursery-grown *S. album* seedlings independent of a host [31]. The nutrient availability unexpectedly did not affect the shoot root ratio or the total count of haustoria, which was contradictory to what Barret and Fox (1997) found with *S. album*.

In support of our hypothesis (ii), we found that the interaction between the nutrient availability and chelated iron significantly increased the shoot height, RCD, dry mass, and chlorophyll index and improved the nutrient status, although the mechanism of effect was different from what was predicted. Rather than chelated iron improving the seedling quality when applied with CRF, it only enhanced the seedling quality in a nutrient-limiting growing environment that did not receive CRF. The foliar Fe concentration was the same for treatments that only received CRF (53.00 ppm ($\pm$5.79)) and for treatments that only received chelated iron (51.92 ppm ($\pm$5.79)), implying that both fertilizers provided adequate levels of Fe, within the range required to avoid deficiency and toxicity symptoms (44–250 ppm) [67]. The nitrogen content increased but the N concentration decreased when chelated Fe was applied in a nutrient-limiting environment. The increased growth and dry mass associated with applying chelated iron in a nutrient-limiting environment resulted in increased N content values. However, the nutrients are diluted across more tissue, which can cause a decrease in the foliar concentration [68,69]. The opposite trend was observed for the Fe concentration, likely because the chelated Fe fertilizer provided sufficient Fe to maintain an increase in Fe concentration despite the increase in dry mass.

The chelated iron treatment only provided a benefit to growth in a nutrient-limiting environment, suggesting that iron may be a limiting factor for growth under these conditions. Fe is required by plants in the greatest quantity of any micronutrient and is necessary for fundamental cellular processes, including chlorophyll biosynthesis, electron transport, and vital enzymatic reactions [53,70]. Fe is the second most abundant mineral in the earth's crust and rarely deficient in soils, although calcareous soil conditions convert plant-available ferrous iron (Fe^{2+}) to the more unavailable form of ferric iron (Fe^{3+}) [69,71,72]. Plants can secrete organic acid compounds and phyto-chelators into the soil to increase Fe availability in calcareous Fe-limiting environments [67,71]. Applying synthetic chelated Fe emulates the natural secretion of phyto-chelates and more efficiently corrects iron deficiencies compared to Fe-compound and natural Fe-complex fertilizers [73]. Chelated iron is widely used in industrial agriculture to correct iron deficiencies and improve yields, although its application in forest restoration systems is largely unstudied [74,75]. Considering that deforestation can decrease the abundance of natural iron chelators in the soil, further investigation into the effects of synthetic iron chelators in silvicultural systems is warranted [76]. Our findings show that chelated iron provided a benefit to growth in a nutrient-limiting nursery environment, although further research is required to see if this effect persists in a field setting. It should be noted that the growing medium that was used (PRO-MIX MP MYCORRHIZAE ORGANIK®) contains endomycorrhizal inoculum PTB297, and endomycorrhiza has been shown to increase the iron uptake by promoting the production of plant-derived iron chelators [77].

The evidence for the influence of the pot host on the seedling quality (hypothesis iii) was largely unsupported by our findings. We found that the pot host only affected the haustoria abundance and Fe concentration, contradicting results from growth trials that demonstrated that co-planting pot hosts of numerous species enhanced the growth of co-planted sandalwood seedlings [28,43]. It is likely that the growth of our sandalwood seedlings did not benefit from the pot host treatment, because of the poor rates of attachment that resulted from the pairings. Although only 54.7% (±5.8) of *A. koa*-paired seedlings and 14.4% (±3.8) of *D. viscosa*-paired seedlings were attached by the end of the experiment, we assumed those that were attached were receiving nutrition from their hosts. Haustoria formation is initiated by chemical signals from the host plant in many hemiparasitic plants, including sandalwood [43,78–80]. Chemical signals from different host plants can provide differential benefits for haustoria formation and root growth prior to being attached to the sandalwood [81]. The suitability and resulting influence of the pot host varies by species, and both the N-fixing and non-N-fixing species have been found to be suitable hosts for *S. album* [28,43,82]. Like studies with other species of sandalwood, we found that a nitrogen-fixing host species produced greater haustoria abundance compared to a non-N-fixing host species, although both had relatively low rates of attachment to the sandalwood [78,83,84]. The duration of our experiment was determined by the prevailing methodology for raising *S. paniculatum* in the nursery, although considering the low percentage of plants that attached to their host, this rearing period may not allow enough time for the sandalwood to establish haustoria connections with the host. A longer rearing time may allow more haustoria connections to form, but may not be practical for nursery cultivation, where an increased rearing time translates to increased costs. Further research is required to elucidate the mechanisms of the haustoria attachment in Hawaiian sandalwood and to determine whether culture practices could be developed to enhance the rate of attachment in nursery propagation. Considering the pot host did not negatively affect the sandalwood development, using a pot host may still be advisable if the host benefits sandalwood field planting performance similar to *S. album*. More research is required to determine the effect of the co-planted nursery host on the field planting success of Hawaiian sandalwood.

5. Conclusions

Hemiparasitic *Santalum* spp. are known to acquire carbon and mineral nutrients autotrophically and heterotrophically. We demonstrated that applying fertilizers significantly improved the growth of *S. paniculatum*, while pairing with a pot host did not affect the growth. The increased nutrient availability from the CRF application consistently improved the growth; however, chelated iron fertilizers only improved the growth in a nutrient-limiting environment where CRF was not applied, indicating iron availability may be a limiting factor to *S. paniculatum* growth in nutrient-poor environments. Applying CRF is an effective and cost-efficient method for improving sandalwood seedling growth and should be integrated into propagation protocols for Hawaiian sandalwood species and would likely enhance the growth of all sandalwood species. The pot host treatment did not affect the growth of the *S. paniculatum* seedlings during nursery propagation. However, the *A. koa*-paired seedlings had more haustoria, so the pot host may provide a benefit following field planting [27]. We suggest that *S. paniculatum* should still be planted with a pot host in the nursery, although further research is required to improve this treatment for Hawaiian sandalwood species. Although our study focused on nursery culture practices specific to the Hawaiian endemic *S. paniculatum*, our results contribute to the scientific knowledge base of propagation practices for the commercially valuable *Santalum* genus and other hemiparasitic species.

Author Contributions: Conceptualization: T.M.S. and D.F.J.; data curation: T.M.S. and D.F.J.; formal analysis: T.M.S. and D.F.J.; funding acquisition: T.M.S. and D.F.J.; investigation: T.M.S. and D.F.J.; methodology: T.M.S. and D.F.J.; project administration: T.M.S. and D.F.J.; resources: T.M.S. and D.F.J.; software: T.M.S. and D.F.J.; supervision: D.F.J.; validation: T.M.S. and D.F.J.; visualization: T.M.S. and D.F.J.; writing—original draft: T.M.S. and D.F.J.; writing—review and editing: T.M.S. and D.F.J. All authors have read and agreed to the published version of the manuscript.

Funding: Funding support was provided by the USDA National Institute of Food and Agriculture McIntire-Stennis Cooperative Forestry Program (IND011535), Purdue University Department of Forestry and Natural Resources, the Tropical Hardwood Tree Improvement and Regeneration Center, Hāloa ʻĀina LLC., the Sloan Indigenous Graduate Partnership, the Fischer Forestry Fund Scholarship, and Hawaii Sandalwood LLC.

Data Availability Statement: Data and code used for analysis will be made available on request from the corresponding author.

Acknowledgments: Thank you to Emily Thyroff, Brianne Innusa, Andrei Toca, Bee Overbey, Caleb Reddick, Aziz Ebrahimi, Sylvia Park, Madeline Montague, Summer Rathfon, Elias Gaffney, Sayon Ghosh, Jeff Stallman, Kylle Roy, Jonathan Shimizu, Rick Meilan, Owen Burney, Irene Sprecher, Nathan Deppe, Asa Suguitan, Chris Brady and Susan Cordell.

Conflicts of Interest: The authors declare no conflict of interest.

References

1. Arunkumar, A.N.; Joshi, G.; Ram, H. Sandalwood: History, Uses, Present Status and the Future. *Curr. Sci.* **2012**, *103*, 1408–1416.
2. Pullaiah, T.; Das, S.C. History of Sandalwood. In *Sandalwood: Silviculture, Conservation and Applications*; Pullaiah, T., Das, S.C., Bapat, V.A., Swamy, M.K., Reddy, V.D., Murthy, K.S.R., Eds.; Springer: Berlin/Heidelberg, Germany, 2021; pp. 9–20.
3. da Silva, T.; Kher, J.A.; Soner, M.M.; Page, D.; Zhang, T.; Nataraj, X.; Ma, M. Sandalwood: Basic Biology, Tissue Culture, and Genetic Transformation. *Planta* **2016**, *243*, 847–887. [CrossRef] [PubMed]
4. Harbaugh, D.T.; Baldwin, B.G. Phylogeny and Biogeography of the Sandalwoods (Santalum, Santalaceae): Repeated Dispersals throughout the Pacific. *Am. J. Bot.* **2007**, *94*, 1028–1040. [CrossRef] [PubMed]
5. Merlin, M.; Thompson, L.; Elevitch, C. Santalum Ellipticum, *S. Freycinetianum*, *S. Haleakalae* and *S. Paniculatum* (Hawaiian Sandalwood). In *Traditional Trees of the Pacific Islands: Their Culture, Environment and Use*; Permanent Agricultural Resources: Honolulu, HI, USA, 2006; pp. 695–714.
6. Harbaugh, D.T.; Oppenheimer, H.L.; Wood, K.R.; Wagner, W.L. Taxonomic Revision of the Endangered Hawaiian Red-Flowered Sandalwoods (*Santalum*) and Discovery of an Ancient Hybrid Species. *Syst. Bot.* **2010**, *35*, 827–838. [CrossRef]
7. Merlin, M.; Vanravenswaay, D. The History of Human Impact on the Genus *Santalum* in Hawai'i. In Proceedings of the Symposium on Sandalwood in the Pacific, Honolulu, HI, USA, 9–11 April 1990; General Technical Report PSW-GTR-122. Pacific Southwest Research Station, Forest Service: Berkeley, CA, USA, 1990; pp. 46–60.
8. Rock, J.F.C. *The Indigenous Trees of the Hawaiian Islands*; College of Hawaii, Territory of Hawaii: Honolulu, HI, USA, 1913.
9. Stemmermann, L. Observations on the Genus *Santalum* (*Santalaceae*) in Hawaii. *Pac. Sci.* **1980**, *34*, 41–54.
10. Price, J.P.; Gon, S.M., III; Jacobi, J.D.; Matsuwaki, D. *Mapping Plant Species Ranges in the Hawaiian Islands: Developing a Methodology and Associated GIS Layers*; USGS: Reston, VA, USA, 2007; Volume 8.
11. Wagner, W.L.; Herbst, D.R. *Manual of the Flowering Plants of Hawaii*, Revised ed.; University of Hawaii Press: Honolulu, HI, USA, 1999.
12. Li, Y.; Mathews, B.W. Effect of Conversion of Sugarcane Plantation to Forest and Pasture on Soil Carbon in Hawaii. *Plant Soil* **2010**, *335*, 245–253. [CrossRef]
13. Parrotta, J.; Wildburger, S.; Mansourian, C. *Understanding Relationships between Biodiversity, Carbon, Forests and People: The Key to Achieving REDD+ Objectives. A Global Assessment Report Prepared by the Global Forest Expert Panel on Biodiversity, Forest Management and REDD+*; IUFRO: Vienna, Austria, 2012; Volume 31.
14. Asquith, A. Alien Species and the Extinction Crisis of Hawaii's Invertebrates. *Endanger. Species Update* **1995**, *12*, 1–8.
15. Howarth, F.G. Impacts of Alien Land Arthropods and Mollusks on Native Plants and Animals in Hawai'i. In *Hawai'i's Terrestrial Ecosystems. Proc. Symposium, 1984*; University of Hawaii Press: Honolulu, HI, USA, 1985; pp. 149–179.
16. Sax, D.F.; Gaines, S.D. Species Invasions and Extinction: The Future of Native Biodiversity on Islands. *Proc. Natl. Acad. Sci. USA* **2008**, *105*, 11490–11497. [CrossRef]
17. Stemmermann, L. Distribution and Status of Sandalwood in Hawai'i. In Proceedings of the Symposium on Sandalwood in the Pacific, Honolulu, HI, USA, 9–11 April 1990; General Technical Report PSW-GTR-122. Pacific Southwest Research Station, Forest Service: Berkeley, CA, USA; Department of Agriculture: Washington, DC, USA, 1990; pp. 62–65.
18. Dlnr, S.H. *Management Plan for the Ahupua'a of Pu'u Wa'awa'a and the Makai Lands of Pu'u Anahulu*; Department of Land and Natural Resources: Honolulu, HI, USA, 2003; pp. 1–82.
19. Scheffel, M. Sandalwood: Current Interest and Activity by the Hawaii Division of Forestry and Wildlife. In Proceedings of the Symposium on Sandalwood in the Pacific, Honolulu, HI, USA, 9–11 April 1990; General Technical Report PSW-GTR-122. Pacific Southwest Research Station, Forest Service: Berkeley, CA, USA, 1990; p. 61.
20. Senock, R.S. *Hāloa 'Āina Forest Management Plan*; On Solid Ground Consulting LLC.: Chico, CA, USA, 2017.
21. Leopold, S. Big Island, Small Planet: Challenges and Failures in Conserving Hawaiian Sandalwood. *HerbalGram.* **2015**, *108*, 60–67.
22. Santha, S.; Dwivedi, C. Properties of Sandalwood Oil. *Anticancer Res.* **2015**, *35*, 3137–3145.
23. Ramanan, S.; George, A.K.; Chavan, S.B.; Kumar, S.; Jayasubha, S. Progress and Future Research Trends on *Santalum Album*: A Bibliometric and Science Mapping Approach. *Ind. Crops Prod.* **2020**, *158*, 112972. [CrossRef]
24. Divakara, B.; Viswanath, S.; Nikhitha, C.; Kumar, S. Forest Research Economics of *Santalum Album* L. Cultivation Under Semiarid Tropics of. *For. Res.* **2018**, *7*, 223.
25. Braun, N.A.; Sim, S.; Kohlenberg, B.; Lawrence, B.M. Hawaiian Sandalwood: Oil Composition of *Santalum Paniculatum* and Comparison with Other Sandal Species. *Nat. Prod. Commun.* **2014**, *9*, 1365–1368. [CrossRef] [PubMed]
26. Duryea, M.L. Nursery Cultural Practices: Impacts on Seedling Quality. In *Forest Nursery Manual: Production of Bareroot Seedlings*; Nijhoff, M., Junk, W., Eds.; The Hague: Corvalis, OR, USA, 1984; pp. 143–164.

27. Radomiljac, A.M. The Influence of Pot Host Species, Seedling Age and Supplementary Nursery Nutrition on *Santalum Album* Linn. (Indian Sandalwood) Plantation Establishment within the Ord River Irrigation Area, Western Australia. *For. Ecol. Manag.* **1998**, *102*, 193–201. [CrossRef]
28. Annapurna, D.; Rathore, T.S.; Joshi, G. Modern Nursery Practices in the Production of Quality Seedlings of Indian Sandalwood (*Santalum Album* L.)—Stage of Host Requirement and Screening of Primary Host Species. *J. Sustain. For.* **2006**, *22*, 33–55. [CrossRef]
29. Struthers, A.R.; Lamont, B.B.; Fox, J.E.D.; Wijesuriya, S.; Struthers, R.; Lamont, B.B.; Fox, J.E.D.; Wijesuriya, S. Mineral Nutrition of Sandalwood (*Santalum Spicatum*). *J. Exp. Bot.* **2022**, *37*, 1274–1284. [CrossRef]
30. Das, C.; Das, S.; Tah, J.; Chandra, B.; Viswavidyalaya, K. Effect of Soil Nutrients on the Growth and Survivility of White Sandal (*Santalum Album* L) In South West Bengal. *Int. J. Curr. Res.* **2020**, *10*, 76264–76267.
31. Barrett, D.R.; Fox, J.E.D. *Santalum Album*: Kernel Composition, Morphological and Nutrient Characteristics of Pre-Parasitic Seedlings under Various Nutrient Regimes. *Ann. Bot.* **1997**, *79*, 59–66. [CrossRef]
32. Radomiljac, A.M.; McComb, J.A.; Shea, S.R. Field Establishment of *Santalum Album* L.—The Effect of the Time of Introduction of a Pot Host (Alternanthera Nana R. Br.). *For. Ecol. Manag.* **1998**, *111*, 107–118. [CrossRef]
33. Neil, P.E. Growing Sandalwood in Nepal—Potential Silvicultural Methods and Research Priorities 1. In Proceedings of the Symposium on Sandalwood in the Pacific, Honolulu, HI, USA, 9–11 April 1990; General Technical Report PSW-GTR-122. Pacific Southwest Research Station, Forest Service: Berkeley, CA, USA, 1990. No. Mathur 1979. pp. 72–75.
34. Peeris, M.K.P.; Senarath, W.T.P.S.K. In Vitro Propagation of *Santalum Album* L. *J. Natl. Sci. Found. Sri Lanka* **2015**, *43*, 265–272. [CrossRef]
35. Annapurna, D.; Rathore, T.S.; Joshi, G. Effect of Container Type and Size on the Growth and Quality of Seedlings of Indian Sandalwood (*Santalum Album* L.). *Aust. For.* **2004**, *67*, 82–87. [CrossRef]
36. Annapurna, D.; Rathore, T.S.; Joshi, G. Refinement of Potting Medium Ingredients for Production of High Quality Seedlings of Sandalwood (*Santalum Album* L.). *Aust. For.* **2005**, *68*, 44–49. [CrossRef]
37. Speetjens, T.M.; Thyroff, E.C.; Yeh, A.K.F.; Jacobs, D.F. Propagation of 'Iliahi (*Santalum Paniculatum* Hook. & Arn.) a Valuable Endemic Hawaiian Sandalwood Species. *Tree Plant. Notes* **2021**, *64*, 33–49.
38. Těšitel, J.; Plavcová, L.; Cameron, D.D. Interactions between Hemiparasitic Plants and Their Hosts: The Importance of Organic Carbon Transfer. *Plant Signal. Behav.* **2010**, *5*, 1072–1076. [CrossRef] [PubMed]
39. Press, M.C. Autotrophy and Heterotrophy in Root Hemiparasites. *Trends Ecol. Evol.* **1989**, *4*, 258–263. [CrossRef] [PubMed]
40. Tennakoon, K.U.; Pate, J.S.; Stewart, G.R. Haustorium-Related Uptake and Metabolism of Host Xylem Solutes by the Root Hemiparasitic Shrub *Santalum Acuminatum* (R. Br.) A. DC. (Santalaceae). *Ann. Bot.* **1997**, *80*, 257–264. [CrossRef]
41. Radomiljac, A.M.; McComb, J.A.; Pate, J.S.; Tennakoon, K.U. Xylem Transfer of Organic Solutes in *Santalum Album* L. (Indian Sandalwood) in Association with Legume and Non-Legume Hosts. *Ann. Bot.* **1998**, *82*, 675–682. [CrossRef]
42. Radomiljac, A.M.; McComb, J.A. Nitrogen-Fixing and Non-Nitrogen-Fixing Woody Host Influences on the Growth of the Root Hemi-Parasite *Santalum Album* L. In Proceedings of the an International Seminar, Bangalore, India, 18–19 December 1998; pp. 54–57.
43. Ouyang, Y.; Zhang, X.; Chen, Y.; Teixeira da Silva, J.A.; Ma, G. Growth, Photosynthesis and Haustorial Development of Semiparasitic *Santalum Album* L. Penetrating into Roots of Three Hosts: A Comparative Study. *Trees Struct. Funct.* **2016**, *30*, 317–328. [CrossRef]
44. Wilkinson, K.M. Propagation Protocol for 'iliahi (*Santalum Freycinetianum*). *Nativ. Plants J.* **2007**, *8*, 248–251. [CrossRef]
45. Hirano, R.T. Propagation of *Santalum*, Sandalwood Tree. In Proceedings of the Symposium on Sandalwood in the Pacific, Honolulu, HI, USA, 9–11 April 1990; General Technical Report PSW-GTR-122. Pacific Southwest Research Station, Forest Service: Berkeley, CA, USA, 1990; Volume 122, pp. 43–45.
46. Vejan, P.; Khadiran, T.; Abdullah, R.; Ahmad, N. Controlled Release Fertilizer: A Review on Developments, Applications and Potential in Agriculture. *J. Control. Release* **2021**, *339*, 321–334. [CrossRef]
47. Cox, D.A. Reducing Nitrogen Leaching-Losses from Containerized Plants: The Effectiveness of Controlled-Release Fertilizers. *J. Plant Nutr.* **1993**, *16*, 533–545. [CrossRef]
48. Alva, A.K. Differential Leaching of Nutrients from Soluble vs. Controlled-Release Fertilizers. *Environ. Manag.* **1992**, *16*, 769–776. [CrossRef]
49. Landis, T.D.; Dumroese, R.K. Using Polymer-Coated Controlled-Release Fertilizers in the Nursery and after Outplanting. *For. Nurs. Notes* **2009**, *Winter 6*, 5–12.
50. Arnott, J.T.; Burdett, A.N. Early Growth of Planted Western Hemlock in Relation to Stock Type and Controlled-Release Fertilizer Application. *Can. J. For.* **1988**, *18*, 710–717. [CrossRef]
51. Dumroese, R.K.; Jacobs, D.F.; Davis, A.S. Inoculating *Acacia Koa* with *Bradyrhizobium* and Applying Fertilizer in the Nursery: Effects on Nodule Formation and Seedling Growth. *HortScience* **2009**, *44*, 443–446. [CrossRef]
52. Clemens, D.F.; Whitehurst, B.M.; Whitehurst, G.B. Chelates in Agriculture. *Fertil. Res.* **1990**, *25*, 127–131. [CrossRef]
53. Marschner, P. *Mineral Nutrition of Higher Plants*, 3rd ed.; Academic Press: Croydon, Australia, 2011.
54. Álvarez-Fernández, A.; García-Marco, S.; Lucena, J.J. Evaluation of Synthetic Iron(III)-Chelates (EDDHA/Fe3+, EDDHMA/Fe3+ and the Novel EDDHSA/Fe3+) to Correct Iron Chlorosis. *Eur. J. Agron.* **2005**, *22*, 119–130. [CrossRef]
55. Shalaby, O.A.E.-S.; Konopinski, M.; Ramadan, M.E.-S. Effect of Chelated Iron and Silicon on the Yield and Quality of Tomato Plants Grown under Semi-Arid Conditions. *Acta Sci. Pol. Hortorum Cultus* **2017**, *16*, 29–40. [CrossRef]

56. Rocha, D.; Santhoshkumar, A.V. Host Plant Influence on Haustorial Growth and Development of Indian Sandalwood (*Santalum Album*). *Indian Sandalwood Compend.* **2022**, 229–244.
57. Baribault, T. *Sandalwood Regeneration and Volume Inventory—Hāloa 'Āina*; Forest Solution Inc.: Kamuela, HI, USA, 2014; Volume I.
58. Friday, J.B.; Keir, M.J.; McMillen, H.L.; Rubenstein, T. Reforestation in the Hawaiian Islands. *Tree Plant. Notes* **2022**, *65*, 34–55.
59. Medeiros, A.C.; Allmen, E.I.V.o.n.; Chimera, C.G. Dry Forest Restoration and Unassisted Native Tree Seedling Recruitment at Auwahi, Maui. *Pac. Sci.* **2014**, *68*, 33–45. [CrossRef]
60. Ammondt, S.A.; Litton, C.M.; Ellsworth, L.M.; Leary, J.K. Restoration of Native Plant Communities in a Hawaiian Dry Lowland Ecosystem Dominated by the Invasive Grass *Megathyrsus Maximus*. *Appl. Veg. Sci.* **2013**, *16*, 29–39. [CrossRef]
61. Soosairaj, S.; Britto, S.J.; Balaguru, B.; Natarajan, D.; Nagamurugan, N. Habitat Similarity and Species Distribution Analysis in Tropical Forests of Eastern Ghats, Tamilnadu. *Trop. Ecol.* **2005**, *46*, 183–191.
62. Mendoza-tafolla, R.O.; Juarez-lopez, P.; Sandoval, M.; Alejo-santiago, G. Estimating Nitrogen and Chlorophyll Status of Romaine Lettuce Using SPAD and at LEAF Readings. *Not. Bot. Horti Agrobot. Cluj-Napoca* **2019**, *47*, 751–756. [CrossRef]
63. Yuan, Z.; Cao, Q.; Zhang, K.; Ata-Ul-Karim, S.T.; Tan, Y.; Zhu, Y.; Cao, W.; Liu, X. Optimal Leaf Positions for SPAD Meter Measurement in Rice. *Front. Plant Sci.* **2016**, *7*, 1–10. [CrossRef] [PubMed]
64. Jones, J.B.; Case, V.W. Sampling, Handling, and Analyzing Plant Tissue Samples. In *Soil Testing and Plant Analysis*; Soil Science Society of America: Madison, WI, USA, 1990; pp. 389–427.
65. Haase, D.L. Morphological and Physiological Evaluations of Seedling Quality. In *National Proceedings: Forest and Conservation Nursery Associations*; U.S. Department of Agriculture Forest Service: Fort Collins, CO, USA, 2007; pp. 3–8.
66. Jacobs, D.F.; Davis, A.S.; Kasten Dumroese, R.; Burney, O.T. Nursery Cultural Techniques Facilitate Restoration of *Acacia Koa* Competing with Invasive Grass in a Dry Tropical Forest. *Forests* **2020**, *11*, 1124. [CrossRef]
67. Welch, R.M. Micronutrient Nutrition of Plants. *Crit. Rev. Plant Sci.* **1995**, *14*, 49–82. [CrossRef]
68. Bauer, G.; Schulze, E.D.; Mund, M. Nutrient Contents and Concentrations in Relation to Growth of *Picea Abies* and *Fagus Sylvatica* along a European Transect. *Tree Physiol.* **1997**, *17*, 777–786. [CrossRef]
69. Pallardy, S.G. *Physiology of Woody Plants*, 3rd ed.; Academic Press: Cambridge, MA, USA, 2008.
70. Samira, R.; Stallmann, A.; Massenburg, L.N.; Long, T.A. Ironing out the Issues: Integrated Approaches to Understanding Iron Homeostasis in Plants. *Plant Sci.* **2013**, *210*, 250–259. [CrossRef]
71. Lindsay, W.L.; Schwab, A.P. The Chemistry of Iron in Soils and Its Availability to Plants. *J. Plant Nutr.* **1982**, *5*, 821–840. [CrossRef]
72. Colombo, C.; Palumbo, G.; He, J.Z.; Pinton, R.; Cesco, S. Review on Iron Availability in Soil: Interaction of Fe Minerals, Plants, and Microbes. *J. Soils Sediments* **2014**, *14*, 538–548. [CrossRef]
73. Abadía, J.; Vázquez, S.; Rellán-Álvarez, R.; El-Jendoubi, H.; Abadía, A.; Álvarez-Fernández, A.; López-Millán, A.F. Towards a Knowledge-Based Correction of Iron Chlorosis. *Plant Physiol. Biochem.* **2011**, *49*, 471–482. [CrossRef]
74. Ilhami Köksal, A.; Dumanoğlu, H.; Güneş, N.T.; Aktaş, M. The Effects of Different Amino Acid Chelate Foliar Fertilizers on Yield, Fruit Quality, Shoot Growth and Fe, Zn, Cu, Mn Content of Leaves in Williams Pear Cultivar (*Pyrus Communis* L.). *Turk. J. Agric. For.* **1999**, *23*, 651–658.
75. Pestana, M.; Varennes, A.D.e.; Faria, E.A. Diagnosis and Correction of Iron Chlorosis in Fruit Trees: A. Review. *Food Agric. Environ.* **2003**, *1*, 46–51.
76. Perry, D.A.; Rose, S.L.; Pilz, D.; Schoenberger, M.M. Reduction of Natural Ferric Iron Chelators in Disturbed Forest Soils. *Soil Sci. Soc. Am. J.* **1984**, *48*, 379–382. [CrossRef]
77. Cress, W.A.; Johnson, G.V.; Barton, L.L. The Role of Endomycorrhizal Fungi in Iron Uptake by Hilaria Jamesii. *J. Plant Nutr.* **1986**, *9*, 547–556. [CrossRef]
78. Bell, T.L.; Adams, M.A.; Rennenberg, H. Attack on All Fronts: Functional Relationships between Aerial and Root Parasitic Plants and Their Woody Hosts and Consequences for Ecosystems. *Tree Physiol.* **2011**, *31*, 3–15. [CrossRef]
79. Tomilov, A.; Tomilova, N.; Yoder, J.I. In Vitro Haustorium Development in Roots and Root Cultures of the Hemiparasitic Plant *Triphysaria Versicolor*. *Plant Cell Tissue Organ. Cult.* **2004**, *77*, 257–265. [CrossRef]
80. Chang, M.; Lynn, D.G. The Haustorium and the Chemistry of Host Recognition in Parasitic Angiosperms. *J. Chem. Ecol.* **1986**, *12*, 561–579. [CrossRef]
81. Li, Z.; Meng, S.; Qin, F.; Wang, S.; Liang, J.; He, X.; Lu, J. Host Root Exudates Initiate a Foraging Preference by the Root Parasite Santalum Album. *Tree Physiol.* **2022**, *43*, 301–314. [CrossRef]
82. Balasubramanian, A.; Prasath, C.N.H.; Radhakrishnan, S.; Sivaprakash, M. Host-Specific Influence on Early Growth and Physiological Attributes of Sandal (*Santalum Album*) Grown in Farmlands. *J. Environ. Biol.* **2014**, *35*, 1013–1020. [CrossRef]
83. Lu, J.K.; Xu, D.P.; Kang, L.H.; He, X.H. Host-Species-Dependent Physiological Characteristics of Hemiparasite *Santalum Album* in Association with N2-Fixing and Non-N2-Fixing Hosts Native to Southern China. *Tree Physiol.* **2014**, *34*, 1006–1017. [CrossRef] [PubMed]
84. Inoue, T.; Yamauchi, Y.; Eltayeb, A.H.; Samejima, H.; Babiker, A.G.T.; Sugimoto, Y. Gas Exchange of Root Hemi-Parasite *Striga Hermonthica* and Its Host *Sorghum Bicolor* under Short-Term Soil Water Stress. *Biol. Plant.* **2013**, *57*, 773–777. [CrossRef]

forests MDPI

Article

Simulation of Episodic Winter Warming on Dehardening of Boreal Forest Seedlings in Northern Forest Nurseries

Mohammed S. Lamhamedi *, Marie-Claude Lambert and Mario Renaud

Direction de la Recherche Forestière, Ministère des Ressources Naturelles et des Forêts 2700, Rue Einstein, Québec, QC G1P 3W8, Canada
* Correspondence: mohammed.lamhamedi@mffp.gouv.qc.ca

Abstract: In recent decades, forest nurseries in eastern Canada have been faced with periods of mild winter weather, delayed snowfall, and low seedling protective snow cover combined with winter rains instead of snowfall. These extreme conditions have resulted in the loss of millions of seedlings, in particular those that overwinter outdoors, probably due to their winter dehardening. The main objective of this study is to simulate different periods of warm weather at the beginning and end of winter and evaluate their effects on the dehardening and growth of *Picea mariana* and *Picea glauca* seedlings in response to different freezing temperatures. Three warming treatments were simulated (control, 1 day, and 3 days of warming at 10 °C) followed by three freezing temperatures (−4, −12, and −20 °C). In winter, regardless of the warming treatment, the seedlings of the two species tolerated the different freezing temperatures without any apparent damage. However, at the end of winter and in the absence of snow cover, the seedlings did not show frost tolerance at −20 °C. On the other hand, the seedlings showed normal growth after undergoing frosts at −4 °C and −12 °C, similar to that observed for control seedlings. Different cultural practices and protection strategies are proposed to improve frost tolerance and reduce the winter loss of seedlings.

Keywords: *Picea mariana*; *Picea glauca*; forest nursery; winter warming; winter freezing; cold hardiness; growth; mineral nutrition

Citation: Lamhamedi, M.S.; Lambert, M.-C.; Renaud, M. Simulation of Episodic Winter Warming on Dehardening of Boreal Forest Seedlings in Northern Forest Nurseries. *Forests* **2022**, *13*, 1975. https://doi.org/10.3390/f13121975

Academic Editor: Antonio Montagnoli

Received: 26 October 2022
Accepted: 16 November 2022
Published: 23 November 2022

1. Introduction

Frost damage is one of the main causes of tree seedling losses in northern forest nurseries [1,2]. Due to the exposure of seedlings to harsh winter conditions in a boreal climate, winter desiccation, root frost, as well as intense early fall and late spring frosts are responsible for a considerable loss of seedlings in forest nurseries. Depending on the intensity of the frost, the estimated annual seedlings losses in each of Québec's forest nurseries vary from 5 to 30% [3]. In Ontario, Canada, evaluations conducted in the same forest nursery over several consecutive growing seasons show that the loss of seedlings caused by frost is very important (16 to 48%) [2].

During recent decades, many forest nursery managers in eastern Canada have been confronted with exceptional inter- and intra-annual variability in environmental variables characterized, for example, by daily warm temperatures above normal in both autumn and winter and late snowfall at the beginning of the winter. In addition, winter rainfall instead of snow combined with drastic changes in temperatures from warm to freezing conditions over two consecutive days and the sometimes absence of a sufficient protective snow cover are predisposing factors for the increase of seedling losses caused by winter frost in forest nurseries. For example, a severe winter frost across the province of Québec, Canada, in 2007 and 2013 caused the loss of more than 7 million and 3.9 million seedlings, respectively [4]. In some nurseries, the percentage of seedlings damaged by winter frost has exceeded 60%. Moreover, the results of a snowmaking project conducted in a forest nursery showed that the absence of a protective snow cover at the beginning of winter alone increases the percentage of seedling mortality depending on the species and varies between 5 and 23% [4]. Recently, the diagnosis of the state of the seedlings produced in the 18 forest nurseries in the province of Québec (Canada) in 2021 revealed that 16 nurseries were affected by root frost, 17 by winter frost, 14 by spring frost, and 9 by fall frost [5]. The winter of 2021, with its climatic extremes, was the most devastating for some northern nurseries in Québec and caused, for example, the loss of nearly 11 million seedlings in a single forest nursery [6].

These losses are strongly related to different cultural practices in forest nurseries [4,7,8], such as wintering seedlings under natural conditions without resorting to the use of cold storage (−2 °C to −4 °C) to protect the seedlings against winter frost. For example, the majority of the 18 forest nurseries (6 government and 12 private) in Québec produced their seedlings in unheated tunnels during the first growing season (1 + 0). The plastic roof of tunnels was removed just before the first snow at the end of the first growing season, around mid or late October, so that the seedlings become buried under the snow during their first winter as well as their second winter before their delivery to the reforestation site [9].

However, with climate change, the frequency of these extreme events in winter will only increase in the years to come. These extreme events involve an increase in the number of days when the temperature is high and relatively long winter thaw periods. Future climate scenarios predict that temperatures will significantly increase at high latitudes during the winter compared with other seasons [10]. This could result in periods of more frequent increase in temperature and rain during winter, followed by a period of rapid freezing, which could negatively affect the survival of plant tissues (roots, stems, buds, needles, etc.), as well as the morphophysiological quality of seedlings produced in forest nurseries. Several forest nursery managers in eastern Canada (Québec) argue that a significant erratic increase in temperature during the winter for a few successive days can directly accelerate the degree of dehardening and decrease frost tolerance (also referred to as frost hardiness or cold hardiness), thereby rendering seedling tissue more vulnerable to freeze injury by subsequent frost events. Under these conditions, the ability of boreal seedlings to survive winter frosts depends not only on the levels of frost tolerance that they can reach at the end of the fall but also on their ability to remain frost-hardy during winter warm spells. However, there is evidence from forest nurseries in the southern United States that seedlings can deharden during winter in response to increasing temperature for as few as three to seven nights [11,12]. Other studies have shown that the frost hardiness of different species can be reduced or lost in a few hours to days in response to the increase in temperature [13–15]. Under such environmental conditions, if seedlings lose their hardening completely, their ability to subsequently reharden can be reduced or may be lost [15,16]. An abrupt and substantial dehardening of 3 to 14 °C has been observed in the current-year foliar of red spruce trees (*Picea rubens* Sarg.) in response to a natural thaw (3 days, 5 to 10 °C) during the winter of 1995 in northeastern North America [17].

Up until now, researchers have intensively emphasized various aspects of the hardening and dehardening of boreal tree seedlings during fall and spring [2,7,15,18]. However, despite a few studies on the effects of artificial warming in winter and spring on frost tolerance of boreal tree seedlings [19–21], the impacts of successive extreme climatic events in winter (e.g., especially a significant erratic increase in temperature followed by frost events) on the dehardening and morphophysiological quality of the whole roots and shoots of the seedlings as well as mineral nutrient concentrations, remain less investigated under forest nursery conditions. Some nursery managers believe that exposing seedlings to mild periods of winter in eastern Canada could reduce foliar nitrogen concentration. This drop in leaf nitrogen concentration could be due to a resumption of physiological activity during the winter thaw at temperatures varying between 3 and 4 °C [22]. Leaf nitrogen concentration is one of the 27 criteria for the payment of nursery managers and qualifying seedlings before they are planted in reforestation sites in Québec, Canada. For example, large seedlings (height $\geq$ 35 cm) must have at least a leaf nitrogen concentration of 1.6% [23].

Black spruce (*Picea mariana* [Mill.] B.S.P.) and white spruce (*Picea glauca* (Moench) Voss.) are the most widespread and used tree species in reforestation programs in North America [24]. The objectives of this study are (i) to assess the frost tolerance of one-year-old black spruce and two-year-old white spruce seedlings at the end of the fall; (ii) to simulate different periods of warm weather at the beginning and end of winter and evaluate their effects on the dehardening and growth of black spruce (1 + 0) and white spruce (2 + 0) seedlings in response to different freezing temperatures; (iii) to compare the nutritional status of black and white spruce seedlings in the fall with those that experienced mild temperatures at the beginning and end of winter; and (iv) to discuss the operational scope of these results in relation to the management of winter protection for seedlings in northern forest nurseries. Thus, to get closer to operational conditions in nurseries and reforestation activities, we used whole seedling freezing tests to quantify different freeze injuries in the entire seedling and evaluate their recovery under optimal growth conditions.

2. Materials and Methods

2.1. Plant Material

In Québec, seedlings are produced in unheated tunnels during the first growing season (May to October). Subsequently, they are transferred outside and spend the winter under the snow (January to April). Depending on the target seedling size and the volume of each cavity/container, the seedlings will be planted in a reforestation site after one year (1 + 0) in the nursery (volume of each cavity/container $\leq$ 110 cm^3) or kept in the nursery (volume of each cavity/container $\geq$ 200 cm^3) for a second growing season (2 + 0) under natural environmental conditions.

Containerized one-year-old black spruce (1 + 0) and two-year-old white spruce (2 + 0) seedlings were grown in two types of containers (67–50 and 25–310, model *IPL*, IPL inc., Saint-Damien, QC, Canada; see Table 1) at the Grandes-Piles governmental forest nursery (latitude: 46°43′54″ N; longitude: 72°42′06″ W, Québec, Canada). The seeds were sown into containers filled with a moist peat/vermiculite growing medium (3/1, v/v; bulk density of 0.084 g/cm^3). Substrate uniformity was closely monitored during potting and seeding. Once an hour (after filling 750–800 containers), a container was removed from the production line for verification of substrate density. Two weeks after germination, the seedlings were thinned to one per cavity. Seedlings were irrigated as needed to maintain the target substrate water content (45%, v/v) with a motorized robot (Aquaboom Harnois model, Québec, Canada) equipped with 22 nozzles and mounted on a ground rail. Water was applied at a pressure of 2.1 bars, and each pass of the robot increased the water content of the substrate by 0.9% (v/v). The coefficient of uniformity of this irrigation system varied between 95 and 98%. The quantities of N, phosphorous (P), and potassium (K) applied per seedling of white spruce (2 + 0) during the growing season were 261 mg, 21 mg, and 61 mg, respectively. On the other hand, the black spruce (1 + 0) seedlings received 16 mg, 5.5 mg, and 9 mg for N, P, and K, respectively. During each fertilization session, seedlings also received micronutrient elements.

In Québec, a short-day treatment is generally applied in forest nurseries towards the end of the growing season (mid-August) to improve hardening processes and frost tolerance. A black polyethylene cover, positioned approximately 40 cm above the shoot tips, was manually installed and removed each day over the seedlings to create a dark period. The short-day treatment was applied for 15 days and consisted of modifying the photoperiod of light/dark to 8 h/16 h. Other details concerning the cultural practices and production of seedlings are described in Lamhamedi et al. [25,26].

To protect the roots of the seedlings against frost at the end of autumn, the Grandes-Piles forest nursery uses a snowmaker system (model Supercrystal, Turbocristal inc., Québec, QC, Canada) to make snow (>45 m^3 of snow per hour) during the day and night when the air temperature is below −8 °C. The snowmaker system is distinguished by a programmable oscillation from 0° to 300°. It is manually adjustable vertically from 0° to 50° and has an application radius of 75 m when environmental conditions are ideal (no wind, etc.). A nursery manager begins by covering the roots with a depth of 5 cm of snow. Then, they ensure total protection of the apical buds of seedlings with at least 5 cm of snow [27,28].

Table 1. Initial morphophysiological characteristics of one-year-old black spruce and two-year-old white spruce seedlings (mean ± SE).

	Black Spruce (1 + 0)	White Spruce (2 + 0)
Container type	Model IPL 67–50	Model IPL 25–310
Number of cavities/container	67	25
Volume/cavity (cm^3)	50	310
Seedling size		
Height (cm)	15.35 ± 0.27	40.25 ± 0.69
Root-collar diameter (mm)	1.65 ± 0.03	5.65 ± 0.12
Dry mass (mg/seedling)		
Shoot	457.83 ± 13.47	8681.81 ± 359.65
Roots	253.92 ± 8.80	1986.79 ± 89.80
Total	711.75 ± 20.72	10,668.60 ± 440.57
Ratio of dry mass to fresh mass (DM/FM, %)	43.04 ± 0.49	46.63 ± 0.49
Shoot nutrient concentrations		
N (%)	1.99 ± 0.05	2.06 ± 0.07
P (%)	0.29 ± 0.01	0.20 ± 0.01
K (%)	0.84 ± 0.02	0.59 ± 0.02
Ca (%)	0.19 ± 0.01	0.23 ± 0.01
Mg (%)	0.14 ± 0.00	0.14 ± 0.00

Growth variables (height and root-collar diameter) were measured at the end of the growing season on subsamples of white and black spruce seedlings (n = 48 seedlings per species, Table 1). These seedlings were randomly sampled from lots of seedlings in forest nurseries (500,000 to 1 million seedlings). Mineral nutrient analyses of seedlings (shoots) were determined on four composite samples (12 seedlings per composite sample per species) at the Laboratoire de chimie organique et inorganique (organic and inorganic chemistry laboratory) of the Direction de la recherche forestière (Québec Forest Research Branch, Québec, QC, Canada), using the methods described in detail in our previous studies [28–30].

The ratio of dry mass (DM) to fresh mass (FM) [(dry mass/fresh mass: DM/FM) × 100] of excised terminal shoot tips (4 cm long) was evaluated at the end of autumn (November 30) after the complete formation of buds and when the chilling sum in eastern Canada (e.g., the number of hours when temperatures < 5 °C) reached at least 200 h [3,31]. For each forest species, this ratio was determined using 4 composite samples with 12 terminal shoot tips per composite sample (Table 1). After weighing the fresh mass of the composite samples, the dry mass was determined after drying for 48 h in an oven at 60 °C.

2.2. *Frost Tolerance of Black and White Spruce Seedlings at the End of the Fall*

In order to determine the degree of hardening and frost tolerance of black (1 + 0) and white (2 + 0) spruce seedlings at the end of the fall, we evaluated various variables, including the relative electrical conductivity (RC) and an Index of Injury (It) of shoots. In addition, to verify the frost tolerance of whole seedlings (roots and shoot) at each of the freezing temperatures tested, we used a bioassay test to assess the survival and growth of the shoots and roots, as well as the bud burst of the main stem and the branches of seedlings.

2.2.1. Electrical Conductivity and Index of Injury of Excised Apical Shoot Tips in Response to Different Artificial Freezing Temperatures

The evaluation of the level of hardening of the shoots of black (1 + 0) and white (2 + 0) spruce seedlings was conducted using the electrolyte leakage method [3,32,33]. For each species, 60 apical shoot tips (4 cm long) randomly sampled from seedlings were rinsed three times with demineralized water and placed in 125 mL Erlenmeyer flasks (3 shoot tips per flask; 20 flasks total). Then, the Erlenmeyer flasks containing the terminal shoot tips were subjected to five temperatures (T0 = 4 °C (control), T1 = −5 °C, T2 = −10 °C, T3 = −15 °C, and T4 = −20 °C). We used 4 replicates (i.e., flasks) for each temperature tested. These temperatures were simulated by a freezer (model T20RS, Tenney environmental inc., Williamsport, PA, USA) equipped with a programming controller (model Versa Tenn II, Union, NJ, USA) with a cooling rate of 2 °C/h [3,34,35].

Freezer programming sequences are described in detail in our previous studies [3,36] and other publications [37,38]. The main steps are stabilizing seedlings at a temperature of 4 °C, lowering temperatures at a rate of 2 °C/h down to 0 °C, maintaining the samples at this temperature for at least 8 h, and starting the freezing cycle at the same cooling rate. Each freezing temperature (−5 °C, −10 °C, −15 °C, and −20 °C) was maintained for one hour, after which four Erlenmeyer flasks were removed from the freezer. These samples were saturated in deionized water overnight at a temperature of 4 °C, then the electrolyte leakage (electrical conductivity measured in μS/cm) of each sample was measured (EC1, μS/cm) using a conductivity meter (model 160, Orion Research Inc., Boston, MA, USA). Subsequently, the samples were placed in an autoclave at a temperature of 121 °C for 15 min to destroy all cells and promote the maximum release of ions. After a full night at 4 °C, the leakage of electrolytes was measured again (EC2, μS/cm).

The relative electrical conductivity (RC, %) for each temperature was calculated using the following formula:

$$\text{RC (\%)} = \frac{(\text{EC1} - B1)}{(\text{EC2} - B2)} \times 100 \tag{1}$$

where $B1$ and $B2$ are optional blanks measured before and after oven-heating to account for possible ion leakage from the control Erlenmeyer flasks containing only deionized water [35].

For a given temperature (t), this index of injury (It) was calculated on a percentage basis by adjusting the RC of frozen samples to the unfrozen controls [39]:

$$\text{I}_\text{t} = \frac{\text{RC}_{\text{frozen(t)}} - \text{RC}_{\text{control}}}{1 - \frac{\text{RC}_{\text{control}}}{100}} \tag{2}$$

2.2.2. Growth of Seedlings, Bud Burst, and Initiation of New Roots in the Case of Whole-Seedling Freeze Treatments

For this experiment, whole seedlings were used to assess the effects of different artificial freezing temperatures (control: 4 °C, −5 °C, −10 °C, −15 °C, and −20 °C) on new root initiation, shoot growth, and bud burst. Unlike other studies that evaluate frost tolerance using the LT_{50}, which corresponds to the temperature at which 50% of the seedlings are considered dead [40], in this study, frost tolerance is defined as the physiological state allowing the seedling to tolerate temperatures below the freezing point without any apparent damage to the various organs of each seedling (buds, needles, branches, main stem, and roots) [3].

For each species and each temperature, 8 whole seedlings (roots and shoot) were used (40 seedlings per species total). Before the freezing tests, seedlings were placed in plastic bags (2 seedlings per bag). Once the seedlings had undergone the various artificial freezing treatments, and after overnight acclimatization at 4 °C, they were transplanted in 2 L pots filled with a standard mixture of peat and vermiculite (*v/v*: 3/1). Because of their different morphological sizes, black spruce seedlings were potted at a rate of 2 seedlings per pot, whereas for white spruce, there was only one seedling per pot. All seedlings were then placed in a greenhouse and randomly distributed in four complete random blocks (2 plants per temperature per block).

Day and night temperatures in the greenhouse were maintained at 25 °C and 18 °C (±2 °C), respectively. The photoperiod was 18 h using supplemental light provided by 400 W high pressure sodium lamps (Lumiponic Inc., Montreal, Québec, QC, Canada). Pots were irrigated during the bioassay to maintain an optimum water content of the substrate (~45%, *v/v*) for seedling growth [27,41]. No fertilizer was applied during the experiment. After 21 days of growth in the greenhouse, the seedlings were removed from pots and the roots were washed. New white roots were separated from the initial roots of each plug. Finally, dry masses of new white roots, total shoot dry mass, and total root dry mass were determined after drying for 48 h in an oven at 60 °C.

2.3. Simulation of Different Periods of Warm Weather and Their Effects on the Dehardening, Growth, and Mineral Nutrition of Black Spruce (1 + 0) and White Spruce (2 + 0) Seedlings at the Beginning and End of Winter

During wintering of one-year-old black spruce (1 + 0) and two-year-old (2 + 0) white spruce seedlings under the snow in eastern Canada, Québec, seedlings of the two species were sampled at the beginning of winter (mid-January 2010) and towards the end of winter (mid-March 2010) when seedlings were not completely covered by snow. Daily maximum, mean, and minimum temperatures, and snow depth accumulation at the Grandes-Piles governmental forest nursery were estimated by linear interpolation using weather data from 8 meteorological stations near the forest nursery and the BioSIM software (version 11.8.6.2) [42] to determine the evolution of environmental variables (maximum, mean, and minimum air temperature, and snow depth accumulation) in winter during the two successive years (2009 and 2010) of this research project and the climatic normals (1991–2020).

These two successive climatic years (2009 and 2010) differed from each other in terms of snow depth accumulation and maximum, average and minimum air temperatures, as well as with the climatic normals (1991–2020) at the Grandes-Piles forest nursery (Figures 1 and 2). In fact, during this experiment, around mid-March 2010 and unlike the roots, most of the shoots of the white spruce and black spruce seedlings were not completely covered by snow (Figure 1b). This was an exceptional year as the snowmelt was early due to certain episodic rainfall.

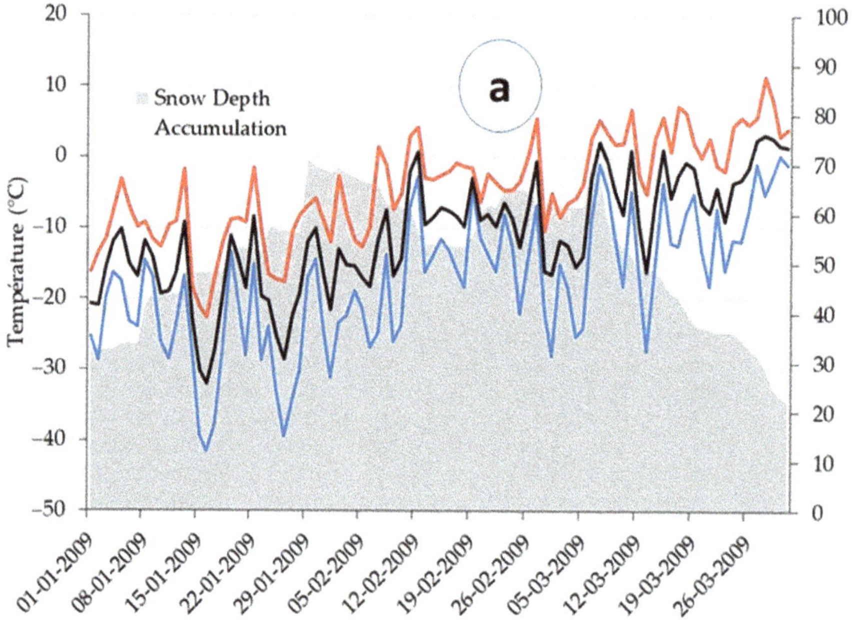

Figure 1. *Cont.*

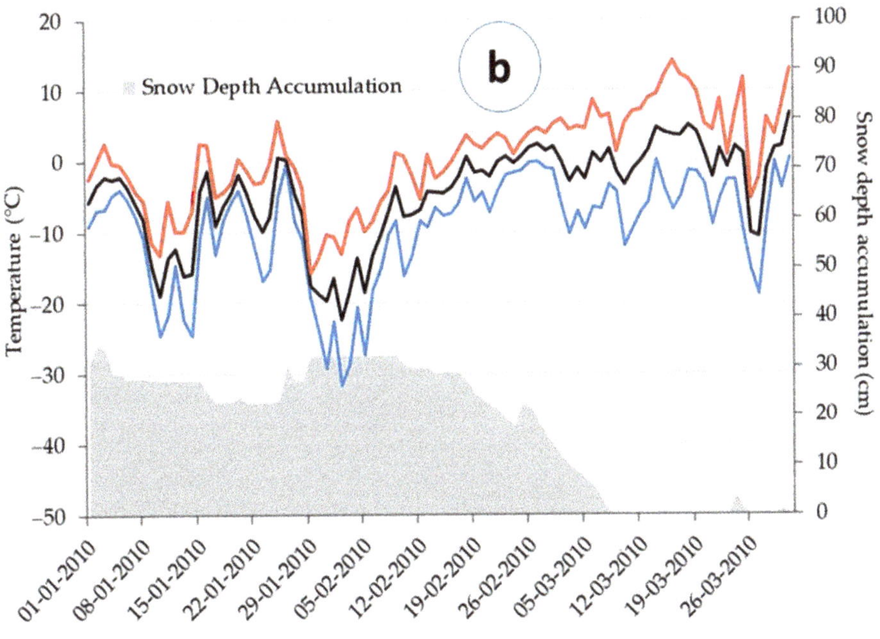

Figure 1. Variation of maximum (red), mean (black), and minimum (blue) daily air temperatures (January-March) at the Grandes-Piles forest nursery (Québec, QC, Canada) in (**a**) 2009 and (**b**) 2010. The areas in gray indicate snow depth accumulation.

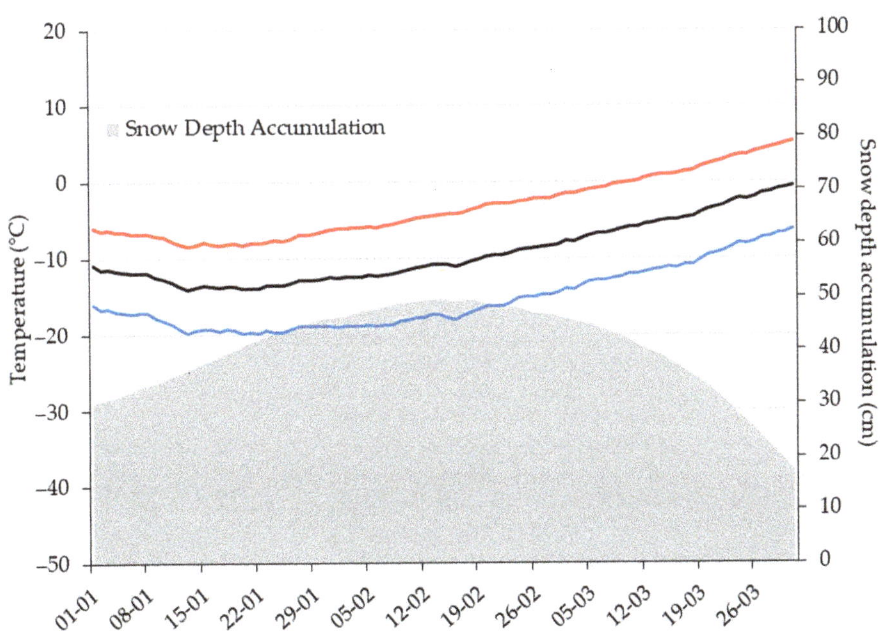

Figure 2. Variation of maximum (red), mean (black), and minimum (blue) daily air temperatures (January-March) for climatic normals (1991–2020) at the Grandes-Piles forest nursery (Québec, QC, Canada). The areas in gray indicate snow depth accumulation.

For each sampling period (mid-January and mid-March) and after having thawed the root plugs at 4 °C in the dark, the seedlings were watered by capillarity to standardize the water content of the peat substrate (55% to 60%, v/v) and the frost kinetics in the root plugs among seedlings of the same species [36,41]. Then, seedlings of the two forest species were subjected to three treatments at the beginning (mid-January) and end of winter (mid-March):

- Treatment 1 (TR1): Control seedlings were kept in a cold room at 4 °C in the dark (thus simulating light conditions under a relatively deep snow cover).
- Treatment 2 (TR2): Seedlings were placed for 1 day in a greenhouse at 10 °C under winter natural photoperiod conditions.
- Treatment 3 (TR3): Seedlings were placed for 3 days in a greenhouse at 10 °C under winter natural photoperiod conditions.

In eastern Canada (Québec), these warming treatments are considered extreme because the seedlings are without any snow cover, and they are subjected to high day/night temperatures (10 °C) for 1 and 3 days. The succession of these extreme environmental conditions applied at the beginning and the end of winter exceeded the normal climatic conditions (Figure 2) and the projected warming scenario (3 °C to 7 °C) for the province of Québec [43].

For each sampling period (mid-January and mid-March), the seedlings of each species from these three treatments were subjected to different freezing temperatures (control 4 °C, −4 °C, −12° C, and −20 °C) using the same freezing routine described in Section 2.2 and in other publications [36–38]. Although rare and extreme, −20 °C was included as a treatment because this temperature represents the minimum threshold to which seedlings could be subjected in the total absence of snow cover (see climatic normals in Figure 2).

To validate the temperature program for each sampling period and to follow the evolution of the temperatures reached both in the aerial parts and in the root plugs of black spruce (50 cm^3) and white spruce (310 cm^3), five temperature probes (Model 107B, Campbell Scientific (Edmonton, AB, Canada) Corp., Edmonton, AB, Canada) were used. For each species, 2 probes were inserted in the center of the root plugs. The remaining probe was placed inside the freezer to monitor the air temperature inside the freezer to which seedlings shoots were subjected during the freezing cycle. A datalogger (model CR10X, Campbell Scientific, Canada Corp., Edmonton, AB, Canada) recorded temperatures every three minutes.

For each forest species and each sampling period (mid-January and mid-March), the growth variables and those associated with frost tolerance were evaluated using the same approach described in Section 2.2. The number of plants used, and variables determined were as follows:

- The ratio of dry mass (DM) to fresh mass (FM) of excised terminal shoot tips (4 cm long) was evaluated using 12 seedlings (3 shoot tips per composite sample, 4 composite samples per treatment). This ratio is used to predict frost tolerance in autumn [3,7].
- Electrical conductivity and an index of injury of excised terminal shoot tips of black spruce and white spruce seedlings in response to different artificial freezing temperatures (28 seedlings total; 3 shoot tips per Erlenmeyer flask per temperature per block distributed in four complete random blocks).
- Artificial freezing of whole seedlings for the evaluation of growth variables (dry shoot dry mass, initial roots and new roots, total dry mass), bud burst and recovery (32 seedlings total per treatment per species at the rate of 2 seedlings per temperature per block distributed in four complete random blocks).

2.4. Mineral Nutrition

The mineral nutrition of white and black spruce seedlings was assessed in the fall, early, and late winter to determine if the nutritional status of the seedlings varied with the duration of the warm temperatures (1 vs. 3 days). Thus, in addition to the characterization of the mineral nutrition in autumn, we also evaluated the mineral nutrition of seedlings that underwent the same treatments (TR1, TR2, and TR3) at the beginning and end of winter. The mineral nutrient concentrations and contents (N, P, K, Ca, and Mg) of the shoots, roots, and whole seedlings of black and white spruce for each treatment were determined using 4 composite samples (10 seedlings per composite sample per block). The mineral content was calculated for each nutrient (concentration multiplied by dry mass) [44]. Mineral nutrient analyses of seedlings were determined at the organic and inorganic chemistry laboratory of the Québec Forest Research Branch using the methods described in our previous studies [28–30].

2.5. Statistical Analyses

An analysis of variance was performed using mixed linear models with the MIXED procedure of SAS/STAT version 14.1 (SAS Institute Inc. Cary, NC, USA, 2015). In all models, the number of degrees of freedom of the denominator for fixed effects tests was calculated using the Kenward-Roger method.

During the various artificial freezing treatments, the seedlings were kept inside the freezer, and they were taken out for one hour after the target temperature was reached. This implied that no randomization was performed when applying the different freezing treatments to assess the different cold tolerance variables. Thus, the statistical model used was a strip-plot, also known as a split-block [45,46]. The model's random part was simplified as described in Bernier-Cardou and Bigras [45] at the 30% threshold, and the significance thresholds for the fixed effects were set at 5%.

To assess the effects of forest species and different artificial freezing temperatures (control 4 °C, −5 °C, −10 °C, −15 °C and −20 °C) on relative electrical conductivity, index of injury, and different growth variables at the end of the fall, forest species, and freezing temperatures, and their interactions were considered fixed effects, while the block and the interaction between the block and factors were considered as random effects.

For frost tolerance in response to different periods of warm weather at the beginning and end of winter, sampling period, treatments at the beginning and end of winter, freezing temperatures (control 4 °C, −4 °C, −12° C and −20 °C), and their interaction were considered as fixed effects, while the block and the interaction between the block and the factors were considered as random effects.

A model with a fixed effects sampling period, treatments at the beginning and end of winter, and their interaction and a random block in the sampling period effect was used to analyze the ratio of dry mass to fresh mass.

To assess the effects of the sampling period and treatments at the beginning and end of winter on nutrient concentrations and the contents of different parts of spruces before freezing, a means model was used to account for the control in autumn measures and analyze the two-way structures with missing treatment combinations. A factor combining the sampling period and treatments was created, and tests and comparisons were made using the contrasts.

For each of the analyses performed, the assumption of normality and homogeneity of variances were graphically checked. To account for the heterogeneity of the variances, the residual variance was weighted according to the variance observed for each forest species for the analysis at the end of the fall. For the other analysis, the models were adjusted to the dataset for the different forest species, since our objective was not to compare the state of the hardening of the seedlings between the two species.

Subsequently, multiple comparisons of the means were performed by forest species when a fixed effect was significant. Multiple comparison thresholds are adjusted with a simulation method available in SAS (SAS Institute Inc. Cary, NC, USA, 2015) to detect significant differences between means, as described by Westfall et al. [47].

3. Results

3.1. Assessment of Frost Tolerance of One-Year-Old Black Spruce (1 + 0) and Two-Year-Old White Spruce (2 + 0) Seedlings at the End of the Fall

At the end of the fall (November 30), no significant difference was observed between the mean relative electrical conductivities and the index of injury of the shoots of the two forest species) in response to the different freezing temperatures tested (Table 2). Thus, for example, the average values of the relative electrical conductivities of black spruce (1 + 0) at temperatures of 4 °C and −20 °C were 3.40 ± 0.20% and 3.58 ± 0.20%, respectively. In the case of white spruce, the averages for these two temperatures (4 °C and −20 °C) were the same (Table 2).

Table 2. Comparison of the adjusted means (±standard error) of the relative electrical conductivity (RC) and the index of injury (It) of white spruce (WS) and black spruce (BS) seedlings in response to different freezing temperatures at the end of autumn (30 November).

		−20 °C	−15 °C	−10 °C	−5 °C	4 °C
RC (%)	WS (2 + 0)	2.48 ± 0.20 a [1]	2.56 ± 0.20 a	2.42 ± 0.20 a	2.75 ± 0.20 a	2.48 ± 0.20 a
	BS (1 + 0)	3.58 ± 0.20 a	3.77 ± 0.20 a	4.09 ± 0.20 a	3.19 ± 0.20 a	3.40 ± 0.20 a
It	WS (2 + 0)	0.00 ± 0.12 a	0.08 ± 0.12 a	−0.06 ± 0.12 a	0.27 ± 0.12 a	
	BS (1 + 0)	0.18 ± 0.35 a	0.37 ± 0.35 a	0.71 ± 0.35 a	−0.23 ± 0.35 a	

[1] Horizontally, the adjusted means (±standard error) followed by similar letters did not show significant differences at the threshold α = 5%, according to a simulation method available in SAS as described by Westfall et al. [47].

The index of injury (It) calculated for each of the forest species in response to the different freezing temperatures tested was always less than one and did not exceed the maximum value of 0.71 for black spruce and 0.27 for white spruce. This indicates that the seedlings of the two species tolerate the different freezing temperatures tested (T1 = −5 °C, T2 = −10 °C, T3 = −15 °C and T4 = −20 °C) at the end of the fall.

3.2. Initiation of New Roots and Bud Break in the Case of Whole-Seedling Freeze Treatments at the End of the Fall

After undergoing the various artificial freezing treatments and after potting the seedlings, followed by a period of 21 days of growth in a greenhouse under optimal environmental conditions, the buds of the main stems and lateral branches of all the spruce seedlings of black spruce (1 + 0) and white spruce (2 + 0) showed bud break (Figures 3 and 4). All the buds and needles showed no sign of browning or mortality, whatever the frost temperature (−5 °C, −10 °C, −15 °C, and −20 °C) (Figures 3 and 4).

For all freezing temperatures tested, all observed seedlings of the two tree species initiated new white roots (Figures 3 and 4, Table 3). However, compared with the control temperature of 4 °C, the frost temperature of −20 °C significantly reduced the new root dry mass in white spruce and black spruce by 44% and 58%, respectively (Table 3). At the end of the bioassay test period (21 days), there was no significant difference in the dry mass of new white roots between the other frost temperatures tested (−5 °C, −10 °C, and −15 °C) and the control temperature (4 °C) (Table 3). With the exception of total and new root dry masses in black spruce, all other growth variables were not significantly affected by the artificial freezing temperatures tested (Table 3).

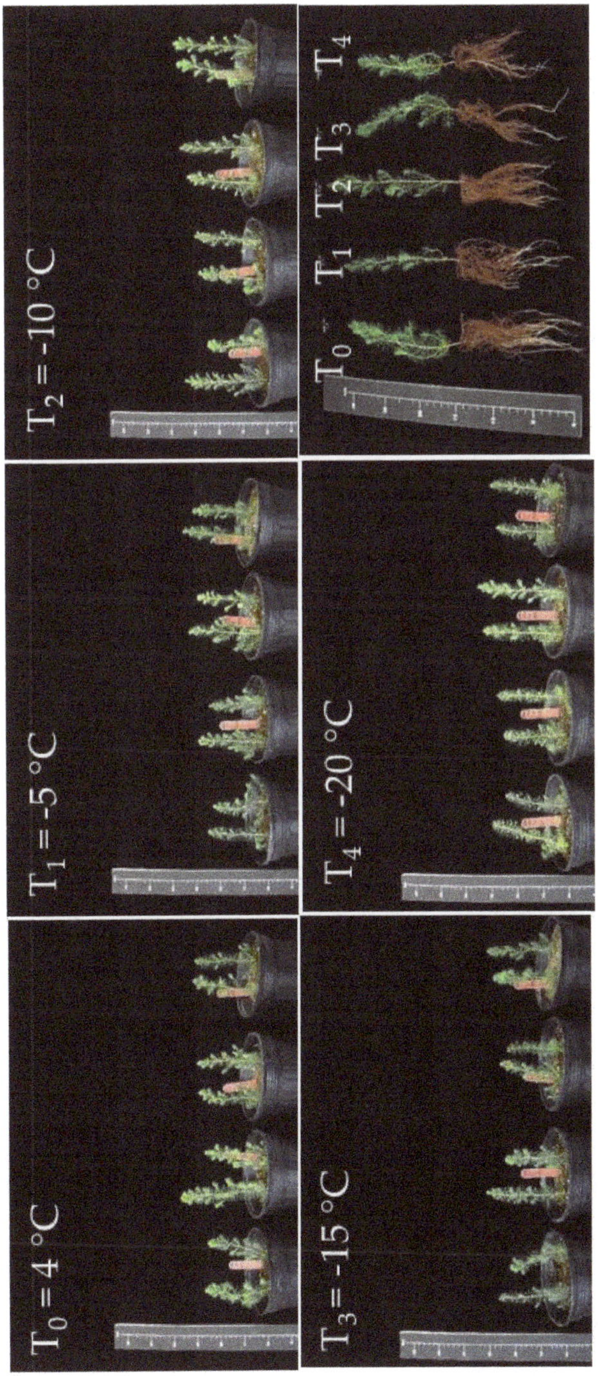

Figure 3. Bud burst of all buds of main stems and lateral branches of all black spruce (1 + 0) seedlings in response to different temperatures (T_0, T_1, T_2, T_3, and T_4) and after 21 days of growth during the recovery period under optimal greenhouse growing conditions. Example of new root initiation in response to the same temperatures and optimal growing conditions.

Figure 4. Bud burst of all buds of main stems and lateral branches of all white spruce (2 + 0) seedlings in response to different temperatures (T_0, T_1, T_2, T_3, and T_4) and after 21 days of growth during the recovery period under optimal greenhouse growing conditions. Example of new root initiation in response to the two freezing temperatures (T_3 and T_4) and optimal growing conditions.

Table 3. Comparison of the adjusted means (± standard error) of different growth variables of white spruce (2 + 0) and black spruce (1 + 0) seedlings at the end of the fall in response to different freezing temperatures and after 21 days of growth during the recovery period under optimal greenhouse growing conditions.

	Black Spruce (1 + 0)					White Spruce (2 + 0)				
	$T_0 = 4\ °C$	$T_1 = -5\ °C$	$T_2 = -10\ °C$	$T_3 = -15\ °C$	$T_4 = -20\ °C$	$T_0 = 4\ °C$	$T_1 = -5\ °C$	$T_2 = -10\ °C$	$T_3 = -15\ °C$	$T_4 = -20\ °C$
Dry mass of new white roots (mg)	37.5 ± 5.1 a [1]	38.3 ± 5.1 a	27.3 ± 5.1 ab	18.4 ± 5.1 ab	15.9 ± 5.1 b	62.4 ± 15.14 a	85.8 ± 15.14 a	79.6 ± 15.14 a	85.1 ± 15.14 a	35.1 ± 15.14 b
Initial root dry mass (mg)	251.0 ± 18.3 a	216.3 ± 18.3 a	236.8 ± 18.3 a	196.9 ± 18.3 a	186.8 ± 18.3 a	1989.5 ± 248.1 a	2330.0 ± 248.1 a	1958.3 ± 248.1 a	2596.4 ± 248.1 a	1987.9 ± 248.1 a
Total root dry mass (mg)	288.5 ± 22.8 a	254.5 ± 22.8 ab	264.0 ± 22.8 ab	215.3 ± 22.8 ab	202.6 ± 22.8 b	2051.9 ± 257.8 a	2415.8 ± 257.8 a	2037.9 ± 257.8 a	2681.5 ± 257.8 a	2023.0 ± 257.8 a
Shoot dry mass (g)	0.67 ± 0.05 a	0.60 ± 0.05 a	0.62 ± 0.05 a	0.53 ± 0.05 a	0.58 ± 0.05 a	10.16 ± 0.98 a	11.57 ± 0.98 a	9.61 ± 0.98 a	11.95 ± 0.98 a	10.16 ± 0.98 a
Total dry mass (g)	0.96 ± 0.06 a	0.86 ± 0.06 a	0.88 ± 0.06 a	0.74 ± 0.06 a	0.79 ± 0.06 a	12.22 ± 1.19 a	13.98 ± 1.19 a	11.65 ± 1.19 a	14.63 ± 1.19 a	12.18 ± 1.19 a

[1] Horizontally, the adjusted means (±standard error) for each species followed by similar letters did not show significant differences at the threshold $\alpha = 5\%$, according to a simulation method available in SAS, as described by Westfall et al. [47].

3.3. Frost Tolerance of Black Spruce (1 + 0) and White Spruce (2 + 0) Seedlings in Response to Different Periods of Warm Weather at the Beginning and End of Winter

3.3.1. Determination of Shoot Cold Tolerance of Black Spruce and White Spruce Using Electrolyte Conductivity Measurements in Response to Different Periods of Warm Weather at the Beginning and End of Winter

The analysis of variance showed that there was a significant effect from the sampling period on the relative electrical conductivity of the shoots of white spruce (p = 0.0427) and a significant interaction between the sampling period and the treatment (p = 0.0409) (Table 4). We also observed a significant effect of date and treatment for the injury index in black spruce (p = 0.0391).

Table 4. Observed probabilities (Pr > F) and degrees of freedom of the fixed effects associated with the analysis of variance of the relative electrical conductivity (RC) and index of injury (It) of shoots of two-year-old white spruce (WS, 2 + 0) and one-year-old black spruce (BS, 1 + 0).

	Relative Electrical Conductivity (%)					Index of Injury				
		WS 2 + 0		BS 1 + 0			WS 2 + 0		BS 1 + 0	
Source of Variation	dln [1]	dld	Pr > F	dld	Pr > F	dln	dld	Pr > F	dld	Pr > F
Period	1	6	0.0427	6	0.0778	1	18	0.9492	18	0.0391
Treatment	2	12	0.1866	12	0.0139	2	18	0.0557	18	0.0391
Period × Treatment	2	12	0.5206	12	0.0409	2	18	0.2919	18	0.7038
Temperature	3	54	0.1098	54	0.4875	2	36	0.2587	36	0.6295
Period × Temperature	3	54	0.5516	54	0.1045	2	36	0.3411	36	0.6515
Treatment × Temperature	6	54	0.2293	54	0.0636	4	36	0.9011	36	0.3974
Period × Treatment × Temperature	6	54	0.5235	54	0.3738	4	36	0.6630	36	0.2057

[1] dln: degrees of freedom of numerator; dld: degrees of freedom of denominator.

The relative electrical conductivity of the shoot of white spruce in mid-March was significantly higher than in mid-January (Table 5). On the other hand, in the case of black spruce, the relative electrical conductivity of the shoots of the control seedlings maintained at 4 °C in the dark was significantly lower than the treatment of exposure for 3 days at 10 °C for the mid-March sampling period (Table 5). The injury index of the shoots of black spruce in mid-March was significantly higher than that of January (Table 5).

Table 5. Comparison [1] of adjusted means of the relative electrical conductivity (RC) and Index of Injury (It) of shoots of two-year-old white spruce and one-year-old black spruce seedlings between sampling date at the beginning (mid-January) and end of winter (mid-March) and three treatments (Control at 4 °C in dark conditions, 1 day at 10 °C and 3 days at 10 °C under the natural photoperiod of the season).

RC (%)	WS 2 + 0	Mid-January	Mid-March				
		2.94 ±0.15 b [2]	3.49 ±0.15 a				
	BS 1 + 0	Mid-January			Mid-March		
		One day	3 days	Control	One day	3 days	Control
		3.24 ±0.23 ab	3.22 ±0.23 ab	3.13 ±0.23 b	3.77 ±0.23 ab	4.25 ±0.23 a	3.38 ±0.23 b
It	BS 1 + 0	One day	3 days	Control			
		0.30 ±0.18 a	0.34 ±0.18 a	−0.28 ±0.18 a			
		Mid-January	Mid-March				
		−0.11 ±0.14 b	0.35 ±0.14 a				

[1] Comparisons are made only when the probabilities are significant. For more details, see Table 4. [2] Horizontally, the adjusted means (±standard error) followed by distinct letters show significant differences at the threshold α = 5%, according to a simulation method available in SAS as described by Westfall et al. [47].

3.3.2. Determination of Whole-Seedling Freezing Temperature Effects on the Growth of New Roots, Shoot, and Bud Breaks in Response to Different Periods of Warm Weather at the Beginning and End of Winter

The monitoring of the different temperatures inside the freezer and in the root plugs of the white and black spruce seedlings is shown in Figure 5. Shoots of the two species were gradually exposed to the different target freezing temperatures (−4 °C, −12 °C, and −20 °C). In the case of root plugs, the cooling kinetics evolved differently depending on the volume of the root plug (50 cm^3 or 310 cm^3) (Figure 5) despite that the substrate water contents were high (55 to 60%, *v/v*) and similar between the two containers (50 cm^3 or 310 cm^3) before placing the seedlings in the freezer. The temperature of −20 °C in the root plugs of white spruce (2 + 0) was reached almost at the same time as in the root plugs of black spruce (50 cm^3), but the transfer kinetics of frost in the two types of root plugs were different (Figure 5).

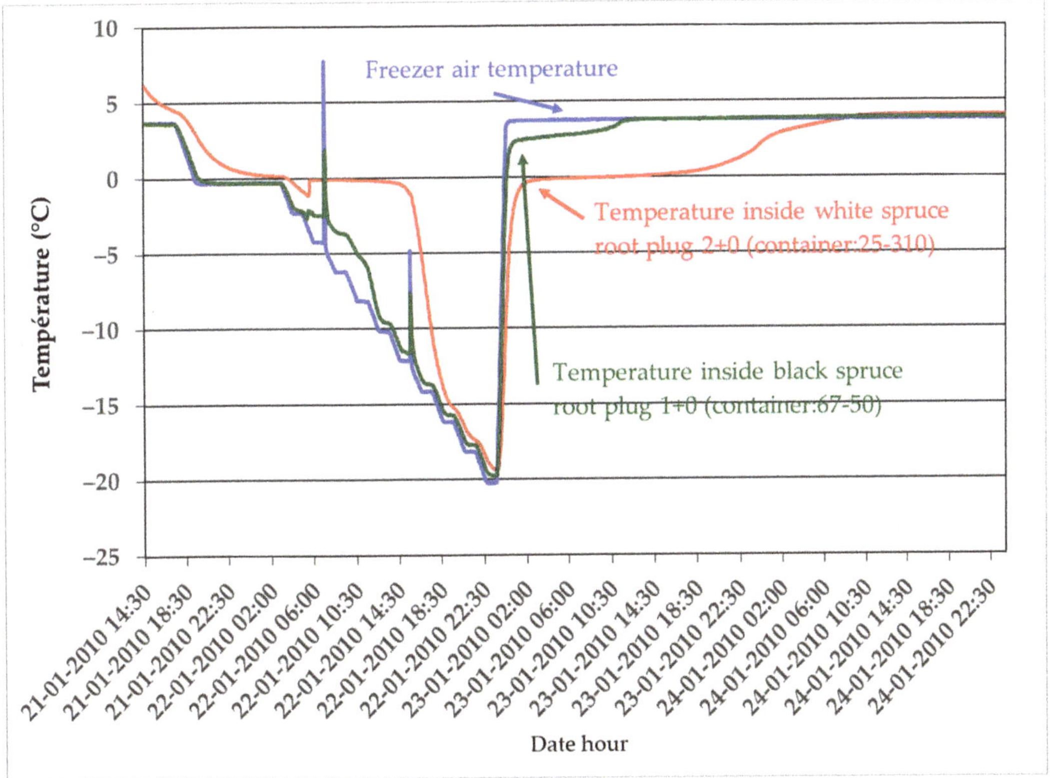

Figure 5. Temperatures registered in the air around the shoots and in the root plugs of black spruce (1 + 0) and white spruce (2 + 0) during artificial frost treatments. The seedlings were removed from the freezer once the target temperature around shoots (−4 °C, −12 °C, and −20 °C) had been reached and maintained for one hour. The peaks in blue indicate when the freezer door was opened to remove the seedlings.

The sampling period had a significant effect on the ratio of dry mass to fresh mass for the white spruce (2 + 0) ($p = 0.0270$) and black spruce (1 + 0) ($p < 0.0001$). On the other hand, the effects of the treatment in relation to the exposure of the seedlings (4 °C in darkness, 1 day, and 3 days exposed to 10 °C) and the interaction sampling period × treatment was not significant ($p > 0.15$). Thus, the ratios determined in mid-March, regardless of the forest species (white spruce: 43.71 ± 0.18; black spruce: 36.67 ± 0.16), were significantly lower than those determined in mid-January (white spruce: 44.33 ± 0.18; black spruce: 38.18 ± 0.16).

After being exposed to the different freezing temperatures and after 21 days under optimal growth conditions in the greenhouse, the analysis of variance of the growth variables showed significant effects of the sampling period × temperature interaction for shoot dry mass, dry mass of new roots for white spruce, total root dry mass for black spruce, and total dry mass of seedling, as well as simple significant effects of temperature for the dry mass of new roots for black spruce and total root dry mass for white spruce and sampling period for the dry mass of new roots for black spruce (Table 6). In the case of white spruce (2 + 0), −20 °C in mid-March significantly reduced the growth of new roots compared to the other freezing temperatures (Figure 6). The comparisons of the means for the other growth variables after the period of recovery under optimal growth greenhouse conditions following the different freezing temperatures are indicated in Table S1.

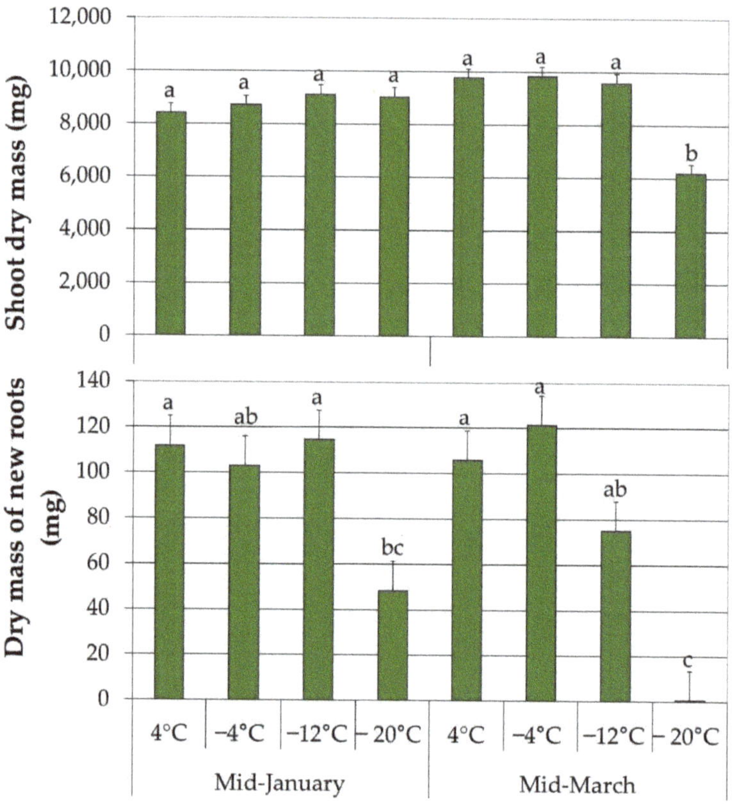

Figure 6. Comparisons of adjusted means (±standard error) of dry masses of shoot and new roots of white spruce (2 + 0) in response to whole-seedling freezing temperatures at the beginning (mid-January) and end of winter (Mid-March). The distinct letters show significant differences at the threshold α = 5%, according to a simulation method available in SAS as described by Westfall et al. [47].

Table 6. Observed probabilities (Pr > F) and degrees of freedom of the fixed effects associated with the analysis of variance of different growth variables of two-year-old white spruce (WS, 2 + 0) and one-year-old black spruce (BS, 1 + 0) seedlings at the beginning (mid-January) and end of winter (mid-March).

		Shoot Dry Mass (mg)				Dry Mass of New Roots (mg)				Total Root Dry Mass (mg)				Total Dry Mass of Seedling (mg)			
		WS 2 + 0		BS 1 + 0		WS 2 + 0		BS 1 + 0		WS 2 + 0		BS 1 + 0		WS 2 + 0		BS 1 + 0	
Source of Variation	**dln** [1]	**dld**	**Pr > F**	**dld**	**Pr > F**	**dld**	**Pr > F**	**dld**	**Pr > F**	**dld**	**Pr > F**	**dld**	**Pr > F**	**dld**	**Pr > F**	**dld**	**Pr > F**
Period	1	18	0.9184	21.4	0.4062	72	0.0401	24	0.0002	18	0.2280	6	0.0762	18	0.7931	20.1	0.8152
Treatment	2	18	0.8173	14.3	0.1117	72	0.5013	48	0.4314	18	0.4471	66	0.1280	18	0.6680	13.6	0.0993
Period × Treatment	2	18	0.6220	14.3	0.2499	72	0.9538	48	0.4529	18	0.2910	66	0.0604	18	0.8633	13.6	0.1743
Temperature	3	54	<0.0001	19.2	0.0004	72	<0.0001	24	0.0002	54	<0.0001	66	0.0005	54	<0.0001	18.6	0.0005
Period × Temperature	3	54	<0.0001	19.2	0.0012	72	0.0429	24	0.1977	54	0.0666	66	0.0006	54	<0.0001	18.6	0.0012
Treatment × Temperature	6	54	0.2458	36.9	0.5266	72	0.3370	48	0.1401	54	0.5086	66	0.8650	54	0.3127	38.2	0.6768
Period × Treatment × Temperature	6	54	0.5690	36.9	0.7818	72	0.4332	48	0.8765	54	0.5230	66	0.9975	54	0.6289	38.2	0.9153

[1] dln: degrees of freedom of numerator; dld: degrees of freedom of denominator.

At the beginning of winter, all the buds of the main stems and branches of the white and black spruce seedlings of these two treatments (1 day at 10 °C and 3 days at 10 °C in the natural photoperiod of the season) showed all the normal bud break similar to the control seedlings (Figure 7). All seedlings also showed normal growth (Figure 7). On the other hand, seedlings sampled in late winter (mid-March) showed that those subjected to −20 °C did not bud break compared with the seedlings of the other warming treatments subjected to other temperatures (4 °C, −4 °C and −12 °C) (Figure 8). A few needles and buds of some of the seedlings subjected to −20 °C died. Thus, the dry shoot masses of seedlings subjected to −20 °C in mid-March were significantly lower than the other combinations of freezing temperatures and the sampling period (Figures 6 and 8).

Figure 7. Normal bud break of the apical buds and lateral branches of white spruce seedlings (2 + 0) in response to their exposure to the three treatments (Control at 4 °C in the dark, 1 day at 10 °C and 3 days at 10 °C under the natural photoperiod of the season) and followed by artificial freezing temperatures (−4 °C, −12 °C and −20 °C) at the beginning of winter (mid-January).

3.3.3. Mineral Nutrition in Autumn, Winter and towards the End of Winter in Response to Warming Treatments

In the case of the shoots of black spruce seedlings (1 + 0), the warming treatments had a significant effect on the concentration of nitrogen (p = 0.0462), potassium (p = 0.0156), calcium (p = 0.0249) and magnesium (p = 0.0196) but no effect on phosphorus concentration (p = 0.4650; Table S2). However, the same treatments had no significant effect on the concentration of all the mineral elements in shoots of white spruce seedlings (2 + 0, p > 0.08; Table S2). Probabilities of significance for root and whole seedling mineral nutrient concentrations and contents are shown in Table S2.

Figure 8. Normal bud break of the apical buds and lateral branches of white spruce seedlings (2 + 0) at the end of winter (mid-March) in response to their exposure to the three warming treatments (control at 4 °C in the dark conditions, 1 day at 10 °C, and 3 days at 10 °C under the natural photoperiod of the season) and followed by artificial freezing temperatures (−4 °C and −12 °C). Conversely, seedlings subjected to the −20 °C freezing temperature and the same warming treatments showed no bud burst.

Additionally, there was no significant difference in shoot nitrogen concentration of the black spruce and white spruce seedlings among the treatments (initial shoot nitrogen concentration in autumn, TR1, TR2 and TR3; Table S3), regardless of the sampling period. Thus, for example, in the case of black spruce, the average shoot nitrogen concentration for the three treatments at the beginning and at the end of winter varied between 1.76 ± 0.05% and 1.96 ± 0.05%. In the case of white spruce, it varied between 1.93 ± 0.07% and 2.06 ± 0.07% (Table S3). The absence of significant difference was also observed in the nitrogen content of the whole seedlings and the roots (Table S3). Thus, warming treatments at the beginning and the end of winter did not induce significant changes in mineral nitrogen nutrition (Table S3).

4. Discussion

Our results revealed that the acquisition of frost tolerance in the fall (Table 2, Figures 3 and 4) combined with the presence of a layer of snow in the forest nursery enabled the seedlings to withstand climatic extremes characterized by mild spells followed by freezing temperatures in winter (Figure 7). On the other hand, the decrease in snow cover depth at the end of winter (Figure 1b) and the exposure of seedlings to periods of realistic mild spells, as predicted by climate change scenarios in eastern Canada [42], followed by freezing temperatures induced the loss of cold hardiness at −20 °C, but not at −4 °C and −12 °C (Figure 8).

The winter survival capacity of seedlings of boreal forest species produced in forest nurseries depends not only on their initial level of cold tolerance at the end of autumn, which is influenced by cultural practices [2,7,8], but also on their ability to maintain a level of cold hardiness in response to mild spells in winter followed by freezing temperatures. To ensure that the seedlings of the two forest species (black spruce 1 + 0 and white spruce 2 + 0) are well hardened at the end of autumn, we measured the relative electrical conductivity and index of injury using 4 cm long apical shoot tips and found no significant difference in all freezing temperatures tested (Table 2). This was also verified using whole seedlings, which were subjected to freezing temperatures followed by a recovery period under optimal environmental greenhouse conditions for growth. All buds of the seedlings of the two forest species showed bud break, good growth of their shoots, and no apparent freezing damage (Figures 3 and 4). These results and observations are similar to those obtained on black and white spruce in autumn in eastern Canada [3,4,36,48].

However, in late autumn, the roots of both forest species survived and showed less growth at −20 °C compared with other freezing temperatures recorded in root plugs. At this time of year (November 30), the −20 °C temperature observed in root plugs was unlikely in eastern Canada [3]. On the other hand, shoots of all white and black spruce seedlings that were subjected to −20 °C were not affected and showed normal budburst and growth (Figures 3 and 4). This clearly indicates that the kinetics of hardening and the tolerance to freezing temperatures of the shoots and roots are not synchronized with a delay in root hardening, as observed in other studies [2,7,8]. This delay in the hardening of the roots of the two tree species in the forest nursery is largely due to the active growth of the root system in late autumn and to the non-freezing temperatures of the root plugs, which are favorable to their growth [7,8]. In fact, after the growth of the shoots has stopped and buds have formed, the dry root mass of white and black spruces doubles in autumn under forest nursery conditions [41,49]. The monitoring of temperatures during a period in autumn (1 September–10 November) in six forest nurseries in Québec showed that the minimum temperatures in the root plugs of white spruce seedlings (1 + 0) rarely dropped below −6 °C, whereas those of the air could reach −12 °C [3].

The improvement in hardening and frost tolerance of black and white spruce seedlings in fall in eastern Canada under forest nursery conditions is the result of the success of different cultural practices, in particular, the control of irrigation [41,49] and fertilization [50] according to the plant growth stages, the accumulation of hardening degree days [48], the imposition of water stress [7,51,52], short-day treatments [2,8,28] to induce bud formation and epicuticular wax formation, and the increase of the ratio of dry mass to fresh mass (DM/FM) of the apical shoots greater than 30% [7,48,53,54]. Our results showed that these ratios reached 43.04% and 46.63% for black spruce and white spruce, respectively (Table 1). By using whole seedlings to assess frost tolerance, it was shown that DM/FM values similar to those obtained in our study indicate that the shoots can tolerate −20 °C without any apparent damage [3,48]. Despite the lack of consensus regarding the effects of high foliar concentrations of mineral nutrients on the cold tolerance of seedlings [7], foliar nitrogen concentrations in the two tree species were optimal in autumn (1.99 to 2.06%, Table S3) and in winter (1.76 to 2.06%, Table S3) and they did not negatively affect the cold tolerance of shoots even for the temperature of −20 °C. These results corroborate those of our previous studies, which showed no negative effect on the optimal foliar nitrogen concentrations (1.63% to 2.41%) of white spruce seedlings (1 + 0) on their frost tolerance in autumn under six forest nursery conditions [3].

Our results showed that winter warming treatments applied in mid-January (1 day at 10 °C or 3 days at 10 °C) did not reduce the cold hardiness of black and white spruce seedlings and these responses are different from those reported in [19] for the same species. These different results could be due to four main factors, including the use of different seed genetic sources [55], cultural practices for growing seedlings [8], their preconditioning to frost [8], and the duration and temperature of warming [56]. Man et al. [19] applied long durations of warming (5, 10, and 15 days) and a high temperature (16 °C), which are far from climatic normals at our forest nursery (Figure 2). Moreover, the warming scenario applied in terms of duration and intensity by Man et al. [19] suggests that seedlings become more susceptible to winter frost. This warming scenario remains improbable in forest nurseries in Québec based on the warming scenario (3 °C to 7 °C) projected by Ouranos for the province of Québec [43]. However, at the end of winter (mid-March), the control seedlings and those of warming treatments subjected to −20 °C did not break bud and did not develop new roots compared with the seedlings of the same treatments subjected to other temperatures (4 °C, −4 °C and −12 °C) (Figure 8). Thereby, the hypothesis that black and white spruce seedlings produced in forest nurseries in eastern Canada rapidly lose their cold hardiness in response to mild winter spells is partially true, especially towards the end of the winter (mid-March). This also clearly indicates that the cold hardiness to −20 °C of white and black spruce control seedlings in mid-March was lost and did not remain constant over the winter (based on mid-January measurements), which corroborates the results from other studies [57,58]. For instance, Tumanov and Krasavtsev [58] showed that under controlled conditions, the maximum level of cold hardiness of northern woody plants could be lost in a few hours when the air temperature increases above 0 °C. The rapid loss of frost tolerance observed in both species at the end of winter when the temperature was −20 °C is probably due to the increase in air temperature (>5 °C) during this exceptional year (Figure 1b), as reported in other studies [59,60]. However, our results are not in agreement with previous studies [61] suggesting that cold hardiness cannot be rapidly lost in late winter to early spring. On the other hand, the survival of the other seedlings of the same treatments, which were subjected to other freezing temperatures (−4 °C and −12 °C), were not negatively affected by these frost levels and the seedlings showed normal growth (bud break, etc.) like the control seedlings (Figure 8). The analysis of climate normals at the Grandes-Piles forest nursery showed that the minimum temperature of −20 °C was rare during the month of March (Figure 2), but for certain exceptional years, the frequency is high (Figure 1a) or rare (Figure 1b). These results suggest that the combination of the early and significant decrease in the depth of snow protecting the seedlings and the periods of warm spells followed by severe freezing temperatures (< −20 °C) at the end of winter or at the beginning of spring during exceptional climatic years will contribute to an increased loss of seedlings in northern forest nurseries.

Significant containerized seedlings losses caused by root frost are very common in forest nurseries in Nordic countries that overwinter outdoors [62]. The absence of initiation of new roots and bud bursts towards the end of winter (mid-March) in the white and black spruce seedlings of the three treatments (Figures 6 and 8) subjected to a temperature of −20 °C is due to root-freezing damages as observed in other studies for the same species [62,63]. Roots damaged by frost cause a significant reduction in the absorption of water and mineral elements and net photosynthesis, as well as survival and growth in reforestation sites [62,63]. This reduction in photosynthesis negatively affects the growth of new roots because the latter is intimately linked to the products of current photosynthates from the shoot in conifer seedlings [64–66]. Carles et al. [66] showed that the growth of new roots during a 21-day period under optimal greenhouse environmental conditions following freezing treatments was negatively related to the proportion of damaged needles and positively linked to the photosynthetic capacity of living needles of white spruce seedlings.

Although several techniques have been developed to determine root freezing damage [62], symptoms of root frost damage are not easily detectable by nursery managers on shoots before bud break. Rapid detection of root frost is of great interest to forest nursery managers to avoid reforestation of damaged seedlings whose survival is compromised.

Conversely, the initiation and growth of new roots, bud breaks, and elongation of main stems and branches are good indicators that integrate both the survival and physiological functioning of the seedling following warming and freezing temperature treatments applied in late fall, early, and late winter (Figures 3, 4 and 6–8). In addition, the absence of visible damage on the shoots in response to warming treatments and freezing temperatures clearly shows the frost tolerance of white and black spruce seedlings. This indicates that the physiological processes (gaz exchange parameters, water and mineral absorption, etc.) of these two spruce species quickly respond during the recovery period following freezing temperatures. Merry et al. [67] showed that the photosynthetic function of white spruce could recover up to three times more rapidly than that of Eastern white pine in response to freezing temperatures during winter.

To avoid any risk associated with climatic extremes in winter, several Nordic nurseries place seedlings in frozen storage during winter (−2 °C to −5 °C) [2,8,68,69]. On the other hand, to reduce the loss of seedlings caused by the different types of frost (early frost, late frost, root frost, winter desiccation, etc.) when seedlings overwinter outside, the nursery managers do not have the choice to use different cultural practices and protection strategies to improve cold hardiness of seedlings. Thus, acquiring and improving the frost tolerance of seedlings in autumn is an essential prerequisite for survival during winter. This improvement in frost tolerance in autumn can be obtained by resorting to a succession of different cultural practices, such as lowering the water content of the substrate from 15% to 20% (*v/v*) to induce water stress [6,40,48,50,51], reducing the nitrogen fertility of the growing media (e.g., black spruce 1 + 0:25 ppm; white spruce 2 + 0:50 ppm) [49], and applying a short-day treatment [6,7,27]. Before the arrival of the first frost temperatures, the water content of the substrate must be increased (50 to 60%, v/v) because moist substrates release heat compared with dry substrates, which offers better protection to the roots against frost. Another cultural practice of overwintering seedlings (1 + 0) is used by directly laying protective covers over seedlings and securing edges [25]. When the first snowfall is late, some nursery managers rent or use their own snowmaking systems to start making artificial snow at an optimum temperature of −8 °C (yield: 45 m^3/h). The protection of seedlings by snow is done in two steps. The first step is to cover the roots of the seedlings by 5 cm to protect most of the seedlings, while the second step is to cover the apical buds of the seedlings by at least 5 cm [25,26]. In the Juniper Forest Nursery in the Province of New Brunswick, Canada, the use of the snowmaking system has become common practice to protect seedlings from winter frost since the early 1990s [70]. With two snowmaking systems, the nursery snows 3.2 hectares with a snow depth of 30 cm (snow quality 3: 370.0 kg/m^3) and almost eliminates seedling losses due to frost. The return on investment of the two snowmaking systems was amortized after 1.6 years [70]. In the province of Québec, forest nurseries use different winter protection techniques, and each forest nursery adopts its protection strategy according to its geographical location, the risk of seedling losses, and economic imperatives [71].

5. Conclusions

The results of this study, conducted at an operational scale in a forest nursery, showed that the application of different durations of realistic mild spells, as predicted by climate change scenarios in eastern Canada, at the beginning of winter followed by freezing temperatures had no effect on hardening, bud break, and growth during the recovery period of white and black spruce seedlings under optimal growth conditions. On the other hand, with a significant decrease in snow cover towards the end of winter (mid-March), only the seedlings of the two species subjected to the freezing temperature of −20 °C were negatively affected, regardless of the warming treatment. Conversely, the other seedlings of the same treatments subjected to the temperatures 4 °C, −4 °C, and −12 °C showed normal growth (bud burst of all buds on apical and lateral branches, root growth, etc.) and no apparent damage. The mild spells applied also did not lead to a very significant decrease, for example, in leaf nitrogen concentrations and content. Different cultural practices and protection strategies are proposed to improve frost tolerance and reduce the winter loss of seedlings.

Supplementary Materials: The following supporting information can be downloaded at: https://www.mdpi.com/article/10.3390/f13121975/s1. Table S1: Comparisons of adjusted means of growth variables of two-year-old white spruce (WS 2 + 0) and one-year-old black spruce (BS, 1 + 0) seedlings between the sampling date at the beginning (mid-January 2010) and end of winter (mid-March 2010) in response to different freezing temperatures; Table S2: Observed probabilities (p-values) and degrees of freedom of the fixed effects related to the different treatments applied before freezing on mineral nutrient concentrations (%) and contents (mg) of different parts (shoot, root, and whole seedling) of two-year-old white spruce (WS 2 + 0) and one-year-old black spruce (BS 1 + 0); Table S3: Comparison of adjusted means between different treatments before freezing related to mineral nutrient concentrations (%) and contents (mg) of different parts (shoot, root, and whole seedling) of two-year-old white spruce (WS, 2 + 0) and one-year-old black spruce (BS, 1 + 0).

Author Contributions: Conceptualization, M.S.L. and M.R.; methodology, M.S.L. and M.R.; supervision, M.R. and M.S.L.; project administration, M.R. and M.S.L.; statistical analyses, M.-C.L. and M.S.L.; data curation, M.S.L., M.R. and M.-C.L.; writing—original draft preparation, M.S.L.; writing—review and editing, M.S.L., M.R. and M.-C.L. All authors have read and agreed to the published version of the manuscript.

Funding: Funding for this research project was provided by the Direction de la recherche forestière (DRF) of the Ministry of Forests, Wildlife and Parks of the Government of Québec as part of project no. 112310038: Determination of the frost tolerance thresholds of seedlings in winter in relation to climatic extremes and development of a model for predicting the evolution of the hardening state of seedlings in forest nurseries (holder: Mohammed S. Lamhamedi).

Data Availability Statement: Not applicable.

Acknowledgments: We sincerely thank Denis Gélinas, Pierre Comtois, and all the staff of the Normandin and Grandes-Piles forest nurseries for their exceptional technical assistance during all phases of this project. We thank all the staff of the organic and inorganic chemistry laboratory of the Direction de la recherche forestière (DRF) for the mineral analyses of the samples and the enriching discussions with the chemical managers (Carol De Blois and Denis Langlois). We thank Steeve Pepin for his editorial comments and Charles Gilles Langlois for his continued support of this project.

Conflicts of Interest: The authors declare no conflict of interest. The funders had no role in the design of the study; in the collection, analyses, or interpretation of data; in the writing of the manuscript, or in the decision to publish the results.

References

1. Bigras, F.J.; Colombo, S.J. (Eds.) *Conifer Cold Hardiness*; Kluwer Academic Publishers: Dordrecht, The Netherlands, 2001; 596p.
2. Colombo, S.J. Frost hardening spruce container stock for overwintering in Ontario. *New For.* **1997**, *13*, 449–467. [CrossRef]
3. Lamhamedi, M.S.; Veilleux, L.; Renaud, M. Élaboration des seuils de tolérance au gel des plants d'épinette blanche (1 + 0) en pépinière forestière selon les régions écologiques du Québec. Gouvernement du Québec, Ministère des Ressources naturelles. *Mémoire De Rech. For.* **2005**, *147*, 50.
4. Lamhamedi, M.S.; Veilleux, L.; Renaud, M.; Desjardins, P. Prédiction et détermination des seuils de tolérance au gel en automne et techniques de protection contre le gel hivernal. In *Production de Plants Forestiers au Québec: La Culture de L'innovation, Proceedings of the Colloque de Transfert de Connaissances et de Savoir-Faire, Carrefour Forêt Innovations, Québec, QC, Canada, 4–6 October 2011*; Colas, F., Lamhamedi, M.S., Eds.; Gouvernement du Québec: Québec, QC, Canada, 2011; pp. 53–64.
5. MFFP. *Insectes, Maladies et Feux Dans les Forêts du Québec en 2021*; Gouvernement du Québec: Québec, QC, Canada, 2022; pp. 1–70.
6. Joly, E.; Les Serres coopératives de Guyenne, Guyenne, Québec, Canada. Personal communication, 2022.
7. Bigras, F.J.; Ryyppö, A.; Lindström, A.; Stattin, E. Cold acclimation and deacclimation of shoots and roots of conifer seedlings. In *Conifer Cold Hardiness*; Bigras, F.J., Colombo, S.J., Eds.; Kluwer Academic Publishers: Dordrecht, The Netherlands, 2001; pp. 57–88.
8. Colombo, S.J.; Menzies, M.I.; O'Reilly, C. Influence of nursery cultural practices on cold hardiness of coniferous forest tree seedlings. In *Conifer Cold Hardiness*; Bigras, F.J., Colombo, S.J., Eds.; Kluwer Academic Publishers: Dordrecht, The Netherlands, 2001; pp. 223–252.
9. Margolis, H.A. Seedling Production and Reforestation in Québec. *J. For.* **1987**, *85*, 39–43.

10. IPCC. *Climate Change 2014: Synthesis Report. Contribution of Working Groups I, II and III to the Fifth Assessment Report of the Intergovernmental Panel on Climate Change*; Core Writing Team, Pachauri, R.K., Meyer, L.A., Eds.; IPCC: Geneva, Switzerland, 2014; pp. 1–151.
11. South, D.B.; Starkey, T.; Enebak, S. *Warm Nighttime Temperatures (with Tents) Affect Terminal Buds of Loblolly Pine Seedlings*; RR 09-04; Southern Forest Nursery Management Cooperative, Auburn University: Auburn, AL, USA, 2009; pp. 1–7.
12. South, D.B.; Starkey, T.; Islam, M.A.; Jacobs, D.F. *Warm Night Time Temperatures Affect the Ability of Loblolly Pine (Pinus taeda) Seedlings to Tolerate Freezing*; RR 08-02; Southern Forest Nursery Management Cooperative, Auburn University: Auburn, AL, USA, 2008; pp. 1–10.
13. Taulavuori, K.M.J.; Laine, K.; Taulavuori, E.B. Artificial deacclimation response of *Vaccinium myrtillus* in mid-winter. *Ann. Bot. Fenn.* **2002**, *39*, 143–147.
14. Cunningham, S.M.; Nadeau, P.; Castonguay, Y.; Laberge, S.; Volenec, J.J. Raffinose and stachyose accumulation, galactinol synthase expression, and winter injury of contrasting alfalfa germplasms. *Crop Sci.* **2003**, *43*, 562–570. [CrossRef]
15. Kalberer, S.R.; Wisniewski, M.; Arora, R. Deacclimation and reacclimation of cold-hardy plants: Current understanding and emerging concepts. *Plant Sci.* **2006**, *171*, 3–16. [CrossRef]
16. Ögren, E. Relationship between temperature, respiratory loss of sugar and premature dehardening in dormant Scots pine seedlings. *Tree Physiol.* **1997**, *17*, 47–51. [CrossRef] [PubMed]
17. Strimbeck, G.R.; Schaberg, P.G.; DeHayes, D.H.; Shane, J.B.; Hawley, G.J. Midwinter dehardening of montane red spruce during a natural thaw. *Can. J. For. Res.* **1995**, *25*, 2040–2044. [CrossRef]
18. Man, R.; Lu, P.; Dang, Q.-L. Cold tolerance of black spruce, white spruce, jack pine, and lodgepole pine seedlings at different stages of spring dehardening. *New For.* **2021**, *52*, 317–328. [CrossRef]
19. Man, R.; Colombo, S.; Lu, P.; Dang, Q.-L. Effects of winter warming on cold hardiness and spring budbreak of four boreal conifers. *Botany* **2016**, *94*, 117–126. [CrossRef]
20. Man, R.; Lu, P.; Dang, Q.-L. Cold hardiness of white spruce, black spruce, jack pine, and lodgepole pine needles during dehardening. *Can. J. For. Res.* **2017**, *47*, 1116–1122. [CrossRef]
21. Pike, C.C.; Warren, J.C.; Montgomery, R.A. Effects of artificial warming during quiescence on budbreak and growth of white spruce, *Picea glauca*. *Can. J. For. Res.* **2017**, *47*, 1538–1545. [CrossRef]
22. Sevanto, S.; Suni, T.; Pumpanen, J.; Grönholm, T.; Kolari, P.; Nikinmaa, E.; Hari, P.; Vesala, T. Wintertime photosynthesis and water uptake in a boreal forest. *Tree Physiol.* **2006**, *26*, 749–757. [CrossRef] [PubMed]
23. MFFP. *Guide de Terrain. Inventaire de Qualification des Plants Résineux Cultivés en Récipients*; Gouvernement du Québec: Québec, QC, Canada, 2019; p. 114.
24. Burns, R.M.; Honkala, B.H. *Silvics of North America: 1. Conifers*; Agriculture Handbook 654; U.S. Department of Agriculture, Forest Service: Washington, DC, USA, 1990; 675p.
25. Lamhamedi, M.S.; Labbé, L.; Margolis, H.A.; Stowe, D.C.; Blais, L.; Renaud, M. Spatial variability of substrate water content and growth of white spruce seedlings. *Soil Sci. Soc. Am. J.* **2006**, *70*, 108–120. [CrossRef]
26. Lamhamedi, M.S.; Renaud, M.; Desjardins, P.; Veilleux, L. Root growth, plug cohesion, mineral nutrition, and carbohydrate content of (1 + 0) *Picea mariana* seedlings in response to a short-day treatment. *Tree Plant. Notes* **2013**, *56*, 35–46.
27. Goulet, S. Utilisation du Canon à Neige en Conditions Opérationnelles. In Proceedings of the Colloque Forestier, Office des Producteurs de Plants Forestiers du Québec, Québec, QC, Canada, 23–24 February 2016; pp. 1–38.
28. Lamhamedi, M.S. Mesures d'adaptation aux changements climatiques: Cas des extrêmes hivernaux dans les pépinières forestières du Québec. Ministère des Ressources naturelles, Direction de la recherche forestière. *Avis Rech. For.* **2012**, *42*, 1–2.
29. Lamhamedi, M.S.; Renaud, M.; Auger, I.; Fortin, J.A. Granular calcite stimulates natural mycorrhization and growth of white spruce seedlings in peat-based substrates in forest nursery. *Microorganisms* **2020**, *8*, 1088. [CrossRef]
30. Lamhamedi, M.S.; Renaud, M.; Auger, I. Granular Calcite: A New Cultural Practice for the Improvement of the Physicochemistry of the Peat Substrate, Growth and Morphophysiological Quality of White Spruce Seedlings in Forest Nursery. *Land* **2021**, *10*, 661. [CrossRef]
31. Lambany, G. La dynamique morphologique et physiologique de plants soumis à un entreposage de longue durée. Gouvernement du Québec, Ministère des Ressources naturelles. *Mémoire Rech. For.* **1994**, *114*, 33.
32. Colombo, S.J.; Raitanen, E.M. Frost hardening in first-year eastern larch (*Larix laricina*) container seedlings. *New For.* **1993**, *7*, 55–61. [CrossRef]
33. Colombo, S.J.; Glreum, C.; Webb, D.P. Frost hardiness testing: An operational manual for use with extended greenhouse culture. Ministry of Natural Resources of Ontario. *For. Res. Rep.* **1984**, *110*, 14.
34. Palta, J.P.; Weiss, L.S. Ice formation and freezing injury: An overview on the survival mechanism and molecular aspects of injury and cold acclimation in herbaceous plants. In *Advances in Plant Cold Hardiness*; Li, P.H., Christersson, L., Eds.; CRC Press, Inc.: Boca Raton, FL, USA, 1993; pp. 143–176.
35. Burr, K.E.; Hawkins, C.D.B.; L'Hirondelle, S.J.; Binder, W.; George, M.F.; Repo, T. Methods for measuring cold hardiness of conifers. In *Conifer Cold Hardiness*; Bigras, F.J., Colombo, S.J., Eds.; Kluwer Academic Publishers: Dordrecht, The Netherlands, 2001; pp. 369–401.
36. Carles, S.; Lamhamedi, M.S.; Stowe, D.C.; Margolis, H.A.; Bernier, P.Y.; Veilleux, L.; Fecteau, B. Frost tolerance of two-year-old *Picea glauca* seedlings grown under different irrigation regimes in a forest nursery. *Scand. J. For. Res.* **2008**, *23*, 137–147. [CrossRef]

37. Lindström, A.; Stattin, E. Root freezing tolerance and vitality of Norway spruce and Scots pine seedlings; influence of strorage duration, storage temperature, and prestorage root freezing. *Can. J. For. Res.* **1994**, *24*, 2477–2484. [CrossRef]
38. Zhu, X.B.; Cox, R.M.; Bourque, C.-P.A.; Arp, P.A. Thaw effect on cold-hardiness parameters in yellow birch. *Can. J. Bot.* **2002**, *80*, 390–398. [CrossRef]
39. Flint, H.L.; Boyce, B.R.; Beattie, D.J. Index of injury—A useful expression of freezing injury to plant tissues as determined by the electrolytic method. *Can. J. Plant Sci.* **1967**, *47*, 229–230. [CrossRef]
40. Tanaka, Y.; Brotherton, P.; Hostetter, S.; Chapman, D.; Dyce SBelanger, J.; Johnson, B.; Duke, S. The operational planting stock quality testing program at Weyerhaeuser. *New For.* **1997**, *13*, 423–437. [CrossRef]
41. Lamhamedi, M.S.; Lambany, G.; Margolis, H.A.; Renaud, M.; Veilleux, L.; Bernier, P.Y. Growth, physiology and leachate losses in *Picea glauca* seedlings (1 + 0) grown in air-slit containers under different irrigation regimes. *Can. J. For. Res.* **2001**, *31*, 1968–1980. [CrossRef]
42. Régnière, J.; St-Amant, R. Stochastic simulation of daily air temperature and precipitation from monthly normals in North America north of Mexico. *Inter. J. Biometeorol.* **2007**, *51*, 415–430. [CrossRef]
43. Ouranos. *Vers l'adaptation. Synthèse des Connaissances sur les Changements Climatiques au Québec. Partie 1: Évolution Climatique au Québec Edition 2015*; Ouranos: Montréal, QC, Canada, 2015; p. 114.
44. Timmer, V.R.; Armstrong, G.; Miller, B.D. Steady-state nutrient preconditioning and early outplanting performance of containerized black spruce seedlings. *Can. J. For. Res.* **1991**, *21*, 585–594. [CrossRef]
45. Bernier-Cardou, M.; Bigras, F.B. The analysis of Cold Hardiness Experiments. In *Conifer Cold Hardiness*; Bigras, F.J., Colombo, S.J., Eds.; Kluwer Academic Publishers: Dordrecht, The Netherlands, 2001; pp. 403–435.
46. Milliken, G.A.; Johnsen, D.E. *Analysis of Messy Data. Vol 1. Designed Experiments*; Van Nostrand Reinhold Company: New York, NY, USA, 1994.
47. Westfall, P.H.; Tobias, R.D.; Russell, D.W. *Multiple Comparisons and Multiple Tests Using SAS®*, 2nd ed.; SAS Institute, Inc.: Cary, NC, USA, 2011.
48. Carles, S.; Lamhamedi, M.S.; Stowe, D.C.; Margolis, H.A. An operational method for estimating cold tolerance thresholds of white spruce seedlings in forest nurseries. *For. Chron.* **2012**, *88*, 448–457. [CrossRef]
49. Lamhamedi, M.S.; Margolis, H.A.; Renaud, M.; Veilleux, L.; Auger, I. Effets de différentes régies d'irrigation sur la croissance, la nutrition minérale et le lessivage des éléments nutritifs des semis d'épinette noire (1 + 0) produits en récipients à parois ajourées en pépinière forestière. *Can. J. For. Res.* **2003**, *33*, 279–291. [CrossRef]
50. Lamhamedi, M.S.; Gagnon, J. *Comment Gérer la Fertilisation et L'irrigation en vue D'améliorer la Tolérance au gel des Plants Produits en Récipients?* Ministère des Ressources Naturelles. Gouvernement du Québec: Québec, QC, Canada, 2002; 15p.
51. Stowe, D.C.; Lamhamedi, M.S.; Margolis, H. Water relations, cuticular transpiration, and bud characteristics of air-slit containerized *Picea glauca* seedlings in response to controlled irrigation regime. *Can. J. For. Res.* **2001**, *31*, 1922–1929. [CrossRef]
52. Bergeron, O.; Lamhamedi, M.S.; Margolis, H.A.; Bernier, P.Y.; Stowe, D.C. Irrigation control and physiological responses of nursery-grown black spruce seedlings (1 + 0) cultivated in air-slit containers. *Hort. Sci.* **2004**, *39*, 599–605. [CrossRef]
53. Calmé, S.; Margolis, H.A.; Bigras, F.J. Influence of cultural practices on the relationship between frost tolerance and water content of containerized black spruce, white spruce, and jack pine seedlings. *Can. J. For. Res.* **1993**, *23*, 503–511. [CrossRef]
54. Colombo, S.J. Bud dormancy status, frost hardiness, shoot moisture content, and readiness of black spruce container seedlings for frozen storage. *J. Am. Soc. Hort. Sci.* **1990**, *115*, 302–307. [CrossRef]
55. Aitken, S.; Hannerz, M. Genecology and gene resource management strategies for conifer cold hardiness. In *Conifer Cold Hardiness*; Bigras, F.J., Colombo, S.J., Eds.; Kluwer Academic Publishers: Dordrecht, The Netherlands, 2001; pp. 23–54.
56. Hänninen, H.; Beuker, E.; Johnsen, Ø.; Leinonen, I.; Murray, M.; Sheppard, L.; Skroppa, T. Impacts of climate change on cold hardiness. In *Conifer Cold Hardiness*; Bigras, F.J., Colombo, S.J., Eds.; Kluwer Academic Publishers: Dordrecht, The Netherlands, 2001; pp. 305–334.
57. Bigras, F.J.; Margolis, H.A. Shoot and root sensitivity of containerized black spruce, white spruce and jack pine seedlings to late fall freezing. *New For.* **1997**, *13*, 29–49. [CrossRef]
58. Tumanov, I.I.; Krasavtsev, O.A. Hardening of northern woody plants by treatment at temperature below 0 °C. *Sov. Plant Physiol.* **1959**, *6*, 654–667.
59. Cannell, M.G.R.; Smith, R.I. Thermal time, chill days and prediction of budburst in *Picea sitchensis*. *J. Appl. Ecol.* **1983**, *20*, 951–963. [CrossRef]
60. Koski, V. The timing of hardening and dehardening of forest trees. *Acta Hort.* **1985**, *168*, 117–124. [CrossRef]
61. Glerum, C. Annual trends in frost hardiness and electrical impedance for seven coniferous species. *Can. J. Plant Sci.* **1973**, *53*, 881–889. [CrossRef]
62. Bigras, F.; Dumais, D. Root-Freezing damage in the containerized nursery: Impact on plantation sites—A review. *New For.* **2005**, *30*, 167–184. [CrossRef]
63. Dumais, D.; Coursolle, C.; Bigras, F.; Margolis, H. Simulated root freezing in the nursery: Effects on the growth and physiology of containerized boreal conifer seedlings after outplanting. *Can. J. For. Res.* **2002**, *32*, 605–615. [CrossRef]
64. Van den Driessche, R. Importance of current photosynthate to new root growth in planted conifer seedlings. *Can. J. For. Res.* **1987**, *17*, 776–782. [CrossRef]
65. Kozlowski, T.T. Carbohydrate sources and sinks in woody plants. *Bot. Rev.* **1992**, *58*, 107–222. [CrossRef]

66. Carles, S.; Lamhamedi, M.S.; Stowe, D.C.; Bernier, P.Y.; Veilleux LMargolis, H.A. Relationships between frost hardiness, root growth potential and photosynthesis of nursery-grown white spruce seedlings. *Ann. For. Sci.* **2011**, *68*, 1303–1313. [CrossRef]
67. Merry, R.; Jerrard, J.; Frebault, J.; Verhoeven, A. A comparison of pine and spruce in recovery from winter stress; changes in recovery kinetics, and the abundance and phosphorylation status of photosynthetic proteins during winter. *Tree Physiol.* **2017**, *37*, 1239–1250. [CrossRef]
68. Wallin, E.; Gräns, D.; Stattin, E.; Verhoef, N.; Mikusiński, G.; Lindström, A. Evaluating methods for storability assessment and determination of vitality status of container grown Norway spruce transplants after frozen storage. *Scan. J. For. Res.* **2019**, *34*, 417–426. [CrossRef]
69. Nadel, R.; Payne, N.D.; Stokes, T.A.; Enebak, S.A. Lifting dates, chilling hours, and storage duration on slash pine seedlings root growth potential, growth, and survival. *Tree Plan. Notes.* **2020**, *63*, 80–90.
70. Turbocristal. *Artificial Snowmaking to Minimize Seedlings Loss in Tree Nursery*; Turbocristal Projet Description: Québec, QC, Canada, 1993; pp. 1–2.
71. MRNF. *Recherche de Nouvelles Méthodes de Protection Hivernale Pour la Production de Plants en Récipient*; Gouvernement du Québec: Québec, QC, Canada, 2009; pp. 1–33.

forests

MDPI

Article

The Importance of Initial Seedling Characteristics in Controlling Allocation to Growth and Reserves under Different Soil Moisture Conditions

Simon M. Landhäusser [1,*,†], Erin T. Wiley [2,†], Kevin A. Solarik [3], Shaun P. Kulbaba [1] and Alexander E. Goeppel [1]

1 Department of Renewable Resources, University of Alberta, Edmonton, AB T6G 2H1, Canada
2 Department of Biology, University of Central Arkansas, Conway, AR 72035, USA; ewiley3@uca.edu
3 National Council for Air and Stream Improvement, Inc. (NCASI), 2000 McGill College Avenue, Montreal, QC H3A 3H3, Canada
* Correspondence: simon.landhausser@ualberta.ca; Tel.: +1-780-492-6381
† These authors contributed equally to this work.

Abstract: After disturbance, forest regeneration and resiliency depend on the ability of seedlings to respond, survive, and grow under a variety of stress conditions, including drought. Despite recent efforts to improve our fundamental knowledge surrounding plant response mechanisms to stress and their application in seedling quality research, initial seedling characteristics are often ignored when exploring seedling responses to stress in field plantings or ecophysiological studies. Here, we explore how initial differences in size, biomass allocation, and non-structural carbohydrate (NSC) storage affect the subsequent partitioning of new biomass, growth potential, and drought response in seedlings of a deciduous broad-leaved (*Populus tremuloides*) and an evergreen coniferous species (*Pinus banksiana*). We exposed seedlings of both species to different growing conditions in their first growing season in order to manipulate the aforementioned seedling characteristics. In a second growing season, we exposed these different seedling types to a subsequent drought stress. While drought reduced both structural growth and NSC storage in all seedling types, the expected shift in allocation favoring roots was only observed in seedling types with initially low root:shoot or root:stem ratios. Overall, we also found that the traits associated with greater growth were quite different between pine and aspen. While larger seedlings led to greater growth in pine, it was the smallest seedling type in aspen with the largest root:stem ratio that produced the most new growth. In aspen, this smaller seedling type was the only one that did not undergo a shift in biomass relative to its initial allometry, suggesting that adjustments in biomass allocation made by other, larger seedling types must have come at the cost of lower growth. In contrast, adjustments in allocation did not appear to negatively impact pine, possibly because reduced root:shoot ratios of larger seedlings did not reduce NSC storage, as it did in aspen. Our results highlight (1) the complexity of how differences in biomass allocation and changes in seedling size may alter storage and the response of species to drought, and (2) the importance of accounting for initial seedling characteristics (both morphological and physiological) when predicting seedling growth and the impacts of environmental stressors.

Keywords: biomass; allocation; reserve; drought; conditioning; structural and non-structural carbohydrates; organs; *Populus tremuloides*; *Pinus banksiana*; seedling; quality

Citation: Landhäusser, S.M.; Wiley, E.T.; Solarik, K.A.; Kulbaba, S.P.; Goeppel, A.E. The Importance of Initial Seedling Characteristics in Controlling Allocation to Growth and Reserves under Different Soil Moisture Conditions. *Forests* **2023**, *14*, 796. https://doi.org/10.3390/f14040796

Academic Editors: Mohammed S. Lamhamedi, Damase P. Khasa and Steeve Pépin

Received: 2 March 2023
Revised: 6 April 2023
Accepted: 9 April 2023
Published: 13 April 2023

1. Introduction

Because of forests' ability to capture and sequester significant quantities of carbon, there are increasing efforts to protect and even expand the amount of forest cover worldwide. However, this goal is challenged by the increasing threat of natural and human-caused disturbances and large-scale forest die-off events, exasperated by climate change [1–7]. In the short-term, forest cover is impacted by the ability of large canopy trees to resist disturbance and cope with post-disturbance conditions [8]. However, the maintenance of forests in perpetuity is ultimately determined by their ability to regenerate successfully [9], which on severely disturbed sites, occurs naturally via recruitment through seed dispersal or via human intervention through artificial planting [10]. Additionally, because the tree seedling stage can be extremely sensitive to environmental changes and prone to much higher mortality rates than mature trees [11–14], forest regeneration may largely depend on the ability of seedlings to survive and grow under a variety of different conditions.

Though seedlings are generally very sensitive to stress, their responses may depend on the seedlings' initial physiological conditions and morphological characteristics—such as growth form, biomass allocation, and carbohydrate or nutrient reserve status—prior to a stress event or at the time of planting [15,16]. This concept has been widely applied in seedling quality research by manipulating the growing conditions in attempts to alter seedling traits that will improve their performance (Reviewed by [17]). For example, seedlings preconditioned with mild stress during their production have shown improved stress tolerance and survival after outplanting (e.g., [8,18–20]). An increasing challenge for forest regeneration efforts is to ensure that establishing seedlings will be able to cope with stressors such as drought. Despite recent efforts to improve our knowledge of how seedling physiology is impacted by drought [21,22] and how species' responses may vary [19,23,24], it is still unclear how dependent a given species' drought response is on the initial physiological and morphological condition of the seedling (but see [25,26]). Thus, better knowledge of how seedling characteristics are influenced by different initial growing conditions—and how these different characteristics may affect subsequent growth and survival—could help us (1) produce seedlings better adapted for different environments or different uses (i.e., target seedling concept [27–29] and (2) better understand and interpret stress responses when executing seedling based studies.

One way that plants commonly cope with stressors such as drought is to alter biomass or resource allocation among organs [30]. According to optimal partitioning theory, plants should favor the partitioning of resources towards organs responsible for capturing the resource that limits growth and survival [31]. For example, many plants growing in light-limiting conditions will allocate proportionally more mass towards the leaves and shoots necessary for the acquisition of light [32,33], while plants growing under water-limiting conditions have consistently shown an increased root mass fraction by either allocating relatively more biomass towards the root system [34–36] or by the shedding of leaves and branches [37–40]. For example, shade-grown seedlings allocated more biomass belowground when exposed to a subsequent drought than seedlings that had been growing in high light conditions [41–43], suggesting that the allocation response to drought depends on the pre-drought conditions the seedling was exposed to. Thus, while allocation changes with environmental conditions are well described, it is still unclear whether *initial* differences in biomass allocation or other physiological traits—caused by differences in genetics and/or previous environmental conditions—can alter how seedlings allocate resources in response to *subsequent* stressors, making these allocation shifts less universal than previously thought. Additionally, if initial biomass allocation differences *do* impact subsequent allocation, it is not known if these different responses lead to differences in seedling performance.

Alterations to biomass allocation may also indirectly affect plant performance and stress responses by affecting the storage of non-structural carbohydrates (NSC). NSC storage provides a source of C that can be remobilized under stress or following disturbances [44–46], with higher concentrations associated with greater growth and survival [26,47–49]. During drought, NSCs can also play an active role in plant hydraulic integrity, as they influence water uptake and turgor maintenance through their role in osmotic adjustment [12,50–52]. Therefore, while increased root biomass allocation in response to drought can be the result of changes in structural root growth, it may also simply reflect increases in organic solutes and NSC reserves [37,50,51,53]. Finally, because species differ in terms of storage patterns between organs [54–56], changes in biomass allocation may have different impacts on storage in different species, which in turn may affect how new biomass is subsequently allocated [24]. Therefore, initial differences and subsequent changes in NSC storage should be considered when evaluating biomass allocation as well considering traits that may improve seedling growth and survival.

In this study, we explore how initial differences in seedling size, biomass allocation, and NSC storage affect the subsequent partitioning of new biomass, growth, and drought response in two common boreal tree species, aspen (*Populus tremuloides* Michx.) and jack pine (*Pinus banksiana* Lamb). Specifically, we ask (1) how do differences in initial characteristics affect growth and storage under well-watered and water-stressed conditions? (2) do the traits associated with higher growth differ between species and growing conditions? and (3) Do allocation shifts in response to drought depend on initial seedling conditions?

2. Materials and Methods

Both aspen and jack pine seeds were obtained from our industry partners and represent a range of populations of open-pollinated seed sources near Fort McMurray, Alberta, Canada (56°73′50″ N; 111°38′01″ W). Approximately 300 seedlings were grown for each species at the Crop Diversification Centre North Facility (Alberta Agriculture and Forestry, Government of Alberta) located in Edmonton, Alberta, Canada (53°64′28″ N; 113°36′13″ W). Aspen was grown in 615A styro-blocks (cavity size: 6 cm diameter 15 cm deep; 340 mL volume), and jack pine was grown in 412A (4 cm diameter and 12 cm deep; 125 mL) (Beaver Plastics Ltd., Acheson, AB, Canada). Cavities were uniformly filled with a commercial growth medium, which was composed of 90% sphagnum peat moss, 10% perlite, and dolomitic limestone (Sun Gro Horticulture Canada Ltd., Seba Beach, AB, Canada). After filling the styro-blocks at a nursery using mechanized equipment which ensured uniformity of the medium mixture, its fill and compaction in each cavity, the growth medium was watered to field capacity (~0.6 g g^{-1} gravimetric soil water content) and seeded. After germination, germinates were treated with a single fertilization of 2 g L^{-1} of a commercial fertilizer (Plants Products Co., Ltd., Leamington, ON, Canada 5-15-5 (N-P-K) containing chelated micronutrients). After three weeks of growth, a second single fertilization of 1 g L^{-1} 10-52-10 was applied to stimulate root growth. Following the seedling establishment phase (first four weeks), seedlings were fertigated twice weekly for 16 weeks (watered and fertilized with a blend of 91 ppm N 77 ppm P and 161 ppm K with chelated micronutrients a blend used by commercial nurseries). The commercial fertilization regime allowed all seedlings to be fertilized in excess throughout the growth period. Figure A1 provides a simplified flow-chart that depicts the different steps used in the execution of this study. Detailed information of each step is provided in the text below.

Once aspen seedlings reached an average height of approximately 20 cm, seedlings were randomly separated into three groups with the intention of creating seedlings with different morphological and physiological characteristics. Based on earlier studies [48,57,58], aspen seedlings were grown in conditions to achieve: (A) high tissue NSC concentrations by artificially restricting seedling height growth using a growth inhibitor during the growing season; (B) a high root-to-stem ratio (RSR) by moving seedlings to outside conditions, and (C) tall seedlings with low RSR and NSC by continually growing seedlings under optimal greenhouse conditions (see also below).

To create three different seedling types with different characteristics in pine, we grew seedlings under three different growing conditions: (A) seedlings were established and grown for 5 months under greenhouse conditions (April to September), (B) seedlings were established at the same time in April as (A), but were moved outdoors between June and September, and (C) seedlings were established from seed in May and grown outside between June and September, representing more natural establishment period.

During the greenhouse phase, seedlings were exposed to average day and night temperatures of 20/16 °C, 50% relative humidity, and 16-17 h of light. After 20 weeks, the greenhouse-grown seedlings were transferred to the outside conditions for bud set and hardening during the fall (10 weeks, between September and November) prior to frozen storage in late November. After all seedlings had hardened outside, they were lifted, placed in plastic bags, and then in waxed cardboard boxes where they were stored frozen at −3 °C for three months. One week prior to the start of the experiment, all seedlings were removed from frozen storage and allowed to slowly thaw in a refrigerator (4 °C). To ensure a comparative distribution of initial seedling sizes within seedling types for the study, 50 seedlings representative of each seedling type and species were selected based on the median height and root collar diameter (RCD) for each seedling type and species. Based on six seedling types (three for each species), a total of about 300 seedlings were selected, 150 for each species. Further, to avoid any potential bias in seedlings selection for the treatment assignment, seedlings of similar size distributions within each seedling type were assigned to the initial pre-treatment measurements and to each subsequent treatment.

2.1. Initial Seedling Measurements

To determine initial seedling characteristics for each seedling type prior to the drought study, a random sample of 10 dormant seedlings was taken from the 50 seedlings in each seedling type. The following variables were measured: seedling height (stem length), RCD, and the dry mass of stems, needles (only in pine), and roots after samples were oven-dried at 100 °C for 1 h and then dried to constant weight at 70 °C (72 h). Additionally, to quantify organ-specific (needle, stem, and root) NSC concentrations all dry samples were first ground using a Wiley Mini-Mill (Thomas Scientific, Swedesboro, NJ, USA) to pass 40 mesh (0.4 mm). Starch and sugar concentrations (% dry mass) for each organ were quantified after water soluble sugars were extracted using 80% ethanol at 90 °C. Soluble sugar concentrations were determined colorimetrically with phenol-sulfuric acid, while the starch in the remaining pellet was digested to glucose using α-amylase and amyloglucosidase (σ) and glucose concentration was colorimetrically quantified using peroxidase-glucose oxidase/o-dianisidine [59,60]. NSC concentrations are presented in this study as the sum of soluble sugars and starch concentrations for each organ.

NSC content (NSC mass) of each organ was calculated based on the total biomass ($biomass_{Total}$) for each organ and its NSC concentration. Whole seedling NSC mass was calculated as the sum of NSC mass of all organs, and whole seedling NSC concentration was calculated as the whole seedling NSC mass divided by total whole seedling biomass (sum of $biomass_{Total}$ of all organs). To distinguish changes in NSC content from structural tissue growth at the organ and whole seedling level, structural biomass ($biomass_{Structural}$) was calculated by subtracting the NSC mass from the $biomass_{Total}$.

2.2. Characterizing and Separating Initial Seedling Conditions

To identify objectively initial differences in seedling conditions, seedling characteristics were analyzed in two ways. First, species-specific multivariate regression tree analyses (MRT) were performed to separate individual seedlings based on their initial seedling characteristics. In short, MRT analyses create dichotomies in a hierarchical manner where seedlings possessing similar characteristics are clustered together, and those that differ are split apart. A Euclidian distance measure was used to determine dissimilarities in our analysis. Final tree selection was determined following 1000 cross-validations, verified using the 1-SE (standard error) rule [14,61]. MRT analyses were conducted using the *mvpart* package 1.6-2 library within the R version 3.3.1 [62]. Second, we also conducted one-way ANOVAs followed by Tukey's Honestly Significant Difference tests in R to test for differences in all measured variables among seedling types within a species.

Within each species, seedling types varied significantly in most variables (See Appendix A Table A1). For aspen, MRT analysis explained 100% of the variation and clearly separated the seedlings by growing conditions. The first split was determined by seedling height, separating the shortest seedlings with the highest RSR (termed *HighRSR* seedling type) were split from the taller seedlings (Figure A2). The second and final split occurred within the taller seedlings (≥24.4 cm, n = 20) and was in response to the NSC concentration, where the seedlings with more than 27.1% total NSC (termed *HighNSC* seedling type) (n = 10) were split from those with less (termed *Tall* seedling type) (Figure A2).

For pine, the MRT analysis also separated seedlings by conditioning treatment relatively well. MRT analysis explained 95% of the total variance (Figure A2). The three seedling types in pine separated mostly by their size rather than NSC or RSR (Table A2). Seedling height provided the primary split, where the tallest seedlings (≥8.7 cm, n = 10), which also had the highest root and shoot biomass (termed *Large* seedling type), were split from shorter seedlings (<8.70 cm, *n* = 20) (Figure A2). The second split occurred in the shorter seedling group, where a stouter seedling type with higher RCD and root mass (≥0.97 g, *n* = 9) (termed *Medium* seedling type) was separated from a seedling type that had a smaller RCD and less root mass (<0.97 g, *n* = 11) (termed *Small* seedling type). Only one seedling was misclassified in this last split (Figure A2).

2.3. Experimental Growing Conditions

Twenty seedlings of each seedling type and species were planted in 2 L square pots (13.7 cm × 13.7 cm × 15.6 cm depth). To achieve similar medium bulk densities among pots and to more accurately assess the water status in each individual pot for all seedlings and drought treatments, each pot was filled with the same amount of medium (1000 g ± 11.9 SD), which was compacted to the same volume in each pot. The planting medium consisted of a 2:1:1 mixture by volume of peat moss (Pro-Moss, Premier Tech Horticulture, Delson, QC, Canada), vermiculite (Grace Specialty Vermiculite, Grace Construction Products, Vancouver, BC, Canada), and clay (MVP, Profile Products LLC, Buffalo Grove, IL, USA). To minimize nutrient limitations during the duration of the experiment, the water had been blended with 2g L^{-1} of 10-52-10 fertilizer containing chelated micronutrients (Plant Products Co., Ltd., Leamington, ON, Canada). The procedure is described in more detail in Galvez et al. [51].

Soil moisture retention curves were developed for this planting medium prior to the experiment to assess soil moisture conditions more accurately during the study. First, bulk samples of the planting medium were oven dried for 48 h at 60 °C to remove most of the water without affecting the properties of the medium. Samples of the medium were then separated into individual bags, where a measured water quantity was added to obtain a predetermined gravimetric water content. The bags were then sealed and stored at 4 °C for 24 h to allow samples to equilibrate. Two subsamples were then taken from each bag and placed on temperature equilibrium plates (maintained at 19 °C; comparable to the temperature in the growth chamber conditions) to have their soil moisture recorded using a WP4C water potential meter (Decagon Devices, Pullman, WA, USA). A soil moisture retention curve was derived by plotting soil water potentials against the corresponding gravimetric water content, and the data were used to fit a retention curve (Figure A3, [63]). The model was then used to generate the corresponding gravimetric water content for the desired soil water potential targets for the respective drought conditions used in this study.

After filling the pots and planting the seedlings, pots were watered to field capacity (gravimetric soil water content of 70%). All seedlings were grown in a controlled growth chamber, where conditions were held constant for day/night time temperatures at 20.5 °C (±1.6 °C)/18 °C (±1.4 °C), an average relative humidity of 47.1% (±7.6%), and 18 h of fluorescent light (PAR 325 µmol m^{-2} s^{-1} at pot level). The experimental design was fully randomized with the individual pots being the experimental unit. Seedlings were widely spaced to avoid neighbor effects, and all seedlings were re-randomized every four weeks to minimize any potential spatial differences in the growth chamber.

2.4. Drought Treatment

Seedlings from each species and seedling type (*n* = 10) were assigned to one of two soil moisture treatments; a well-watered (control) and dry (drought). Target moisture conditions were chosen based on xylem vulnerability curves that had been developed for both aspen [64] and jack pine [65]. Target water potentials for the drought treatments was chosen for both species at 50% loss of stem hydraulic conductivity (P50). Since catastrophic embolisms (run-away cavitation) can occur past the P50, care was taken to ensure that seedlings did not exceed a shoot water potential of −2.4 MPa for aspen and −1.4 MPa for pine. Gravimetric soil water content was 41.6%, with a corresponding soil water potential of −1.37 MPa.

To replicate more natural soil drying conditions, a gradual dry-down of the pots was implemented by watering the pots with half of the amount of water lost during a 24 h period until target soil moisture conditions were reached [50]. Target mass for gravimetric soil water content was reached after about 20 days in both aspen and pine. Once the target weights were reached, a constant soil water content was maintained. To do so, pots were weighed daily to determine daily water loss and re-watered to the target weight. Control treatments were watered to field capacity twice weekly and weighed daily to verify target weights for the duration of the study.

2.5. Post-Drought Seedling Measurements

After 16 weeks (all seedlings had terminated shoot expansion), seedlings were destructively sampled, and the same variables were measured as described for the *initial* seedlings (Tables A1 and A2) to determine the *final* total and structural biomass and the mass and concentration of NSC (Tables A3 and A4). Prior to harvest, the average midday shoot water potentials were measured in all seedlings. Aspen control seedlings had an average shoot water potential of −0.71 MPa compared to drought with −1.43 MPa, while pine seedlings had a shoot water potential of −0.65 MPa and −1.4 MPa for control and drought seedlings, respectively. For the *final* measurement, aspen leaf mass was added, as it was not available at the initial harvest when aspen seedlings were dormant. Organ-level change in structural biomass (i.e., *structural growth*) was estimated for each seedling by subtracting the estimated *initial* structural biomass from the structural biomass measured at the final harvest. For aspen, initial structural biomass estimates were obtained in the following way. To estimate initial structural biomass, linear regressions models for each seedling type were fit first using the *initial* harvest seedlings, with structural biomass as the dependent variable and either height, RCD, or both as the independent variables. Second, the regression model with the highest R^2 for each seedling type was then used to estimate the initial structural mass of the experimental aspen seedlings, using their *initial* height and/or RCD measurements. Stem structural mass for aspen was best predicted by a combination of root collar diameter and height, whereas root structural mass of *HighNSC* seedlings was best predicted by RCD alone. Root structural biomass of *HighRSR* and *Tall* seedlings was not significantly predicted by either variable, so the *average* root structural biomass of the initially harvested seedlings was used as the estimate for the experimental seedlings. For pine, only the *average* structural biomass from the initial harvest for each seedling type was used as an estimate of *initial* biomass for each experimental seedling. Whole seedling growth estimates were obtained by summing organ-level estimates. Changes in reserve storage were also estimated by subtracting the seedling-type specific average *initial* NSC concentration from each experimental seedling's final NSC concentration. Negative values indicate a net remobilization.

Finally, we assessed the relative allocation of new growth (i.e., new structural biomass) between roots, stems or leaves/needles for the different seedling types in response to drought and well-watered conditions by dividing the organ-level growth by the whole plant growth (i.e., total increase in structural biomass) over the experimental period. Seedlings with negative values of organ-level growth or whole seedling growth were excluded from these analyses.

2.6. Data Analysis

Data were analyzed by separate ANOVAs for each species and followed a 3 × 2 full-factorial design with three seedling types and two drought treatment levels: control and drought. Data that did not meet the assumptions of homogeneity of variance were ln- or square-root transformed. In some cases, transformations did not fix heteroscedasticity, but removal of a single outlier did. To analyze allocation of new growth in pine, however, there was much higher variance in the drought relative to the control treatments. We therefore analyzed the effect of seedling-type separately for the control and drought treatments, and we tested the main effect of drought across seedling types using Welch's *t*-test, which does not assume equal variance.

Finally, because biomass allocation can vary with changes in plant size, we tested for shifts in biomass allocation during the experiment relative to initial allocation patterns by comparing the allometric relationships between root and stem or root and shoot biomass [30]. Differences in biomass allocation between *initial* harvest seedlings, drought, and control seedlings at the final harvest would be manifested either as a difference in the slopes or in the intercepts of the allometric relationships. For each seedling type, we first ran an ANOVA with ln-transformed root biomass as the dependent variable, ln-transformed stem or shoot biomass as the covariate, a treatment effect (initial harvest, drought or control seedlings), and the interaction between covariate and treatment. When the interaction term was not significant ($p > 0.10$), we removed it from the model and used an ANCOVA to test for a treatment effect (i.e., differences in intercept). For one seedling type, the interaction between covariate and treatment was significant (i.e., differences in slopes), and so we tested for differences in slopes between treatments using 95% confidence intervals for the estimated slope parameters for each treatment. Data were analyzed using R (version 4.1.2), and post hoc comparisons were made using Tukey's HSD test ($\alpha = 0.10$).

3. Results

3.1. Aspen

Overall, the *HighRSR* seedling-type tended to grow more in total and aboveground structural biomass than the other seedling types, particularly under well-watered conditions (Figure 1). *HighNSC* seedlings had significantly less leaf and stem growth than both other seedling types under control conditions, but differences became smaller under drought when *HighRSR* seedlings grew more leaves and stem tissue that the other two seedling types (both seedling type × drought effects: $p < 0.001$). However, root growth did not differ between seedling types under well-watered or drought conditions ($p > 0.40$ for both seedling types; seedling type × drought effects). For all seedling types, growth at the seedling and organ levels was reduced by drought (all $p < 0.001$; Table 1, Figure 1).

Table 1. Summary of analysis of variance for change in structural biomass (i.e., structural growth (ΔBiomass) and change in NSC concentrations ($\Delta NSC_{concentration}$) of aspen and pine seedlings by seedling type and drought. Significant effects ($p \leq 0.10$) are indicated in bold, $n = 10$ per seedling type/drought treatment combination, except where noted.

Species	Organ	Response Variable	Seedling Type	Drought	Seedling Type × Drought	Error
		DF	2	*1*	2	*54 or 53*
Aspen	Seedling	ΔBiomass *	0.0379	**<0.0001**	0.21	
		$\Delta NSC_{concentration}$	**<0.0001**	**0.0018**	0.36	
	Leaves	Δ Biomass *	**<0.0001**	**<0.0001**	**0.0007**	
		$\Delta NSC_{concentration}$	0.34	0.28	0.37	
	Stem	Δ Biomass *	0.0006	**<0.0001**	**<0.0001**	
		$\Delta NSC_{concentration}$ *	**<0.0001**	0.17	0.27	
	Roots	ΔBiomass *	0.61	**<0.0001**	0.92	
		$\Delta NSC_{concentration}$	**<0.0001**	**0.0017**	0.40	
Pine	Seedling	ΔBiomass **	**<0.0001**	**<0.0001**	**0.0174**	
		$\Delta NSC_{concentration}$ †	**0.0423**	**<0.0001**	0.72	
	Needles	ΔBiomass **	**<0.0001**	**<0.0001**	**0.0271**	
		$\Delta NSC_{concentration}$ *	**<0.0001**	**<0.0001**	0.20	
	Stem	ΔBiomass *	**<0.0001**	**<0.0001**	**0.0027**	
		$\Delta NSC_{concentration}$	**<0.0001**	**0.0032**	0.41	
	Roots	ΔBiomass **	**<0.0001**	**<0.0001**	**0.0135**	
		$\Delta NSC_{concentration}$ †	**0.0001**	**<0.0001**	0.59	

* Indicates that data were ln transformed. ** Indicates data were square-root transformed. † One Medium pine outlier from control treatment removed from analysis.

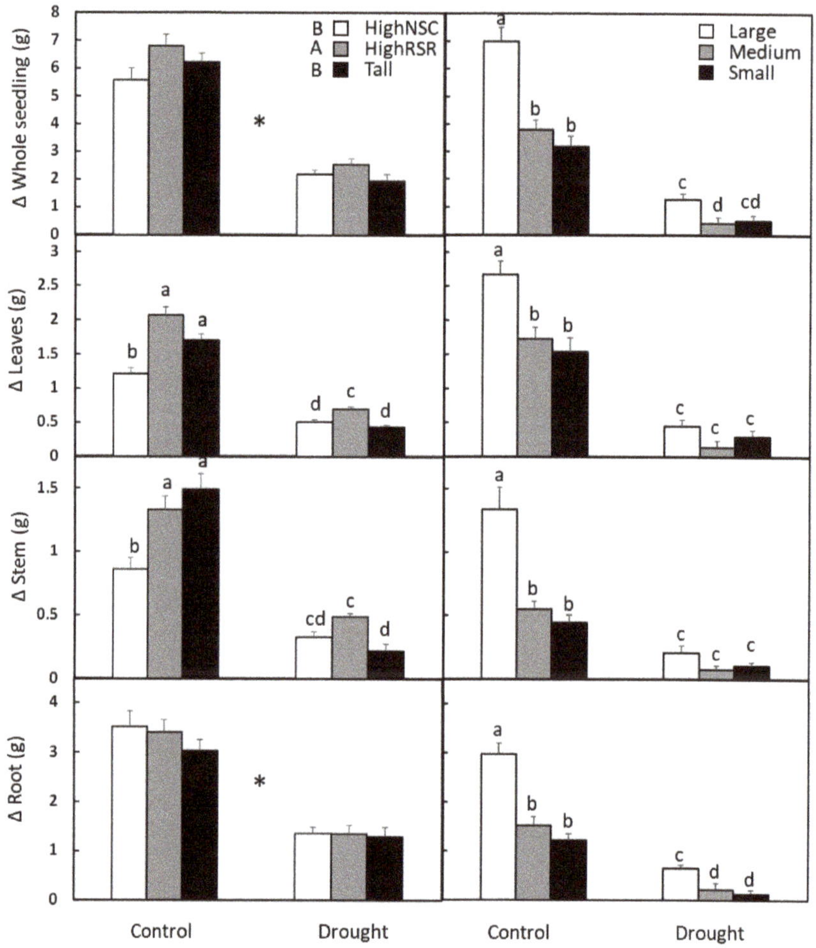

Figure 1. Average (±SE) changes in structural biomass (i.e., growth (ΔBiomass)) at the whole seedling and organ level for three aspen (LEFT column) and three pine (RIGHT column) seedling types under well-watered (control) and dry (drought) conditions. Species and organs were analyzed separately. An * and/or different letters on legend labels indicate significant differences among means of main effects (drought, seedling type), while different letters above bars indicate significant differences among all treatment combination means (Table 1; Tukey's HSD, $p \leq 0.10$).

With a few exceptions, NSC concentrations declined in all aspen seedling types at the whole seedling and organ level over the experimental period (Table 1; Figure 2). The reduction in whole seedling NSC concentrations was significantly greater in seedlings experiencing drought and driven by a significant NSC reduction in the root system rather the stem (Table 1). *HighNSC* and *HighRSR* seedlings, which initially had the highest NSC concentrations (29.6 and 26.9%, respectively), had an approximately four times greater reduction in NSC concentrations than *Tall* seedlings, which had initially much lower NSC concentration (19.7%; Table A1). Leaf concentration did not indicate any treatment or seedling type effects. *HighNSC* and *Tall* seedlings had small positive or negative changes in stem NSC concentrations, while *HighRSR* seedling generally showed the largest reduction in stem NSC concentration (Figure 2).

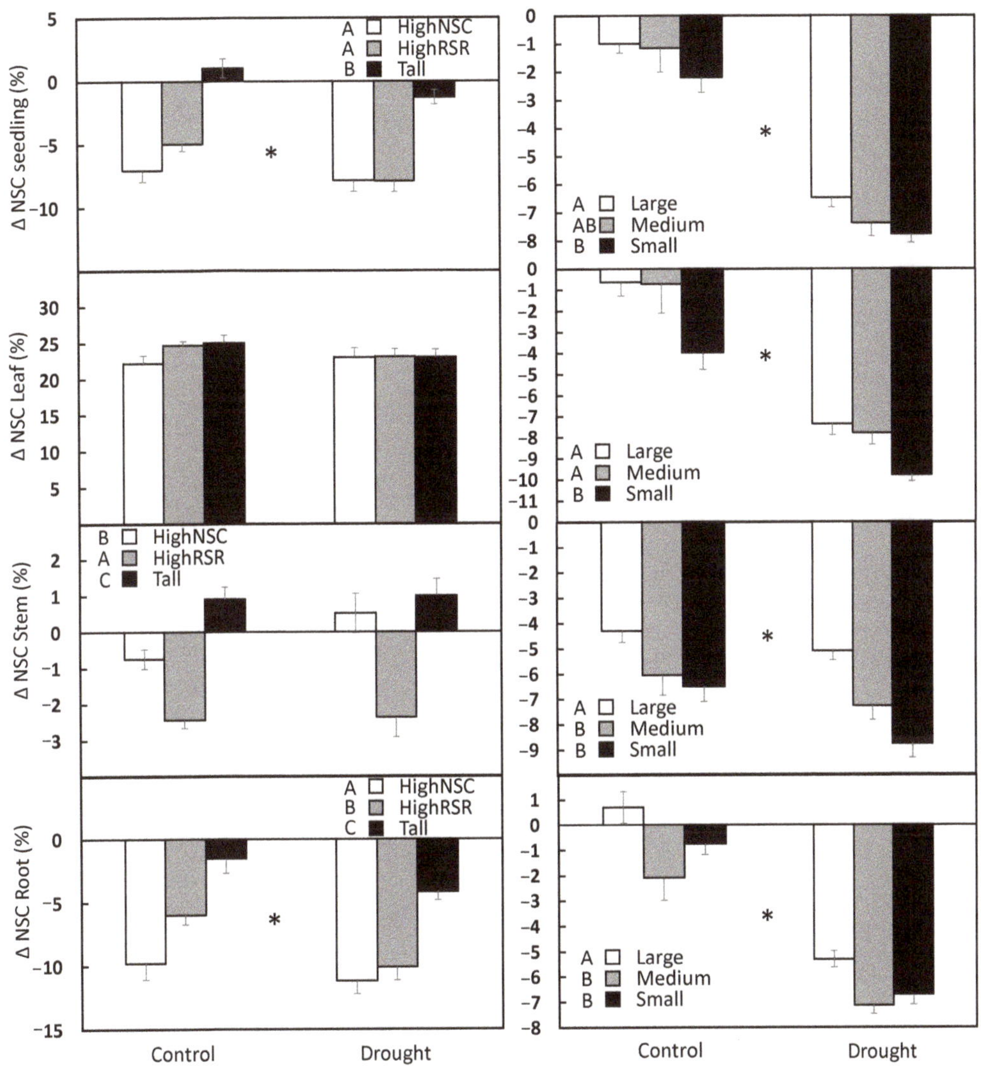

Figure 2. Average (±SE) changes in $NSC_{concentration}$ ($\Delta NSC_{concentration}$) at the organ and whole seedling level for aspen (LEFT column) and pine (RIGHT column) seedling types under drought and control conditions. Species and organs were analyzed separately. An * and/or different letters on legend labels indicate significant differences among means of main effects (drought, seedling type), while different letters above bars indicate significant differences among all treatment combination means (Table 1; Tukey's HSD, $p \leq 0.10$).

The three aspen seedling types allocated new structural biomass differently among organs (Figure 3, $p < 0.05$ for seedling type effect), and only the *Tall* seedlings showed the expected increase in relative allocation toward root growth in response to drought (Table 3; drought × seedling effects: $p < 0.001$). Under drought, *Tall* seedlings significantly decreased allocation of new structural biomass to stems and increased allocation to roots, but neither *HighRSR* nor *HighNSC* seedlings significantly altered their allocation to roots or stems when compared with seedlings in well-watered conditions (Figure 3). Allocation to leaf mass was not affected by drought; however, *HighRSR* seedlings allocated overall more new structural biomass to leaves than *HighNSC* seedlings (Figure 3). Under well-watered conditions,

HighNSC seedlings allocated significantly more to roots and less to leaves compared to the other two seedling types, though under drought, both *Tall* seedlings allocated more to roots and less to stems compared to *HighRSR* seedlings.

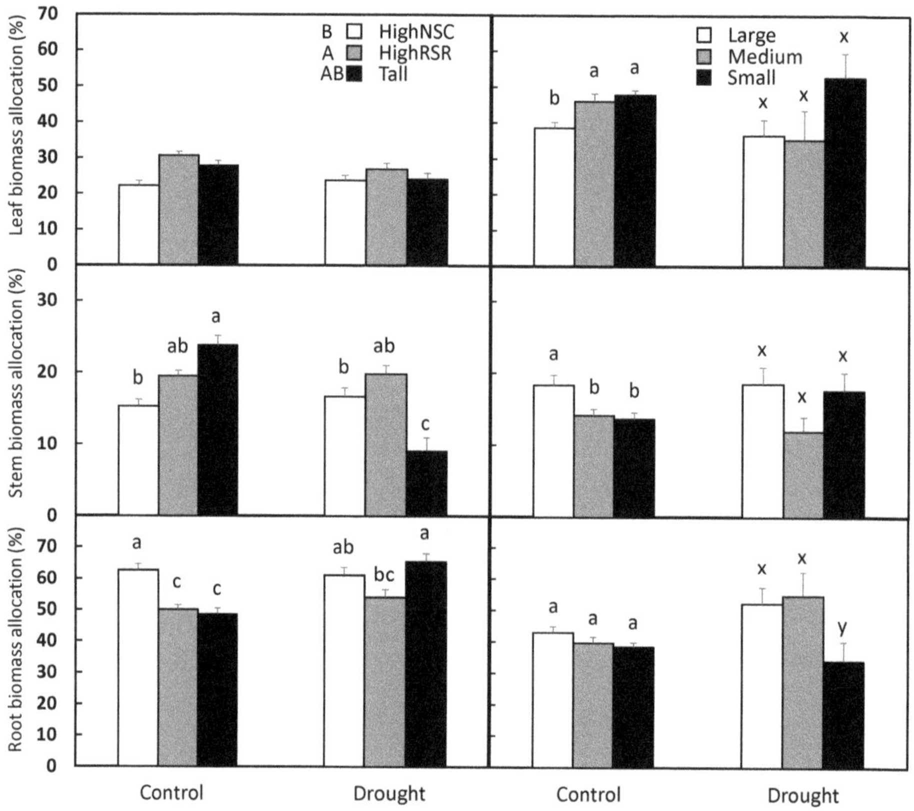

Figure 3. Average (SE) partitioning of new structural biomass (i.e., growth) to leaves, stems and roots in three aspen (LEFT column) and three pine (RIGHT column) seedling types under well-watered (control) and dry (drought) conditions. Species and organs were analyzed separately. Different letters on legend labels indicate significant differences among means of main effects (seedling type), while different letters above bars indicate significant differences among all treatment combination means for aspen (Table 3) and differences in seedling types in the control and drought separately for pine (Table 2) (Tukey's HSD, $p \leq 0.10$).

Table 2. ANOVA summaries for effects of seedling type on relative allocation of new structural biomass (%) for pine. Due to unequal variance between drought and control treatments, the effect of seedling type was tested using oneway ANOVAs for control and drought treatments separately. The main effect of drought was tested using Welch's *t*-test across all seedling types. Significant effects ($p \leq 0.10$) are indicated in bold.

Species	Organ	Seedling Type (for Control)	*n*	Seedling Type (for Drought)	*n*	Main Drought Effect	*n*
	DF	2		2		1	
Pine	Needles	**0.0016**	10	0.12	9,6,8	0.56	30,23
	Stem	**0.0075**	10	0.20	8,5,8	0.46	30,21
	Roots	0.19	10	**0.0633**	10,6,7	**0.0940**	30,23

Table 3. Two-way ANOVA summary for effects of seedling type and drought on relative allocation of new structural biomass (%) for aspen. Significant effects ($p \leq 0.10$) are indicated in bold.

Species	Organ *DF*	Seedling Type 2	Drought 1	Seedling Type × Drought 2	Error
Aspen	Leaves *	**0.0008**	0.102	0.122	53
	Stem **	**0.0116**	**0.0001**	**<0.0001**	52
	Roots *	**0.0002**	**0.0007**	**0.0003**	53

* one outlier from HighRSR, drought treatment removed. ** one outlier from Tall, drought treatment and one from High NSC, drought removed.

Allometric analysis of root versus shoot final biomass$_{Total}$ indicates that *Tall* and *HighNSC* seedlings' final biomass allocation was significantly altered (adjusted) from the initial conditions, displaying an increase in root mass for a given stem biomass (Figure 4; treatment effect: $p < 0.001$). These shifts were generally greater under control conditions than under drought, perhaps in part due to the smaller amount of new biomass that seedlings added under drought, as the final biomass allocation reflects the combination of *initial* differences in biomass allocation and differences in how *new* biomass (including NSC) was partitioned. In contrast, *HighRSR* seedlings did not differ in final biomass allocation regardless of soil moisture conditions (treatment effect: $p = 0.24$).

3.2. Pine

For pine, *Large* seedlings had significantly greater structural growth than *Medium* and *Small* seedlings for both control and drought conditions (Figure 1, Table 1). These patterns were largely mirrored at the organ level, as well. *Large* seedlings had significantly greater root and needle growth than *Small* and *Medium* seedlings (with the exception of needle growth under drought in *Small* seedlings). *Large* seedlings also had greater stem growth under control conditions, but this difference was not significant under drought (Figure 1). As seen for aspen, drought significantly reduced biomass growth of all organs in all pine seedling types (Figure 1, drought effects: $p < 0.001$).

Overall, NSC concentrations declined at the whole seedling and organ level from the initial concentrations (Figure 2, Table 1). These NSC declines were significantly larger under drought for all seedling types in the roots and needles, but were not significant in the stem for any seedling type individually (Figure 2). Generally, *Large* seedlings, which had the lowest initial whole seedling NSC concentrations (Table A2), had smaller organ-level decreases in NSC concentrations compared to the *Medium* and *Small* seedlings. *Medium* and *Small* seedlings had similar declines, except that under control conditions *Medium* seedlings had a significantly greater decline in the roots while *Small* seedlings had a greater decline in the needles (Figure 2).

In contrast to aspen, there were fewer differences in the relative allocation of new structural biomass to different organs in pine (Figure 3, Table 2). Under control conditions, *Large* seedlings allocated relatively more new biomass to stems and less to needles compared to *Medium* and *Small* seedlings (Figure 3). The drought had limited impacts on the partitioning of new biomass with only a marginally significant increase in root biomass allocation across seedling types (Table 2, drought effect: $p = 0.094$). Under drought, the different seedling types did not strongly differ, except that *Small* seedlings allocated marginally less new biomass to roots than *Large* and *Medium* seedlings (Figure 3, Table 2).

Finally, the allometric relationships indicate that *Large* and *Small* pine seedling types varied in their final biomass allocation from their initial allocation (Figure 4). *Large* seedlings had an altered relationship between shoot and root biomass compared to the initial seedlings, regardless of soil moisture conditions, with a smaller increase in root biomass for a given increase in stem biomass (i.e., covariate × treatment effect: $p = 0.060$). *Small* seedling allometry also differed between initial and final treatments (treatment effect: $p = 0.071$), with droughted seedlings having slightly lower root biomass for a given shoot biomass (Figure 4). *Medium* seedlings, however, did not show any difference in allometry between drought and controls or between initial and final harvests (treatment effect: $p = 0.93$).

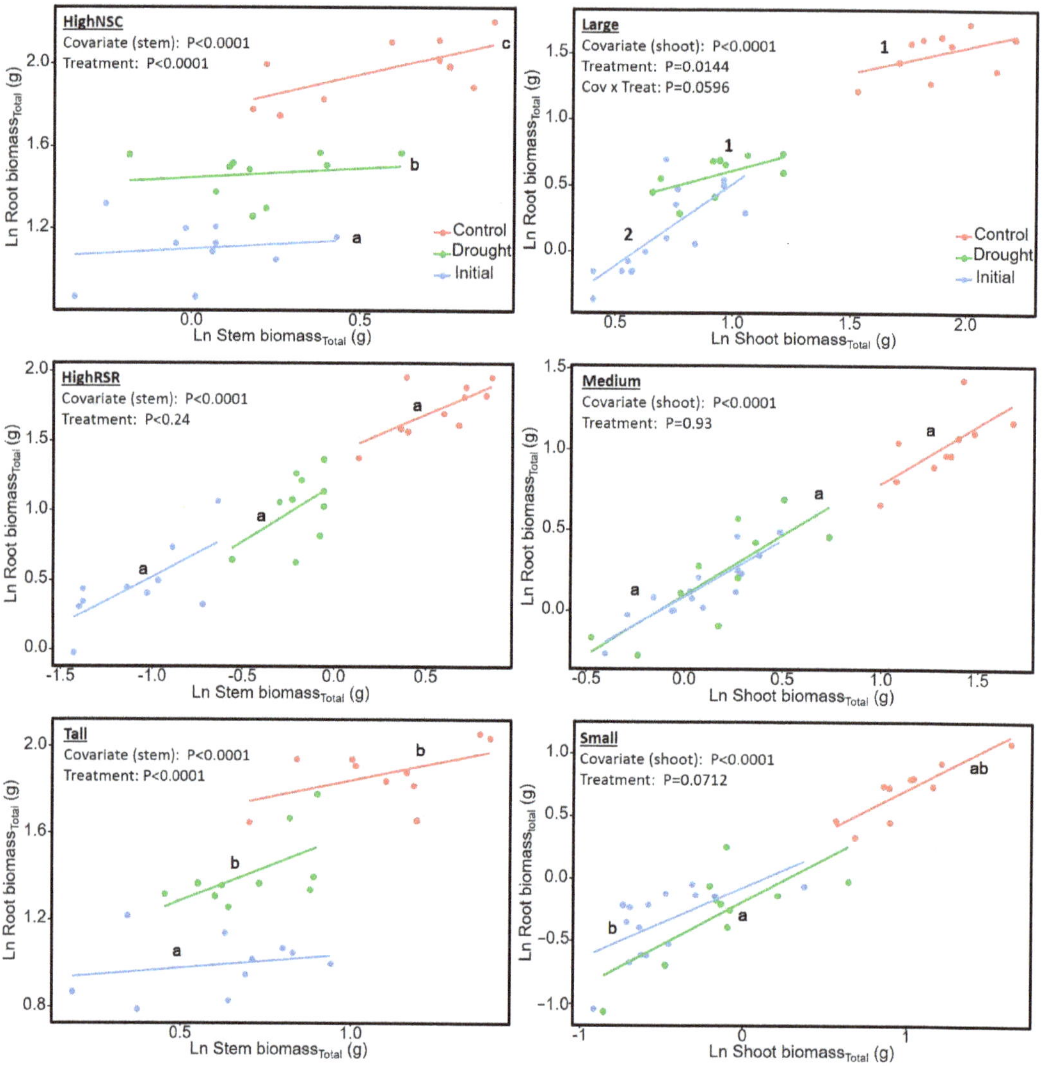

Figure 4. Allometric relationships between root and stem (excl. leaves) Biomass$_{Total}$ (aspen, LEFT column) and root and shoot (incl. needles) Biomass$_{Total}$ (pine, RIGHT column) for three different seedling types at the initial harvest, after well-watered (control) and dry (drought) conditions. Relationships between root and stem biomass in aspen are shown, as seedlings at the initial harvest were leafless. Lines represent individual linear regressions for each treatment and seedling type combination. Significant treatment or covariate × treatment effects indicate differences in biomass allocation between drought, control and initial seedlings. Different letters represent significant differences in intercepts according to Tukey's HSD ($p < 0.10$). For *Large* pine, there was a significant covariate × treatment interaction; different numbers therefore represent significant differences in slope according to 95% confidence intervals.

4. Discussion

In both species, the different conditioning treatments produced seedling types that differed in size and in morphological and physiological characteristics. Initial biomass allocation and NSC concentrations were particularly different for aspen, which subsequently influenced how new biomass was allocated between organs in the following growing season, and ultimately affected their amount of new growth. For aspen, the *HighRSR* seedlings—the smallest seedling type—were the most productive in terms of whole seedling growth across well-watered and drought conditions (Figure 1), demonstrating that larger seedlings are not always better. This greater growth likely stems from their initially higher root:stem biomass ratio. During the experiment, both *HighNSC* and *Tall* seedlings significantly altered their root-stem allometric trajectory from their initial trajectory, increasing root mass for a given stem mass (Figure 4). In contrast, *HighRSR* seedlings remained on the same allometric trajectory during the experimental period, even under drought conditions, suggesting that this pre-conditioning treatment produced aspen seedlings with a more optimal organ balance. With an already greater root allocation, *HighRSR* seedlings could partition relatively more to the new growth of leaves (Figure 3), which would increase C gain, supporting the current and future growth of other organs as well. Thus, while growing conditions can produce taller or larger seedlings, if these conditions also alter aspen allometry, it may require seedlings to make allocation adjustments later that will incur a cost and reduce overall growth potential and potentially establishment success.

While both *Tall* and *HighNSC* aspen seedling types generally altered their allometry during the experiment by increasing root biomass allocation (Figure 4), the drivers underlying these adjustments likely differ. *Tall* seedings had the lowest initial NSC concentrations and exhibited relatively small NSC reductions, suggesting their root growth likely relied more on current photosynthates. In addition, in response to the higher water demand, the allocation to roots increased further with drought, potentially requiring greater demand from current photosynthates. In contrast, *HighNSC* seedlings—with the highest initial root NSC concentrations—had the greatest reduction in root reserve concentrations during the experimental period (Figure 2); it therefore seems likely that a large portion of this remobilized NSC was utilized to support new root growth. Thus, we suggest that the allometric shift favoring root biomass in this seedling type was driven by their greater resource supply for root growth. This is supported by the fact that *HighNSC* seedlings with a higher root:stem ratio still partitioned significantly more biomass to root growth than *Tall* seedlings under well-watered conditions, despite having less water demand because of lower leaf area. Root NSC storage may be largely constrained for use in the root system, explaining these seedlings' lower aboveground growth despite the large remaining reserve pool belowground. The potential for root reserves to be largely restricted from remobilization to other organs has been suggested in other studies [66,67] (Hart et al. in prep.), supporting the idea that NSC pools in different organs are regulated independently [45] (Fermaniuk et al. in prep.). Alternatively, because *HighNSC* seedlings were conditioned by the application of a growth inhibitor, it is also possible that there were residual hormonal effects of this application which might have continued to restrict aboveground growth, leading to greater C supply belowground (and/or restricted use of root reserves aboveground). Such a constraint could also explain why *HighNSC* seedlings ended up with a higher root mass for a given stem mass than both other seedling types under well-watered conditions (Figure A4). A longer-lasting aboveground sink limitation effect could make this form of seedling conditioning problematic, particularly in sites with strong competition; however, residual effects of this treatment have not been observed in earlier outplanting studies [48,58,68].

Optimal partitioning theory predicts that under well-watered conditions, plants should preferentially allocate resources to aboveground organs responsible for carbon capture [69], but under drought, allocation should shift to favor root biomass to increase water extraction [70]. While increased root allocation in response to drought has been commonly reported in many plants [30,37,70], we found that only the response of *Tall* seedlings—which had the lowest initial root-to-stem ratio—was consistent with the optimal partitioning theory. In *Tall* seedlings, the increased allocation to root growth came at the expense of stem growth, consistent with the results of a meta-analysis for severe drought impacts on biomass allocation [30]. In contrast, *HighRSR* and *HighNSC* seedling types did not significantly alter their biomass partitioning in response to water stress (Figure 3). This lack of response is likely due to their 'preconditioning', which initially increased their root:stem ratios, perhaps allowing them to maintain higher water potentials and avoid or reduce the severity of the physiological impacts of soil drought. This would also be consistent with the findings of Poorter et al. [30], who found root biomass allocation to increase only under severe but not mild drought. Though we did not test the effect of drought on biomass allocation directly in pine, we did see similar trends: allocation of new growth to roots increased in all seedling types except the *Small* pine seedlings, which initially had the highest root: shoot ratios (Figure 3). Therefore, initial seedlings characteristics, particularly in plastic species, must be considered carefully when evaluating seedling responses to growing conditions in the field or experimental studies, as these initial seedling conditions can affect how individuals respond to stressors, or if they respond at all.

Unlike aspen, the largest pine seedling type generally had the greatest growth in the next growing season, although this advantage disappeared under drought conditions. The *Large* pine seedling type grew more than smaller seedling types under well-watered conditions, consistent with many studies that indicate taller or larger seedlings have higher outplanting survival and growth under non-adverse conditions [47,58,71–73]. The greater growth of *Large* seedlings may be the combined result of initially higher leaf area driving photosynthetic gain as well as initially higher NSC mass, suggesting that differences in pool size of available reserves rather than tissue NSC concentrations can be more important determinants of growth potential or survival [41,66,72,74,75]. However, when faced with drought, all pine seedling types showed a large decline in new growth, and the differences in growth among the three seedling types became much smaller. This is consistent with findings that the potential advantages of large size (e.g., taller seedlings) are significantly reduced or even reversed when water becomes limiting, particularly in pine species [71,76–78].

In further contrast to aspen, we found that alterations to organ biomass allometry in pine did not incur a cost for new growth, likely because the main organ where reserves are stored differs between species. As in aspen, we observed significant shifts in the allometry of some pine seedling types (*Large* and *Small*; Figure 4). However, unlike aspen, the *Medium* pine seedlings that did *not* exhibit an allometric shift, but instead tended to grow less, not more (Figure 1). Thus, shifting allometry was not associated with reduced growth in pine seedling types as in aspen. This difference between pine and aspen may derive from the difference in where the long-term reserves are stored in deciduous and evergreen species [23,24,67]. For both species, an initially larger size came at the expense of a higher root:stem or root:shoot ratio. In aspen, because the roots are a major location of stored reserves (i.e., higher concentrations; Table A1), the increase in initial size during conditioning came at the expense of their NSC storage. In contrast, needles of evergreen conifer seedlings tend to be the major NSC storage site, often containing more than half of the total NSC pool (e.g., [79,80]). Here, 50%–60% of NSC reserves in pine were located in needles, and both stem and needles maintained higher NSC concentrations than roots. Thus, producing a larger, taller pine seedling did not cause a reduction in storage. On the contrary, *Large* seedlings had about double the NSC mass of *Small* seedlings with higher root:shoot ratios (Table A2). Ultimately, this suggests that the cost of reducing root allocation is likely lower in pine than in aspen, though this cost likely increases under drought. Therefore, because species differ in terms of how storage pools are distributed among organs, the type of initial seedling characteristics (and biomass allocation) that affect subsequent growth and survival should be expected to vary between species and functional types [23,72].

Changes in seedling allometry that impact storage pools may be particularly important during stressful periods, when seedlings may have to increase their reliance on reserves.

During drought, both aspen and pine seedlings had significantly lower NSC, with reduced concentrations in the roots of aspen and declines in all organs of pine. As photosynthesis likely declined under drought [50,81], these larger NSC declines may have resulted from an increased reliance of growth, respiration, and/or constitutive defenses on stored NSC [32,80,82–84]. Alternatively, the drought may have reduced the allocation of assimilates to refilling of reserves (primary growth had ceased) relative to well-watered conditions. Whatever the cause, the larger reductions in reserves may limit the ability of these seedlings to recover from a successive stress or impact their ability to survive the winter [51,85]. Therefore, seedling types such as *HighNSC* aspen that maintained higher NSC concentrations at the end of the experiment may still have an advantage in the longer-term, despite their lower aboveground growth in the short-term. In addition, the different location of storage between pine and aspen may also lead to differences in their ability to utilize reserves under drought. If phloem malfunction limits carbohydrate redistribution belowground [86–88], then the reliance of pine on needle NSC storage might make pine roots particularly sensitive to severe water limitation, whereas root function in aspen may be preserved longer because of their preferential storage belowground.

Finally, our study suggests that while both species can be manipulated to alter initial characteristics, aspen may exhibit greater phenotypic plasticity and thus a greater range of responses to pre-conditioning treatments than jack pine, as the allometric adjustments between seedling types—though not statistically compared between species—appeared much larger for aspen. Additionally, while initial differences in seedling characteristics had profound effects on size and biomass allocation (i.e., root-shoot balance) in both species, these changes were associated with much greater plasticity in physiological attributes (here NSC concentrations) in aspen than in pine. The larger seedling-type differences in aspen are consistent with other studies that suggest physiological attributes, including photosynthetic and leaf hydraulic traits of deciduous angiosperms, display more phenotypic plasticity than evergreen conifers [32,89–91]. Plasticity in physiological traits associated with changes in size and allometry should therefore be considered, as conditioning for larger seedling size may be associated with maladaptive physiological changes such as reduced storage in some species. Recognizing these types of taxonomic differences in response to initial growing conditions is essential when assessing seedling survival and growth after outplanting, and when using seedlings for ecophysiological studies and research.

5. Conclusions

We found that initial differences in seedling characteristics can substantially alter carbon allocation. The expected allocation shift with drought favoring roots was only observed when root:shoot or root:stem ratios were initially low. Additionally, we found that the traits associated with greater growth were quite different between pine and aspen. While larger seedlings had greater growth in pine, the smallest seedling type with the highest root:stem ratio grew most in aspen. In aspen, this smaller seedling type was the only one that did not undergo a shift in biomass relative to its initial allometry, suggesting that adjustments in biomass allocation made by other, larger seedling types may have come at the cost of reduced growth. In contrast, adjustments in allocation did not appear to negatively impact pine, possibly because reduced root:shoot ratios did not reduce NSC storage, as it did in aspen. We conclude that differences in biomass allocation and size can alter storage and seedling drought response; initial seedling characteristics should therefore be considered when predicting the effects of environmental stressors.

Author Contributions: Authors significantly contributed to the conception, design of the work and/or the acquisition, analysis, and interpretation of data. Conceptualization, S.M.L.; Methodology and data collection S.M.L., S.P.K., A.E.G.; Formal Analysis, E.T.W., K.A.S.; Investigation, K.A.S., A.E.G.; Resources, S.M.L.; Writing—Original Draft Preparation, E.T.W., S.M.L., K.A.S., S.P.K. and A.E.G.; Writing—Review and Editing, E.T.W. and S.M.L.; Visualization, E.T.W., S.M.L., K.A.S., S.P.K. and A.E.G.; Supervision, S.M.L.; Project Administration, S.M.L.; Funding Acquisition, S.M.L. All authors have read and agreed to the published version of the manuscript.

Funding: We thank the Natural Sciences and Engineering Research Council of Canada (NSERC), Syncrude Canada Ltd., Suncor Energy, Capital Power Corporation, and Shell Canada (IRCPJ 378901-13) for funding the research.

Data Availability Statement: The data presented in this study are available on request from the corresponding author.

Acknowledgments: We owe gratitude to Pak Chow for the non-structural carbohydrate analyses. Thank you also to Fran Leishman, Eckehart Marenholtz, and David Galvez for their enormous help with the set up and maintenance of the study.

Conflicts of Interest: The authors declare no conflict of interest.

Appendix A

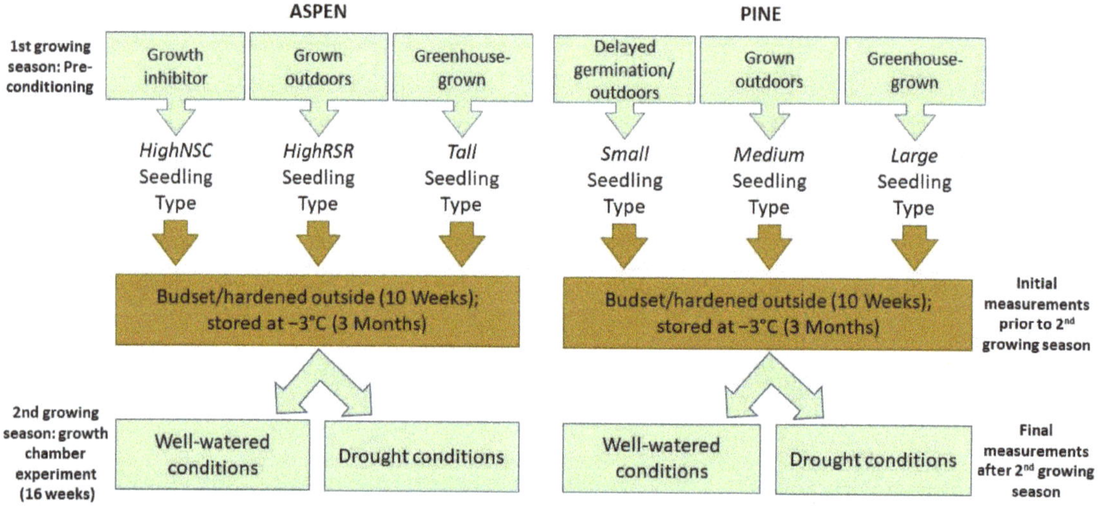

Figure A1. Flow-chart showing a simplified summary and timeline for the whole study; more details on the different steps are provided in the Section 2. The top row depicts the different seedling preconditioning treatments for aspen (**left**) and pine (**right**) creating three different seedling types for each species. The naming of each of the seedling types is based on the dominant seedling characteristic separating the seedling types from each other (see also Figure A2). All seedlings were hardened, dormant and stored frozen prior to exposing them to drought and well-watered conditions over a second growing season. Seedling characteristics were measured for all seedling types before and after the second growing season.

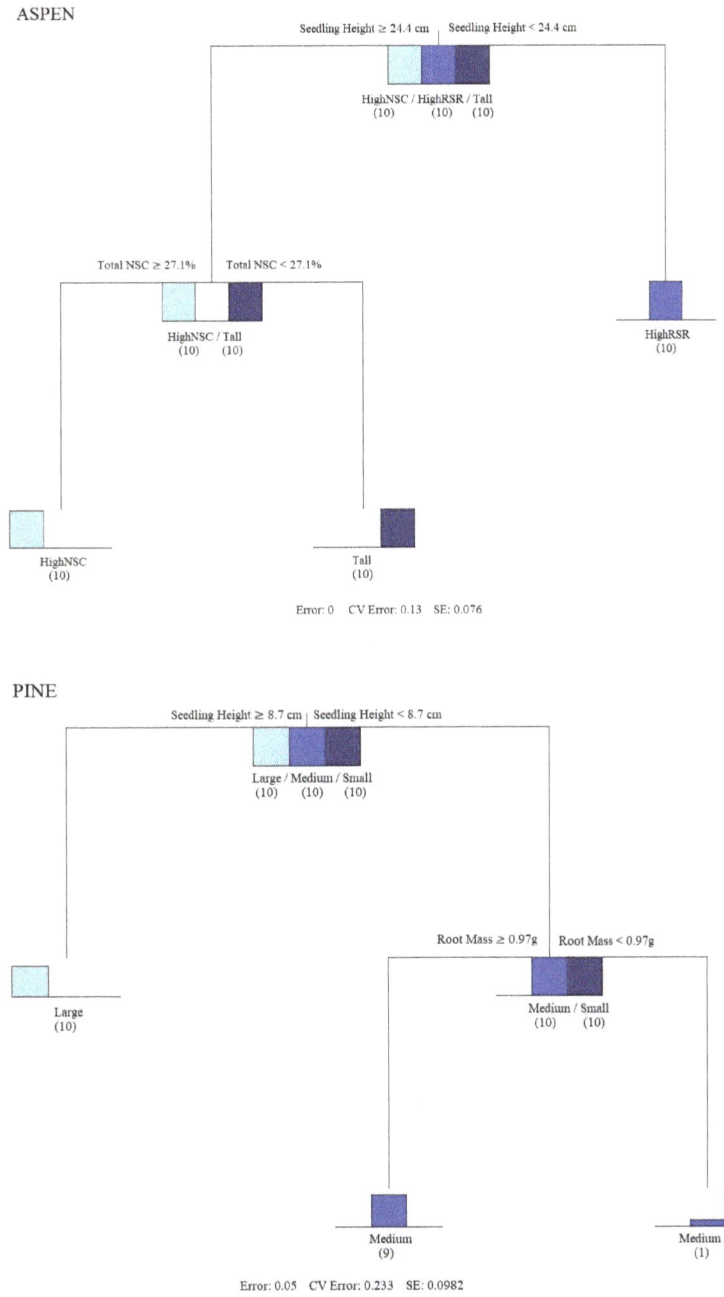

Figure A2. Multiple Regression Tree (MRT) analysis based on the initial aspen seedling characteristics (**top**) to determine the contribution of variables determining each seedling type. This tree explained 100% of the total variance, and the vertical depth of each split is proportional to the variation explained. The numbers below each of the splits correspond to the number of seedlings by corresponding seedling type. MRT analysis (**bottom**) of initial pine seedling characteristics that determine the contributing factors in determining seedling type. This tree correctly explained 95% of the total variance, where the vertical depth of each split is proportional to the variation explained.

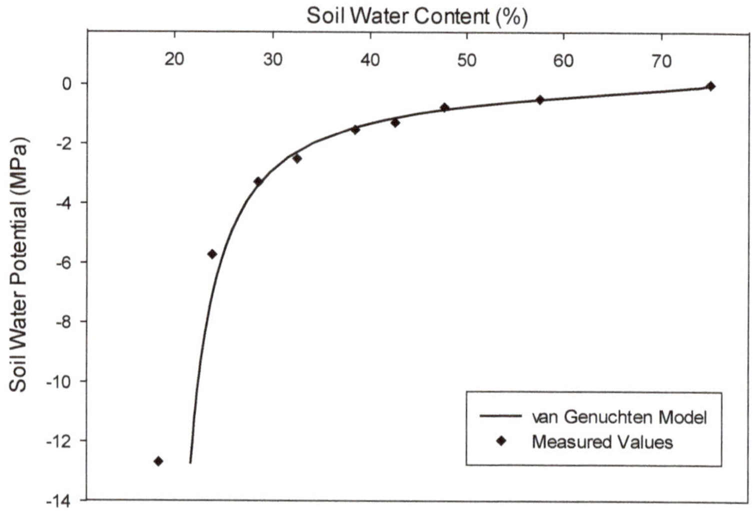

Figure A3. Soil moisture retention curve for the growing medium. Measured values were obtained using a WP4C dewpoint potentiometer, and the van Genuchten model was used for curve fitting. This allowed for the estimation of the required soil water content required to achieve the drought-like conditions.

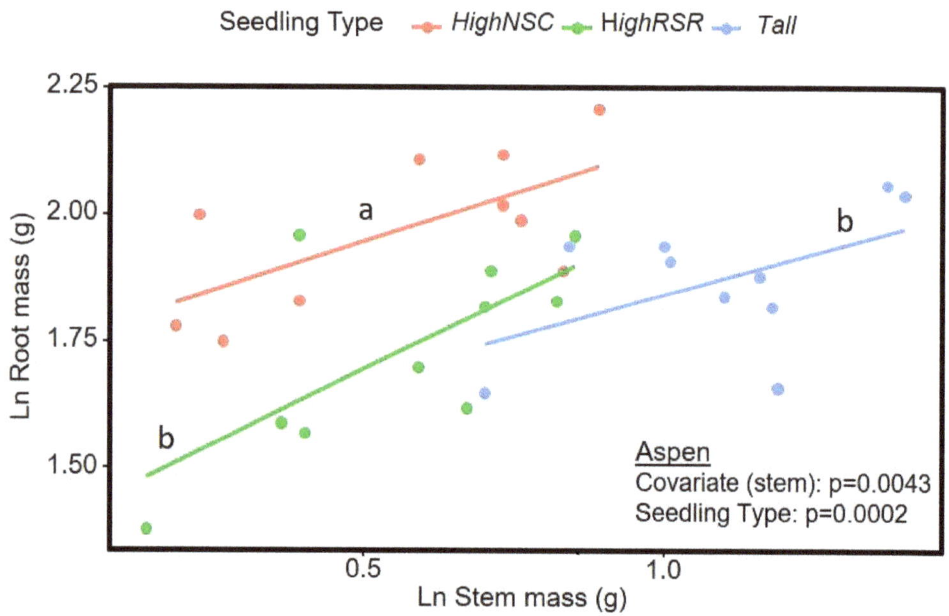

Figure A4. Allometric relationships between final root and stem $Biomass_{Total}$ for aspen seedling types under well-watered conditions (control). Lines represent individual linear regressions for each treatment and seedling type combination. Differences between seedling types were tested using ANCOVA, after testing for homogeneity of slopes. Different letters represent significant differences in intercepts between seedling types according to Tukey's HSD (all $p < 0.005$).

Table A1. Summary of *initial* measurement means (±SE) for aspen by seedling type. Different letters indicate significant differences among seedling types using Tukey-HSD post hoc tests ($p \leq 0.10$, n = 15).

	Measurement	Seedling Type		
		HighNSC	*HighRSR*	*Tall*
	Height (cm) *	32.35 (0.79) b	15.90 (0.61) c	51.78 (1.26) a
	Root Collar Diameter (mm) *	3.91 (0.08) b	3.03 (0.07) c	4.76 (0.11) a
	Root Mass (g)	3.04 (0.13) a	1.64 (0.17) b	2.72 (0.12) a
	Stem Mass (g) *	1.05 (0.08) b	0.35 (0.03) c	1.89 (0.14) a
	Whole Seedling Mass (g)	4.09 (0.16) a	1.99 (0.19) b	4.61 (0.19) a
	Root Stem Ratio (g g^{-1}) *	3.04 (0.25) b	4.83 (0.31) a	1.51 (0.13) c
NSC concentration (%)	Root Sugar	24.17 (0.92) a	21.92 (0.60) a	18.24 (0.74) b
	Root Starch	9.79 (0.46) a	7.49 (0.87) b	5.78 (0.77) b
	Root NSC	33.96 (0.99) a	29.41 (1.36) b	24.01 (1.34) c
	Stem Sugar *	16.52 (0.43) a	15.27 (0.51) a	12.72 (0.29) b
	Stem Starch	0.21 (0.03) a	0.22 (0.03) a	0.26 (0.02) a
	Stem NSC *	16.73 (0.43) a	15.48 (0.49) a	12.98 (0.29) b
	Whole Seedling NSC	29.55 (0.76) a	26.94 (1.19) a	19.67 (1.08) b
NSC mass (g)	Root NSC	1.03 (0.05) a	0.48 (0.06) b	0.66 (0.06) c
	Stem NSC *	0.18 (0.01) b	0.05 (0.006) c	0.24 (0.02) a
	Whole Seedling	1.20 (0.05) a	0.53 (0.06) c	0.90 (0.05) b
Structural biomass (g)	Root	2.01 (0.10) a	1.16 (0.11) b	2.06 (0.08) a
	Stem *	0.87 (0.06) b	0.29 (0.03) c	1.65 (0.12) a
	Whole Seedling	2.88 (0.12) b	1.45 (0.13) c	3.71 (0.18) a

* Data were ln-transformed prior to analysis.

Table A2. Summary of *initial* measurement means (±SE) for pine by seedling type. Different letters indicate significant differences among seedling types using Tukey's HSD post hoc tests ($p \leq 0.10$, n = 15).

	Measurement	Seedling Type		
		Large	*Medium*	*Small*
	Height (cm) *	14.35 (0.62) a	5.45 (0.28) b	5.52 (0.34) b
	Root Collar Diameter (mm)	2.73 (0.07) a	2.49 (0.13) b	1.98 (0.06) c
	Root Mass (g) *	1.20 (0.10) a	1.17 (0.06) a	0.73 (0.05) b
	Stem Mass (g) *	0.48 (0.03) a	0.26 (0.02) b	0.13 (0.01) c
	Needle Mass (g)	1.56 (0.08) a	0.86 (0.06) b	0.51 (0.06) c
	Whole Seedling Mass (g) *	3.24 (0.19) a	2.29 (0.13) b	1.36 (0.10) c
	Root:Shoot Ratio	0.59 (0.04) c	1.07 (0.03) b	1.20 (0.08) a
NSC concentration (%)	Root Sugar *	4.21 (0.10) a	4.57 (0.22) a	4.43 (0.17) a
	Root Starch	6.12 (0.54) b	7.57 (0.34) a	7.56 (0.36) a
	Root NSC	10.33 (0.57) b	12.14 (0.46) a	12.00 (0.39) a
	Stem Sugar	13.87 (0.55) b	15.77 (0.53) a	16.47 (0.42) a
	Stem Starch	0.38 (0.07) b	1.15 (0.10) a	1.13 (0.11) a
	Stem NSC	14.26 (0.58) b	16.92 (0.59) a	17.60 (0.44) a
	Needle Sugar	16.65 (0.31) b	16.74 (0.30) b	18.23 (0.38) a
	Needle Starch	0.15 (0.01) a	0.11 (0.01) b	0.11 (0.01) a b
	Needle NSC	16.63 (0.41) b	16.63 (0.29) b	18.16 (0.52) a
	Whole Seedling NSC	14.03 (0.33) b	14.47 (0.27) a	14.85 (0.21) a
NSC mass (g)	Root NSC	0.123 (0.012) a	0.141 (0.007) a	0.088 (0.007) b
	Stem NSC *	0.070 (0.007) a	0.043 (0.004) b	0.023 (0.002) c
	Needle NSC *	0.263 (0.017) a	0.146 (0.011) b	0.092 (0.011) c
	Whole Seedling NSC *	0.457 (0.032) a	0.330 (0.018) b	0.203 (0.016) c
Structural biomass (g)	Root *	1.08 (0.09) a	1.03 (0.06) a	0.64 (0.04) b
	Stem *	0.41 (0.03) a	0.22 (0.02) b	0.11 (0.01) c
	Needle	1.29 (0.06) a	0.71 (0.05) b	0.41 (0.05) c
	Whole Seedling *	2.78 (0.16) a	1.96 (0.11) b	1.16 (0.08) c

* Data were ln-transformed prior to analysis.

Table A3. Summary of *final* measurement means (±SE) for aspen seedlings by seedling type and drought interaction. $Biomass_{Total}$ is the sum of structural and NSC mass. Different letters within a row indicate significant differences according to Tukey's HSD ($p \leq 0.10$).

Organ	Variable	Control			Drought		
		HighNSC	*HighRSR*	*Tall*	*HighNSC*	*HighRSR*	*Tall*
Whole Seedling	$Biomass_{Total}$ *	10.6 (0.56) a	10.3 (0.55) a	11.9 (0.51) a	6.3 (0.22) b	4.61 (0.25) c	6.81 (0.34) b
	$Biomass_{Structural}$ *	8.2 (0.38) ab	8.0 (0.31) b	9.4 (0.29) a	4.9 (0.16) c	3.72 (0.18) d	5.56 (0.25) c
	NSC_{Mass} *	2.4 (0.15) a	2.3 (0.14) a	2.5 (0.18) a	1.4 (0.07) b	0.89 (0.07) c	1.26 (0.08) b
	$NSC_{Concentration}$	22.5 (0.9) a	22.0 (0.5) ab	20.7 (0.8) abc	21.7 (0.9) ab	19.0 (0.8) bc	18.4 (0.6) c
Leaves	$Biomass_{Total}$ *	1.57 (0.13) b	2.7 (0.15) a	2.28 (0.13) a	0.66 (0.05) d	0.91 (0.05) c	0.57 (0.04) d
	$Biomass_{Structural}$ *	1.21 (0.09) b	2.1 (0.12) a	1.70 (0.09) a	0.51 (0.03) d	0.70 (0.03) c	0.44 (0.02) d
	NSC_{Mass} *	0.4 (0.05) b	0.68 (0.04) a	0.58 (0.05) a	0.15 (0.02) cd	0.21 (0.02) c	0.13 (0.02) d
	$NSC_{Concentration}$	22.2 (1.1) a	24.7 (0.6) a	25.0 (1.1) a	23.1 (1.5) a	23.1 (1.1) a	23.0 (1.1) a
Stem	$Biomass_{Total}$ *	1.80 (0.15) b	1.80 (0.13) b	3.06 (0.20) a	1.26 (0.09) c	0.83 (0.04) d	2.06 (0.10) b
	$Biomass_{Structural}$ *	1.51 (.12) b	1.56 (0.11) b	2.63 (0.17) a	1.04 (0.07) c	0.72 (0.03) d	1.77 (0.09) b
	NSC_{Mass} *	0.3 (0.02) b	0.23 (0.02) b	0.42 (0.03) a	0.22 (0.02) b	0.11 (0.01) c	0.29 (0.02) b
	$NSC_{Concentration}$	15.9 (0.3) a	13.1 (0.2) b	13.9 (0.3) b	17.3 (0.5) a	13.1 (0.5) b	14.0 (0.5) b
Root	$Biomass_{Total}$ *	7.3 (0.35) a	5.75 (0.34) b	6.57 (0.28) ab	4.35 (0.15) c	2.87 (0.22) d	4.19 (0.25) c
	$Biomass_{Structural}$ *	5.5 (0.32) a	4.39 (0.25) a	5.09 (0.22) a	3.36 (0.13) b	2.30 (0.16) c	3.35 (0.20) b
	NSC_{Mass} *	1.7 (0.10) a	1.36 (0.10) ab	1.48 (0.11) a	0.99 (0.05) bc	0.57 (0.06) d	0.84 (0.06) c
	$NSC_{Concentration}$	24.3 (1.3) a	23.4 (0.7) ab	22.5 (1.2) ab	22.8 (1.0) ab	19.3 (1.1) b	19.9 (0.6) b

* Data were ln-transformed prior to analysis.

Table A4. Summary of final measurement means (±SE) seedling for pine seedlings by seedling type and drought interaction. $Biomass_{Total}$ is the sum of structural and NSC mass. Different letters within a row indicate significant differences according to Tukey's HSD ($p < 0.10$).

Organ	Source of Variation	Control			Drought		
		Large	*Medium*	*Small*	*Large*	*Medium*	*Small*
Whole Seedling	$Biomass_{Total}$ *	11.2 (0.56) a	6.59 (0.40) b	4.92 (0.44) bc	4.40 (0.22) c	2.55 (0.25) d	1.75 (0.18) e
	$Biomass_{Structural}$ *	9.78 (0.49) a	5.76 (0.35) b	4.30 (0.39) b	4.07 (0.20) b	2.37 (0.23) c	1.63 (0.17) d
	$Mass_{NSC}$ *†	1.46 (0.08) a	0.87 (0.10) b	0.62 (0.06) b	0.33 (0.02) c	0.18 (0.02) d	0.12 (0.01) e
	$NSC_{Concentration}$ †	13.0 (0.3) a	13.3 (0.9) a	12.6 (0.5) a	7.6 (0.3) b	7.1 (0.5) b	7.1 (0.3) b
Needles	$Biomass_{Total}$ *	4.75 (0.25) a	2.91 (0.20) b	2.22 (0.25) bc	1.91 (0.11) c	0.93 (0.11) d	0.72 (0.10) d
	$Biomass_{Structural}$ *	3.97 (0.19) a	2.44 (0.17) b	1.90 (0.21) b	1.73 (0.10) b	0.84 (0.10) c	0.66 (0.09) c
	NSC_{Mass} *	0.78 (0.07) a	0.47 (0.06) b	0.32 (0.05) b	0.18 (0.02) c	0.09 (0.01) d	0.06 (0.01) d
	$NSC_{Concentration}$ *	16.2 (0.6) a	16.1 (1.4) a	14.4 (0.8) a	9.4 (0.5) b	9.1 (0.6) b	8.6 (0.3) b
Stem	$Biomass_{Total}$ *	1.94 (0.19) a	0.85 (0.07) b	0.61 (0.07) c	0.68 (0.06) b	0.31 (0.04) c	0.23 (0.03) c
	$Biomass_{Structural}$ *	1.75 (0.18) a	0.76 (0.06) b	0.55 (0.06) b	0.61 (0.06) b	0.28 (0.03) c	0.21 (0.03) c
	NSC_{Mass} *	0.19 (0.01) a	0.09 (0.01) b	0.07 (0.01) b	0.06 (0.01) b	0.03 (0.01) c	0.02 (0.01) d
	$NSC_{Concentration}$	10.0 (0.4) a	10.9 (0.8) a	11.1 (0.6) a	9.2 (0.4) a	9.7 (0.6) a	8.8 (0.6) a
Roots	$Biomass_{Total}$ *	4.55 (0.23) a	2.82 (0.20) b	2.09 (0.15) bc	1.81 (0.08) c	1.31 (0.13) d	0.81 (0.08) e
	$Biomass_{Structural}$ *	4.06 (0.22) a	2.56 (0.18) b	1.85 (0.13) bc	1.72 (0.08) d	1.24 (0.13) d	0.76 (0.08) e
	NSC_{Mass} †*	0.50 (0.03) a	0.28 (0.03) b	0.23 (0.02) b	0.09 (0.01) c	0.06 (0.01) cd	0.04 (0.004) d
	$NSC_{Concentration}$ †	11.0 (0.6) a	10.1 (0.9) a	11.3 (0.4) a	5.0 (0.3) b	5.0 (0.3) b	5.3 (0.4) b

* Data were ln-transformed prior to analysis. † One *Medium* seedling from control treatment excluded for analysis.

References

1. Adams, H.D.; Guardiola-Claramonte, M.; Barron-Gafford, G.A.; Villegas, J.C.; Breshears, D.D.; Zou, C.B.; Troch, P.A.; Huxman, T.E. Temperature Sensitivity of Drought-Induced Tree Mortality Portends Increased Regional Die-off under Global-Change-Type Drought. *Proc. Natl. Acad. Sci. USA* **2009**, *106*, 7063–7066. [CrossRef] [PubMed]
2. Allen, C.D.; Macalady, A.K.; Chenchouni, H.; Bachelet, D.; McDowell, N.; Vennetier, M.; Kitzberger, T.; Rigling, A.; Breshears, D.D.; Hogg, E.H.; et al. A Global Overview of Drought and Heat-Induced Tree Mortality Reveals Emerging Climate Change Risks for Forests. *For. Ecol. Manag.* **2010**, *259*, 660–684. [CrossRef]

3. Anderegg, W.R.L.; Trugman, A.T.; Badgley, G.; Anderson, C.M.; Bartuska, A.; Ciais, P.; Cullenward, D.; Field, C.B.; Freeman, J.; Goetz, S.J.; et al. Climate-Driven Risks to the Climate Mitigation Potential of Forests. *Science* **2020**, *368*, eaaz7005. [CrossRef]
4. Beier, C.; Beierkuhnlein, C.; Wohlgemuth, T.; Penuelas, J.; Emmett, B.; Körner, C.; de Boeck, H.; Christensen, J.H.; Leuzinger, S.; Janssens, I.A.; et al. Precipitation Manipulation Experiments—Challenges and Recommendations for the Future. *Ecol. Lett.* **2012**, *15*, 899–911. [CrossRef] [PubMed]
5. Clark, J.S.; Iverson, L.; Woodall, C.W.; Allen, C.D.; Bell, D.M.; Bragg, D.C.; D'Amato, A.W.; Davis, F.W.; Hersh, M.H.; Ibanez, I.; et al. The Impacts of Increasing Drought on Forest Dynamics, Structure, and Biodiversity in the United States. *Glob. Chang. Biol.* **2016**, *22*, 2329–2352. [CrossRef]
6. Cobb, R.C.; Ruthrof, K.X.; Breshears, D.D.; Lloret, F.; Aakala, T.; Adams, H.D.; Anderegg, W.R.L.; Ewers, B.E.; Galiano, L.; Grünzweig, J.M.; et al. Ecosystem Dynamics and Management after Forest Die-off: A Global Synthesis with Conceptual State-and-Transition Models. *Ecosphere* **2017**, *8*, e02034. [CrossRef]
7. Royer, P.D.; Cobb, N.S.; Clifford, M.J.; Huang, C.-Y.; Breshears, D.D.; Adams, H.D.; Villegas, J.C. Extreme Climatic Event-Triggered Overstorey Vegetation Loss Increases Understorey Solar Input Regionally: Primary and Secondary Ecological Implications. *J. Ecol.* **2011**, *99*, 714–723. [CrossRef]
8. Hodgson, D.; McDonald, J.L.; Hosken, D.J. What Do You Mean, 'Resilient'? *Trends Ecol. Evol.* **2015**, *30*, 503–506. [CrossRef]
9. Grubb, P.J. Control of Forest Growth and Distribution on Wet Tropical Mountains: With Special Reference to Mineral Nutrition. *Annu. Rev. Ecol. Syst.* **1977**, *8*, 83–107. [CrossRef]
10. Macdonald, S.E.; Landhäusser, S.M.; Skousen, J.; Franklin, J.; Frouz, J.; Hall, S.; Jacobs, D.F.; Quideau, S. Forest Restoration Following Surface Mining Disturbance: Challenges and Solutions. *New For.* **2015**, *46*, 703–732. [CrossRef]
11. Hanson, P.J.; Todd, D.E.; Amthor, J.S. A Six-Year Study of Sapling and Large-Tree Growth and Mortality Responses to Natural and Induced Variability in Precipitation and Throughfall. *Tree Physiol.* **2001**, *21*, 345–358. [CrossRef] [PubMed]
12. Lloret, F.; Peñuelas, J.; Ogaya, R. Establishment of Co-Existing Mediterranean Tree Species under a Varying Soil Moisture Regime. *J. Veg. Sci.* **2004**, *15*, 237–244. [CrossRef]
13. Poorter, L.; Hayashida-Oliver, Y. Effects of Seasonal Drought on Gap and Understory Seedlings in a Bolivian Moist Forest. *J. Trop. Ecol.* **2000**, *16*, 481–498. [CrossRef]
14. Solarik, K.A.; Messier, C.; Ouimet, R.; Bergeron, Y.; Gravel, D. Local Adaptation of Trees at the Range Margins Impacts Range Shifts in the Face of Climate Change. *Glob. Ecol. Biogeogr.* **2018**, *27*, 1507–1519. [CrossRef]
15. Kozlowski, T.T.; Pallardy, S.G. Acclimation and Adaptive Responses of Woody Plants to Environmental Stresses. *Bot. Rev.* **2002**, *68*, 270–334. [CrossRef]
16. Niinemets, Ü. Responses of Forest Trees to Single and Multiple Environmental Stresses from Seedlings to Mature Plants: Past Stress History, Stress Interactions, Tolerance and Acclimation. *For. Ecol. Manag.* **2010**, *260*, 1623–1639. [CrossRef]
17. Grossnickle, S.C.; MacDonald, J.E. Why Seedlings Grow: Influence of Plant Attributes. *New For.* **2018**, *49*, 1–34. [CrossRef]
18. Moler, E.R.V.; Nelson, A.S. Perspectives on Drought Preconditioning Treatments With a Case Study Using Western Larch. *Front. Plant Sci.* **2021**, *12*, 741027. [CrossRef] [PubMed]
19. Sloan, J.L.; Burney, O.T.; Pinto, J.R. Drought-Conditioning of Quaking Aspen (Populus Tremuloides Michx.) Seedlings During Nursery Production Modifies Seedling Anatomy and Physiology. *Front. Plant Sci.* **2020**, *11*, 557894. [CrossRef] [PubMed]
20. Villar-Salvador, P.; Ocaña, L.; Peñuelas, J.; Carrasco, I. Effect of Water Stress Conditioning on the Water Relations, Root Growth Capacity, and the Nitrogen and Non-Structural Carbohydrate Concentration of Pinus Halepensis Mill. (Aleppo Pine) Seedlings. *Ann. For. Sci.* **1999**, *56*, 459–465. [CrossRef]
21. Dietze, M.C.; Sala, A.; Carbone, M.S.; Czimczik, C.I.; Mantooth, J.A.; Richardson, A.D.; Vargas, R. Nonstructural Carbon in Woody Plants. *Annu. Rev. Plant Biol.* **2014**, *65*, 667–687. [CrossRef] [PubMed]
22. Hoch, G. Carbon Reserves as Indicators for Carbon Limitation in Trees. In *Progress in Botany: Vol. 76*; Lüttge, U., Beyschlag, W., Eds.; Progress in Botany; Springer International Publishing: Cham, Switzerland, 2015; pp. 321–346. ISBN 978-3-319-08807-5.
23. Piper, F.I.; Fajardo, A. Foliar Habit, Tolerance to Defoliation and Their Link to Carbon and Nitrogen Storage. *J. Ecol.* **2014**, *102*, 1101–1111. [CrossRef]
24. Tomlinson, K.W.; van Langevelde, F.; Ward, D.; Bongers, F.; da Silva, D.A.; Prins, H.H.T.; de Bie, S.; Sterck, F.J. Deciduous and Evergreen Trees Differ in Juvenile Biomass Allometries Because of Differences in Allocation to Root Storage. *Ann. Bot.* **2013**, *112*, 575–587. [CrossRef] [PubMed]
25. O'Brien, M.J.; Burslem, D.F.R.P.; Caduff, A.; Tay, J.; Hector, A. Contrasting Nonstructural Carbohydrate Dynamics of Tropical Tree Seedlings under Water Deficit and Variability. *New Phytol.* **2015**, *205*, 1083–1094. [CrossRef]
26. O'Brien, M.J.; Leuzinger, S.; Philipson, C.D.; Tay, J.; Hector, A. Drought Survival of Tropical Tree Seedlings Enhanced by Non-Structural Carbohydrate Levels. *Nat. Clim. Chang.* **2014**, *4*, 710–714. [CrossRef]
27. Dumroese, K.; Landis, T.D.; Pinto, J.; Haase, D.L.; Wilkinson, K.W.; Davis, A.S. Meeting Forest Restoration Challenges: Using the Target Plant Concept. *Reforesta* **2016**, *1*, 37–52. [CrossRef]
28. Landis, T.D. The Target Plant Concept-a History and Brief Overview. In *National Proceedings: Forest and Conservation Nursery Associations—2010*; Proc. RMRS-P-65; Riley, L.E., Haase, D.L., Pinto, J.R., Eds.; USDA Forest Service, Rocky Mountain Research Station: Fort Collins, CO, USA, 2011; Volume 65, pp. 61–66.

29. Rose, R.; Campbell, S.J.; Landis, T.D. *Target Seedling Symposium: Proceedings, Combined Meeting of the Western Forest Nursery Associations*; Gen. Tech. Rep. RM-GTR-200; U.S. Department of Agriculture, Forest Service, Rocky Mountain Forest and Range Experiment Station: Fort Collins, CO, USA, 1990; Volume 200. [CrossRef]
30. Poorter, H.; Niklas, K.J.; Reich, P.B.; Oleksyn, J.; Poot, P.; Mommer, L. Biomass Allocation to Leaves, Stems and Roots: Meta-Analyses of Interspecific Variation and Environmental Control. *New Phytol.* **2012**, *193*, 30–50. [CrossRef]
31. Bloom, A.J.; Chapin, F.S.; Mooney, H.A. Resource Limitation in Plants-An Economic Analogy. *Annu. Rev. Ecol. Syst.* **1985**, *16*, 363–392. [CrossRef]
32. Landhäusser, S.M.; Lieffers, V.J. Photosynthesis and Carbon Allocation of Six Boreal Tree Species Grown in Understory and Open Conditions. *Tree Physiol.* **2001**, *21*, 243–250. [CrossRef]
33. Niinemets, Ü. Photosynthesis and Resource Distribution through Plant Canopies. *Plant Cell Environ.* **2007**, *30*, 1052–1071. [CrossRef]
34. Buckley, T.N.; John, G.P.; Scoffoni, C.; Sack, L. The Sites of Evaporation within Leaves. *Plant Physiol* **2017**, *173*, 1763–1782. [CrossRef]
35. Claeys, H.; Inzé, D. The Agony of Choice: How Plants Balance Growth and Survival under Water-Limiting Conditions. *Plant Physiol.* **2013**, *162*, 1768–1779. [CrossRef]
36. Valladares, F.; Gianoli, E.; Gómez, J.M. Ecological Limits to Plant Phenotypic Plasticity. *New Phytol.* **2007**, *176*, 749–763. [CrossRef] [PubMed]
37. Hart, A.T.; Merlin, M.; Wiley, E.; Landhäusser, S.M. Splitting the Difference: Heterogeneous Soil Moisture Availability Affects Aboveground and Belowground Reserve and Mass Allocation in Trembling Aspen. *Front. Plant Sci.* **2021**, *12*, 654159. [CrossRef]
38. Martínez-Vilalta, J.; Cochard, H.; Mencuccini, M.; Sterck, F.; Herrero, A.; Korhonen, J.F.J.; Llorens, P.; Nikinmaa, E.; Nolè, A.; Poyatos, R.; et al. Hydraulic Adjustment of Scots Pine across Europe. *New Phytol.* **2009**, *184*, 353–364. [CrossRef] [PubMed]
39. Rood, S.B.; Patiño, S.; Coombs, K.; Tyree, M.T. Branch Sacrifice: Cavitation-Associated Drought Adaptation of Riparian Cottonwoods. *Trees* **2000**, *14*, 248–257. [CrossRef]
40. Sterck, F.J.; Martínez-Vilalta, J.; Mencuccini, M.; Cochard, H.; Gerrits, P.; Zweifel, R.; Herrero, A.; Korhonen, J.F.J.; Llorens, P.; Nikinmaa, E.; et al. Understanding Trait Interactions and Their Impacts on Growth in Scots Pine Branches across Europe. *Funct. Ecol.* **2012**, *26*, 541–549. [CrossRef]
41. Canham, C.D.; Kobe, R.K.; Latty, E.F.; Chazdon, R.L. Interspecific and Intraspecific Variation in Tree Seedling Survival: Effects of Allocation to Roots versus Carbohydrate Reserves. *Oecologia* **1999**, *121*, 1–11. [CrossRef] [PubMed]
42. Galiano, L.; Martínez-Vilalta, J.; Eugenio, M.; Granzow-de la Cerda, Í.; Lloret, F. Seedling Emergence and Growth of Quercus Spp. Following Severe Drought Effects on a Pinus Sylvestris Canopy. *J. Veg. Sci.* **2013**, *24*, 580–588. [CrossRef]
43. Suarez, M.L.; Kitzberger, T. Recruitment Patterns Following a Severe Drought: Long-Term Compositional Shifts in Patagonian Forests. *Can. J. For. Res.* **2008**, *38*, 3002–3010. [CrossRef]
44. Chapin, F.S.; Schulze, E.; Mooney, H.A. The Ecology and Economics of Storage in Plants. *Annu. Rev. Ecol. Syst.* **1990**, *21*, 423–447. [CrossRef]
45. Hartmann, H.; Adams, H.D.; Hammond, W.M.; Hoch, G.; Landhäusser, S.M.; Wiley, E.; Zaehle, S. Identifying Differences in Carbohydrate Dynamics of Seedlings and Mature Trees to Improve Carbon Allocation in Models for Trees and Forests. *Environ. Exp. Bot.* **2018**, *152*, 7–18. [CrossRef]
46. Maguire, A.J.; Kobe, R.K. Drought and Shade Deplete Nonstructural Carbohydrate Reserves in Seedlings of Five Temperate Tree Species. *Ecol. Evol.* **2015**, *5*, 5711–5721. [CrossRef] [PubMed]
47. Landhäusser, S.M.; Pinno, B.D.; Lieffers, V.J.; Chow, P.S. Partitioning of Carbon Allocation to Reserves or Growth Determines Future Performance of Aspen Seedlings. *For. Ecol. Manag.* **2012**, *275*, 43–51. [CrossRef]
48. Landhäusser, S.M.; Rodriguez-Alvarez, J.; Marenholtz, E.H.; Lieffers, V.J. Effect of Stock Type Characteristics and Time of Planting on Field Performance of Aspen (Populus Tremuloides Michx.) Seedlings on Boreal Reclamation Sites. *New For.* **2012**, *43*, 679–693. [CrossRef]
49. Signori-Müller, C.; Oliveira, R.S.; de Barros, F.V.; Tavares, J.V.; Gilpin, M.; Diniz, F.C.; Zevallos, M.J.M.; Yupayccana, C.A.S.; Acosta, M.; Bacca, J.; et al. Non-Structural Carbohydrates Mediate Seasonal Water Stress across Amazon Forests. *Nat. Commun.* **2021**, *12*, 2310. [CrossRef]
50. Galvez, D.A.; Landhäusser, S.M.; Tyree, M.T. Root Carbon Reserve Dynamics in Aspen Seedlings: Does Simulated Drought Induce Reserve Limitation? *Tree Physiol.* **2011**, *31*, 250–257. [CrossRef]
51. Galvez, D.A.; Landhäusser, S.M.; Tyree, M.T. Low Root Reserve Accumulation during Drought May Lead to Winter Mortality in Poplar Seedlings. *New Phytol.* **2013**, *198*, 139–148. [CrossRef]
52. Rennenberg, H.; Loreto, F.; Polle, A.; Brilli, F.; Fares, S.; Beniwal, R.S.; Gessler, A. Physiological Responses of Forest Trees to Heat and Drought. *Plant Biol.* **2006**, *8*, 556–571. [CrossRef]
53. Kobe, R.K.; Iyer, M.; Walters, M.B. Optimal Partitioning Theory Revisited: Nonstructural Carbohydrates Dominate Root Mass Responses to Nitrogen. *Ecology* **2010**, *91*, 166–179. [CrossRef]
54. Barbaroux, C.; Breda, N. Contrasting Distribution and Seasonal Dynamics of Carbohydrate Reserves in Stem Wood of Adult Ring-Porous Sessile Oak and Diffuse-Porous Beech Trees. *Tree Physiol.* **2002**, *22*, 1201–1210. [CrossRef] [PubMed]
55. Hoch, G.; Richter, A.; Körner, C. Non-Structural Carbon Compounds in Temperate Forest Trees. *Plant Cell Environ.* **2003**, *26*, 1067–1081. [CrossRef]

56. Uscola, M.; Villar-Salvador, P.; Gross, P.; Maillard, P. Fast Growth Involves High Dependence on Stored Resources in Seedlings of Mediterranean Evergreen Trees. *Ann. Bot.* **2015**, *115*, 1001–1013. [CrossRef]
57. Kelly, J.W.G.; Landhäusser, S.M.; Chow, P.S. The Impact of Light Quality and Quantity on Root-to-Shoot Ratio and Root Carbon Reserves in Aspen Seedling Stock. *New For.* **2015**, *46*, 527–545. [CrossRef]
58. Le, K.D.; Schreiber, S.G.; Landhäusser, S.M.; Schoonmaker, A.L. Manipulating Aspen (Populus Tremuloides) Seedling Size Characteristics to Improve Initial Establishment and Growth on Competitive Sites. *Scand. J. For. Res.* **2020**, *35*, 29–45. [CrossRef]
59. Chow, P.S.; Landhäusser, S.M. A Method for Routine Measurements of Total Sugar and Starch Content in Woody Plant Tissues. *Tree Physiol.* **2004**, *24*, 1129–1136. [CrossRef]
60. Landhäusser, S.M.; Chow, P.S.; Dickman, L.T.; Furze, M.E.; Kuhlman, I.; Schmid, S.; Wiesenbauer, J.; Wild, B.; Gleixner, G.; Hartmann, H.; et al. Standardized Protocols and Procedures Can Precisely and Accurately Quantify Non-Structural Carbohydrates. *Tree Physiol.* **2018**, *38*, 1764–1778. [CrossRef]
61. De'ath, G. Multivariate Regression Trees: A New Technique for Modeling Species–Environment Relationships. *Ecology* **2002**, *83*, 1105–1117. [CrossRef]
62. R Development Core Team. *R: A Language and Environment for Statistical Computing*; R Foundation for Statistical Computing: Vienna, Austria, 2017.
63. van Genuchten, M.T. A Closed-Form Equation for Predicting the Hydraulic Conductivity of Unsaturated Soils. *Soil Sci. Soc. Am. J.* **1980**, *44*, 892–898. [CrossRef]
64. Cai, J.; Tyree, M.T. The Impact of Vessel Size on Vulnerability Curves: Data and Models for within-Species Variability in Saplings of Aspen, Populus Tremuloides Michx. *Plant Cell Env.* **2010**, *33*, 1059–1069. [CrossRef]
65. Schoonmaker, A.L.; Hacke, U.G.; Landhäusser, S.M.; Lieffers, V.J.; Tyree, M.T. Hydraulic Acclimation to Shading in Boreal Conifers of Varying Shade Tolerance. *Plant Cell Env.* **2010**, *33*, 382–393. [CrossRef] [PubMed]
66. Wiley, E.; King, C.M.; Landhäusser, S.M. Identifying the Relevant Carbohydrate Storage Pools Available for Remobilization in Aspen Roots. *Tree Physiol.* **2019**, *39*, 1109–1120. [CrossRef] [PubMed]
67. Wyka, T.P.; Karolewski, P.; Żytkowiak, R.; Chmielarz, P.; Oleksyn, J. Whole-Plant Allocation to Storage and Defense in Juveniles of Related Evergreen and Deciduous Shrub Species. *Tree Physiol.* **2016**, *36*, 536–547. [CrossRef] [PubMed]
68. Schott, K.M.; Pinno, B.D.; Landhäusser, S.M. Premature Shoot Growth Termination Allows Nutrient Loading of Seedlings with an Indeterminate Growth Strategy. *New For.* **2013**, *44*, 635–647. [CrossRef]
69. Jordan, M.-O.; Habib, R. Mobilizable Carbon Reserves in Young Peach Trees as Evidenced by Trunk Girdling Experiments. *J. Exp. Bot.* **1996**, *47*, 79–87. [CrossRef]
70. Oberhuber, W.; Gruber, A.; Lethaus, G.; Winkler, A.; Wieser, G. Stem Girdling Indicates Prioritized Carbon Allocation to the Root System at the Expense of Radial Stem Growth in Norway Spruce under Drought Conditions. *Environ. Exp. Bot.* **2017**, *138*, 109–118. [CrossRef]
71. Tuttle, C.L.; South, D.B.; Golden, M.S.; Meldahl, R.S. Relationship between Initial Seedling Height and Survival and Growth of Loblolly Pine Seedlings Planted during a Droughty Year. *South. J. Appl. For.* **1987**, *11*, 139–143. [CrossRef]
72. Villar-Salvador, P.; Puértolas, J.; Cuesta, B.; Peñuelas, J.L.; Uscola, M.; Heredia-Guerrero, N.; Rey Benayas, J.M. Increase in Size and Nitrogen Concentration Enhances Seedling Survival in Mediterranean Plantations. Insights from an Ecophysiological Conceptual Model of Plant Survival. *New For.* **2012**, *43*, 755–770. [CrossRef]
73. Navarro, R.M.; Retamosa, M.J.; Lopez, J.; del Campo, A.; Ceaceros, C.; Salmoral, L. Nursery Practices and Field Performance for the Endangered Mediterranean Species Abies Pinsapo Boiss. *Ecol. Eng.* **2006**, *27*, 93–99. [CrossRef]
74. Myers, J.A.; Kitajima, K. Carbohydrate Storage Enhances Seedling Shade and Stress Tolerance in a Neotropical Forest. *J. Ecol.* **2007**, *95*, 383–395. [CrossRef]
75. Wiley, E.; Hoch, G.; Landhäusser, S.M. Dying Piece by Piece: Carbohydrate Dynamics in Aspen (Populus Tremuloides) Seedlings under Severe Carbon Stress. *J. Exp. Bot.* **2017**, *68*, 5221–5232. [CrossRef] [PubMed]
76. Tuttle, C.L.; South, D.B.; Golden, M.S.; Meldahl, R.S. Initial Pinustaeda Seedling Height Relationships with Early Survival and Growth. *Can. J. For. Res.* **1988**, *18*, 867–871. [CrossRef]
77. Tan, W.; Hogan, G.D. Physiological and Morphological Responses to Nitrogen Limitation in Jack Pine Seedlings: Potential Implications for Drought Tolerance. *New For.* **1997**, *14*, 19–31. [CrossRef]
78. Trubat, R.; Cortina, J.; Vilagrosa, A. Nutrient Deprivation Improves Field Performance of Woody Seedlings in a Degraded Semi-Arid Shrubland. *Ecol. Eng.* **2011**, *37*, 1164–1173. [CrossRef]
79. Furze, M.E.; Huggett, B.A.; Aubrecht, D.M.; Stolz, C.D.; Carbone, M.S.; Richardson, A.D. Whole-Tree Nonstructural Carbohydrate Storage and Seasonal Dynamics in Five Temperate Species. *New Phytol.* **2019**, *221*, 1466–1477. [CrossRef]
80. Schoonmaker, A.L.; Hillabrand, R.M.; Lieffers, V.J.; Chow, P.S.; Landhäusser, S.M. Seasonal Dynamics of Non-Structural Carbon Pools and Their Relationship to Growth in Two Boreal Conifer Tree Species. *Tree Physiol.* **2021**, *41*, 1563–1582. [CrossRef]
81. Pinto, J.R.; Marshall, J.D.; Dumroese, R.K.; Davis, A.S.; Cobos, D.R. Photosynthetic Response, Carbon Isotopic Composition, Survival, and Growth of Three Stock Types under Water Stress Enhanced by Vegetative Competition. *Can. J. For. Res.* **2012**, *42*, 333–344. [CrossRef]
82. Hillabrand, R.M.; Gordon, H.; Hynes, B.; Constabel, C.P.; Landhäusser, S.M. Populus Root Salicinoid Phenolic Glycosides Are Not Mobilized to Support Metabolism and Regrowth under Carbon Limited Conditions. *Tree Physiol.* **2023**, tpad020. [CrossRef] [PubMed]

83. Najar, A.; Landhäusser, S.M.; Whitehill, J.G.A.; Bonello, P.; Erbilgin, N. Reserves Accumulated in Non-Photosynthetic Organs during the Previous Growing Season Drive Plant Defenses and Growth in Aspen in the Subsequent Growing Season. *J. Chem. Ecol.* **2014**, *40*, 21–30. [CrossRef]
84. Wiley, E.; Rogers, B.J.; Hodgkinson, R.; Landhäusser, S.M. Nonstructural Carbohydrate Dynamics of Lodgepole Pine Dying from Mountain Pine Beetle Attack. *New Phytol.* **2016**, *209*, 550–562. [CrossRef]
85. Bréda, N.; Huc, R.; Granier, A.; Dreyer, E. Temperate Forest Trees and Stands under Severe Drought: A Review of Ecophysiological Responses, Adaptation Processes and Long-Term Consequences. *Ann. For. Sci.* **2006**, *63*, 625–644. [CrossRef]
86. Hartmann, H.; Ziegler, W.; Trumbore, S. Lethal Drought Leads to Reduction in Nonstructural Carbohydrates in Norway Spruce Tree Roots but Not in the Canopy. *Funct. Ecol.* **2013**, *27*, 413–427. [CrossRef]
87. Sala, A.; Piper, F.; Hoch, G. Physiological Mechanisms of Drought-Induced Tree Mortality Are Far from Being Resolved: Letters. *New Phytol.* **2010**, *186*, 274–281. [CrossRef] [PubMed]
88. Sevanto, S.; Mcdowell, N.G.; Dickman, L.T.; Pangle, R.; Pockman, W.T. How Do Trees Die? A Test of the Hydraulic Failure and Carbon Starvation Hypotheses. *Plant Cell Env.* **2014**, *37*, 153–161. [CrossRef]
89. Brodribb, T.J.; Holbrook, N.M.; Hill, R.S.; Brodribb, T.J.; Holbrook, N.M.; Hill, R.S. Seedling Growth in Conifers and Angiosperms: Impacts of Contrasting Xylem Structure. *Aust. J. Bot.* **2005**, *53*, 749–755. [CrossRef]
90. Carnicer, J.; Barbeta, A.; Sperlich, D.; Coll, M.; Penuelas, J. Contrasting Trait Syndromes in Angiosperms and Conifers Are Associated with Different Responses of Tree Growth to Temperature on a Large Scale. *Front. Plant Sci.* **2013**, *4*, 409. [CrossRef]
91. Givnish, T.J. Adaptive Significance of Evergreen vs. Deciduous Leaves: Solving the Triple Paradox. *Silva Fenn.* **2002**, *36*, 703–743. [CrossRef]

forests

MDPI

Article

Effects of Seedling Size, Stock Type, and Mechanical Site Preparation Method on Initial Survival and Growth of Japanese Larch (*Larix kaempferi*) Seedlings

Hisanori Harayama [1,*], Ikutaro Tsuyama [2], Mitsutoshi Kitao [2], Takeshi Yamada [1], Naoyuki Furuya [2], Hajime Utsugi [1] and Shozo Sasaki [3]

1 Forestry and Forest Products Research Institute, Matsunosato 1, Tsukuba 305-8687, Japan
2 Hokkaido Research Center, Forestry and Forest Products Research Institute, Hitsujigaoka 7, Toyohira-ku, Sapporo 062-8516, Japan
3 KITARIN Lab, 2-16-1101 Minami 13-jo Nishi 1-chome, Chuo-ku, Sapporo 064-0913, Japan
* Correspondence: harahisa@ffpri.affrc.go.jp

Abstract: It is important to understand the characteristics of Japanese larch (*Larix kaempferi*) seedlings that allow them to grow vigorously after planting and quickly exceed the height of surrounding vegetation, resulting in lower weeding costs. Seven stock types, including bareroot and container-grown seedlings, were planted in two plots with different mechanical-site-preparation (MSP) methods and evaluated for survival, height, and root collar diameter (RCD) for four consecutive years. Three-year-old bareroot seedlings, which were one year older and larger than normal, had low survival rates in the mulcher MSP. Initial seedling height significantly differed among the seven stock types, while almost no significant differences were observed after four growing seasons. Model analyses showed that initial seedling height and RCD had a significant effect on seedling height after planting until the second growing season, while the effect of planted seedling age and plot became increasingly significant after the third growing season. The difference in seedling type, bareroot versus container-grown seedlings, had no effect on the seedling height during the four growing seasons after planting. A decision tree analysis suggests that the seedlings with sufficiently large RCD and young age, regardless of seedling type, can grow taller than surrounding vegetation more quickly.

Keywords: bareroot seedlings; container-grown seedlings; height; root collar diameter; seedling age

Citation: Harayama, H.; Tsuyama, I.; Kitao, M.; Yamada, T.; Furuya, N.; Utsugi, H.; Sasaki, S. Effects of Seedling Size, Stock Type, and Mechanical Site Preparation Method on Initial Survival and Growth of Japanese Larch (*Larix kaempferi*) Seedlings. *Forests* **2023**, *14*, 784. https://doi.org/10.3390/f14040784

Academic Editors: Mohammed S. Lamhamedi, Damase P. Khasa and Steeve Pépin

Received: 3 March 2023
Revised: 30 March 2023
Accepted: 7 April 2023
Published: 11 April 2023

1. Introduction

Seedling quality is one of the most important factors in determining the success of afforestation [1–4]. Morphological traits, such as seedling height and root collar diameter (RCD), the root system, and the ratio of aboveground to belowground biomass allocation (i.e., top:root ratio), are associated with initial field performance of survival and growth after planting [1,2,5]. These traits can vary widely between the two typical seedling types, namely bareroot and container-grown seedlings, and can also be altered by the nursery practices in each seedling type [4,6–8].

Japanese larch (*Larix kaempferi* (Lamb.) Carrière) is an important forestry tree species in cool temperate regions of Japan [9,10]; it can have a rapid initial growth in the regions in Japan as well as in Europe and North America [11–13]. The planted area of Japanese larch in Japan has rapidly increased in recent years, from approximately 5000 ha year^{-1} from 2012 to 2017 to approximately 9000 ha year^{-1} in 2021, and the annual number of shipped planting seedlings has correspondingly increased from approximately 10 million to 16 million. Only bareroot seedlings were previously planted in Japan. However, with the aging and decreasing number of forestry workers and seedling producers, the production of container-grown seedlings started in 2012 in response to the expectation that container-grown seedlings will extend the planting period and reduce the labor required for seedling production, and the proportion of container-grown seedlings in planted seedlings has rapidly increased from 3% in 2018 to 19% in 2021. Container-grown seedlings are produced by dozens of relatively small nurseries, using a variety of cultivation methods. In Japanese larch, container-grown seedlings are smaller in size than bareroot seedlings, but they sell for 1.3 to 2 times the price. It has been reported that the small container-grown seedlings do not grow taller than bareroot seedlings after planting [14,15]. Some studies have recently been conducted on methods to produce high-quality container-grown seedlings; however, knowledge regarding what characteristics contribute to the effective performance of container-grown Japanese larch seedlings after planting is still insufficient [16–19].

The performance of the planted seedlings can also be influenced by the reforestation site conditions, such as soil moisture, nutrients, and light, as well as by competing vegetation [1,2]. Japanese larch has low shade tolerance and is therefore susceptible to significant loss of survival and growth, due to competition from surrounding vegetation [14,20]. Meanwhile, the vegetation growth is also influenced by site environmental conditions [21]. The influence of environmental conditions and vegetation growth on seedling performance must be considered when evaluating seedling quality based on the survival and growth in planting trials.

Mechanical site preparation (MSP) is a major technique for vegetation control in afforestation sites [22,23]. Vegetation in Japan grows intensively, with humid and warm temperatures during the growing season, often exceeding 200 cm in height in summer. The use of herbicides in forests is discouraged; thus, vegetation release with motor-manual brush cutters is typically performed yearly for four years after planting in Japanese larch plantations [24]. This method increases the cost of reforestation and is thus a major factor in the low reforestation rate, of approximately 30%, after harvest [25]. Site preparation is still typically performed manually mainly due to the steep slopes of many reforestation sites in Japan; however, MSP has been increasingly applied to sites with gentle slopes in recent years in the expectation of inhibiting vegetation growth and thus reducing vegetation release costs [26].

Excavators with a 0.5 m^3 bucket are the most commonly used machines for MSP in Japan. The bucket is typically used to pile up logging residues at the edge of the reforestation area by dragging and pushing them with the claws of the bucket and scraping the topsoil to remove vegetation by the roots throughout the reforestation area. Oya et al. (2021) found that bucket MSP was more effective in suppressing vegetation and reducing the number of weeding years than manual site preparation. However, surface soil removal by bucket MSP negatively affected the height growth of planted Japanese larch and Japanese cedar (*Cryptomeria japonica*) seedlings, probably due to reduced soil nutrients [26]. Meanwhile, the use of excavators with forestry mulchers has been tested in recent years for reforestation in logging areas by forest harvester machines, which generate large amounts of logging residue [27]. The mulcher shreds the vegetation and logging residues of branches and stumps through a fixed blade attached to a high-speed rotating drum, and the shredded debris mulches the ground surface, potentially inhibiting vegetation development as much as, or more than, the bucket MSP [28]. The mulcher MSP slightly disturbs the soil; thus, this mulcher is expected to have less of the negative effect on tree growth [29] observed with the bucket MSP. However, its effect on the growth of planted seedlings has only been partially elucidated.

Here, a total of seven stock types, container-grown Japanese larch seedlings from five nurseries and two types of bareroot seedlings from another nursery using their own growing practices, were planted in a bucket and mulcher MSP plot in a general reforestation site in Hokkaido, Japan. Their survival and growth were evaluated annually for four consecutive growing seasons. The objective of this study was to determine what characteristics allow planted seedlings, especially container-grown seedlings, to grow taller than the surrounding vegetation more quickly, thus reducing weeding costs, while considering the effects of plot factors, including MSP methods. Thus, the effects of planting seedling characteristics (such as initial height, initial RCD, seedling type (bareroot or container-grown), and seedling age) and plot (including MSP) on planting seedling height, were quantitatively evaluated using generalized linear mixed models and decision tree analysis.

2. Materials and Methods

2.1. Study Site and Stock Types

The study site was located in a town-owned forest in Shimokawa Town, Hokkaido, northern Japan (44°18′ N, 142°41′ E, 200–240 m elevation). The 1991–2020 averages recorded at the nearest weather station (44°18′ N, 142°37′ E, 140 m elevation) approximately 3 km from the study site revealed the following: the annual mean air temperature was 5.4 °C, the monthly mean daily maximum and minimum air temperatures were 25.0 °C in August and −8.8 °C in January, the annual precipitation was 965 mm, and the deepest snowfall was 116 cm, in March.

In February 2017, 220 m^3 ha^{-1} of timber from a 1.7 ha 50-year-old *Abies sachalinensis* plantation with snow cover was cleared and logged by harvesters and transported by forwarders. The slope of the site was relatively gentle, with an average slope angle of 10–16°. The site was divided into two plots, and MSP was conducted using a 0.5 m^3-capacity excavator bucket (approximately 0.6 ha) and an excavator mulcher (approximately 1.1 ha), in early May 2017 after snowmelt (Figure 1). The mulcher MSP plot was located on the same slope at an elevation of approximately 225–240 m, while the bucket MSP plot was located on a lower slope at an elevation of 200–225 m. In the excavator bucket MSP (Figure S1A), the bucket was used to drag, push, and pile the logging residues, such as tree tops and branches, around the reforestation area and remove vegetation and scrape the soil surface (Figure S1B). In the MSP with the excavator mulcher (MINI-BMS 125, Seppi, Italy), the logging residues and vegetation were crushed by the mulching head (Figure S1C). The mulcher also shredded stumps that obstructed the path of the excavator. Consequently, 93% of the ground surface was mulched with crushed woody debris (Figure S1D), which ranged in height from 1 cm to 13 cm (mean 5 cm ± 1 cm standard error, median 1–2 cm) throughout the mulcher MSP plot [30]. Due to site and forest-management constraints, it was not possible to place the two MSP plots on the same slope or to establish replications.

After the MSP, seedlings of seven stock types were planted in the mulcher MSP plot, and four of the seven stock types were planted in the bucket MSP plot at 2.58 m spacing (1503 seedlings ha^{-1}). The number of prepared seedlings varied widely among the stock types; thus, five of the seven stock types with low seedling numbers were not planted in the bucket MSP plot (Figure 1 and Table 1). Of the seven stock types, two were two-year-old (bareroot (1+1)) and three-year-old (bareroot (1+2)) bareroot seedlings from a single nursery in central Hokkaido. The five remaining stock types were container-grown seedlings varying in seedling age, container type (volume, density, and with/without side slits), and nursery (Table 1).

LIECO 390 (0+1) was sown directly into LIECO 15 blue containers (LIECO GmbH & Co KG, Kalwang, Austria) with a cell volume of 390 mL and density of 198 cells m^{-2} filled with peat-moss-based growing media, and grown in containers for one year by a nursery in western Hokkaido. JFA300.C (1+2) and JFA300.E (1+2) were grown by a nursery in central and eastern Hokkaido, respectively, in JFA300 containers (Zenbyouren, Tokyo, Japan) of 300 mL volume and 178 cells m^{-2} density. Seedlings grown in a field for one year after sowing were transplanted to JFA300 containers and grown in these containers for two years. HRO200 (1+1) was grown by a nursery in northeastern Hokkaido in HRO200 containers (Dobyouso, Hokkaido, Japan) with a volume of 200 mL and a density of 112 cells m^{-2}. Seedlings were also grown in the field for one year after sowing, and then transplanted and grown in HRO200 containers for one year. BCC150 (0+2) was grown by a nursery in northern Hokkaido in a 150 mL capacity BCC SideSlit Cell 150 container (BCC, Landskrona, Sweden) at a density of 362 cells m^{-2} in FlexiFrame 77 (BCC, Landskrona, Sweden). Seedlings were transplanted into the container immediately after germination, and grown for two years. All containers used coconut-husk-based growing media except for LIECO 390 (0+1), which used peat-moss-based media. Containers other than JFA300 had side slits to prevent root rolling, while JFA300 had no side slits, but ribs on the inside of the cell wall. The label of the container-grown seedling indicated the product name and cell volume, followed by the two numbers in parentheses, which indicated the number of years grown in the field after sowing and in the container. Except for LIECO390 (0+1) and HRO200 (1+1), different seed lots were used to produce bareroot and container-grown seedlings (Table 1). Seed orchards are not well developed in Japan. Therefore, the seeds were collected from common plantations in each town, and were not genetically verified.

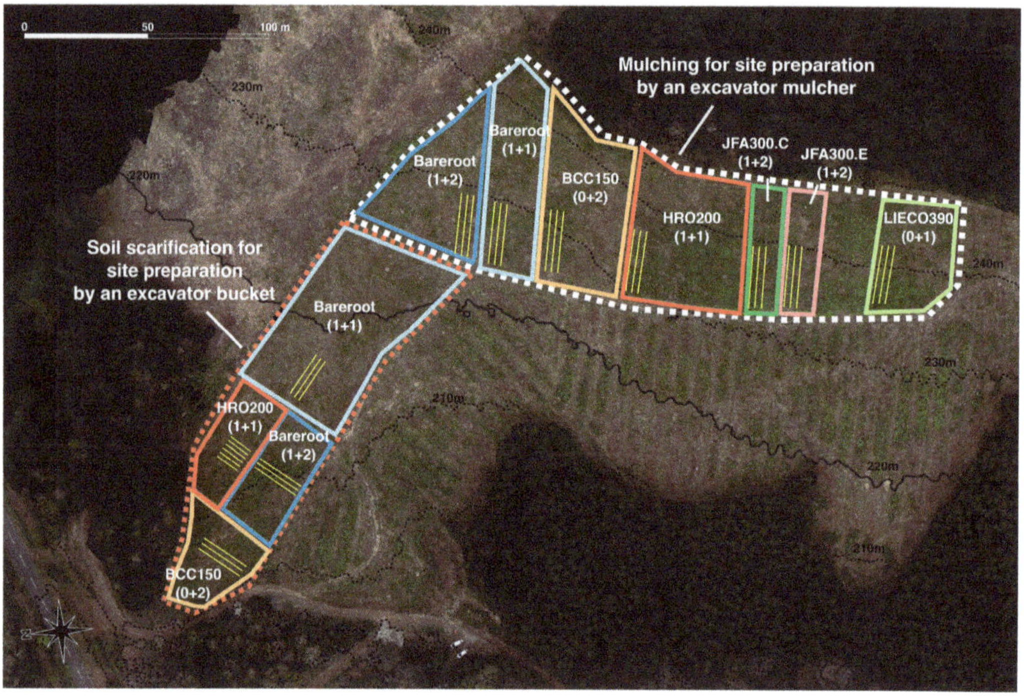

Figure 1. Overview of the study site. Seven stock types were planted in the mulcher site preparation plot and four stock types were planted in the bucket site preparation plot in areas enclosed by solid lines of different colors, blue: bareroot (1+2), light blue: bareroot (1+1), red: HRO200 (1+1), orange: BCC150 (0+2), green: JFA300.C (1+2), pink: JFA300.E (1+2), and light green: LIECO390 (0+1). Yellow lines indicate the investigated planting rows. Details of stock types are shown in Table 1. The photograph was taken by S. Sasaki.

2.2. Measurement of the Dry Mass of the Planted Seedlings

Nine seedlings of each stock type were taken to the laboratory in plastic bags to prevent desiccation. The seedlings were then cut into stems, branches, leaves, and thick and fine roots based on 2 mm diameter, dried in a 70 °C oven to constant mass, and measured for dry mass (accuracy 1 mg). The top:root and top:fine root ratios were calculated by dividing the top mass (stems + branches) by the root mass (thick + fine roots) or by the fine root mass, respectively.

2.3. Field Measurement

In each stock type of each MSP plot, 3 adjacent rows with 10 planting seedlings each were selected, to investigate seedling survival and growth, except for HRO200 (1+1) in the bucket MSP plot, wherein 6 rows of 5 seedlings each were selected ($n = 30$ per stock type in each MSP plot, Figure 1). Seedling height (accuracy 1 cm) and RCD (accuracy 0.1 mm, the average of two cross-measurements) were measured at planting in mid-May 2017, after the first and second growing seasons (GS) in mid-October 2017 and 2018, and after the third and fourth GS in mid-May 2020 and 2021. Survival, animal damage, and miscutting during brush-cutter weeding (conducted in the summer of the second and third GS) were also recorded. Height:RCD ratio and annual (i.e., during one GS) growth of height and RCD were calculated. Relative growth rates (RGRs) of height and RCD during each of the first to fourth GS and for all the four GS after planting were calculated as follows [31]:

$$\text{RGR} = (\ln(\text{dimension after the GS}) - \ln(\text{dimension before the GS}))/\text{GS}, \quad (1)$$

where dimension was height or RCD and ln() was the natural logarithm.

Table 1. Summary of planting stock types.

Name [1]	Bareroot (1+1)	Bareroot (1+2)	LIECO390 (0+1)	JFA300.C (1+2)	JFA300.E (1+2)	HRO200 (1+1)	BCC150 (0+2)
Type	Bareroot	Bareroot	Container	Container	Container	Container	Container
Seedling age [2]	1 + 1	1 + 2	0 + 1	1 + 2	1 + 2	1 + 1	0 + 2
Volume of cell (mL)	-	-	390	300	300	200	150
Density in container (cells m^{-2})	-	-	198	178	178	112	362
Side slit	-	-	Yes	No	No	Yes	Yes
Product name of container	-	-	LIECO 15 blue	JFA300	JFA300	HRO200	BCC SideSlit cell 150
Nursery location in Hokkaido	Central	Central	Western	Central	Eastern	Northeastern	Northern
Seed collection year and town	2013 Ashoro	2011 Yuubetsu	2015 Aibetsu	2015 Asahikawa	2015 Bihoro	2015 Aibetsu	Unknown
Planted plot	Both	Both	Mulcher	Mulcher	Mulcher	Both	Both

[1] The label of a container-grown seedling comprises the product name and the cell capacity. The number in parentheses indicates the seedling age. [2] Seedling age is expressed as the number of years at sowing site + years at transplanting site. The latter number represents years in the field nursery and in the container after transplanting for bareroot and container seedlings, respectively.

Many seedlings in the study area suffered from feeding damage by wild voles, and died between the first and second GS during the winter of 2017–2018, increasing the difficulty in continuing the study with only the initially selected seedlings. Therefore, in order to be able to continuously compare average and annual growth, an additional 5–20 seedlings for each stock type were selected in each preparation area for seedlings close to the row to be measured at the end of the second GS, to facilitate the investigation of 21 or more seedlings per stock type per MSP method. Animal damage and miscutting also occurred during the other periods. Consequently, the number of seedlings for which growth could be calculated was 21–30, 4–26, 17–31, and 11–29 after the first, second, third, and fourth GS, respectively, per stock type in each MSP plot.

2.4. Statistical Analyses

All statistical analyses were performed using R (Version 4.1.2, Vienna, Austria) [32]. For the field measurement data, seedlings affected by animal damage or miscutting were excluded from analyses to assess intrinsic seedling growth potential.

Linear models (LMs) with Tukey–Kramer post hoc comparisons using the "glht" function from the "multcomp" package [33] were used to evaluate differences in organ dry mass (such as stem, branches, leaves, roots, and thick and fine roots) and organ allocation (such as top:root and top:fine-root ratios) among all stock types and among stock types of container-grown seedlings. The methods were also used to evaluate differences in seedling height, RCD, and height:RCD ratio at planting and after each first to fourth GS, annual height and RCD growth, and RGR of height and RCD during each first to fourth GS among four stock types in the bucket MSP plot and seven stock types in the mulcher MSP plot (i.e., a total of 11 treatments). A general linear model with binomial distribution and logit-link function followed by post hoc comparisons was used for survival after a GS among the total 11 treatments, using the "brglm2" [34] and "eemeans" [35] packages. The significance level for post hoc comparisons was set at $\alpha = 0.05$.

To quantify the effect of seedling characteristics on post-planting seedling height, linear mixed models (LMMs) were constructed for surviving seedlings in both MSP plots, with height at the end of each first-through-fourth GS as the objective variable, seedling characteristics such as initial height, initial RCD, seedling type as bareroot or container-grown, and seedling age as explanatory variables, and plot as a random effect, using the "lmer" function of the "lme4" package [36]. In addition, to quantify the effects of seedling characteristics as well as plots with different MSP, LMMs were constructed for the data of four stock types (bareroot (1+1), bareroot (1+2), HRO200 (1+1), and BCC150 (0+2)) planted in both MSP plots, with seedling height after each growing season as the objective variable and seedling characteristics (initial height, initial RCD, and seedling type as bareroot or container-grown), as well as plot as explanatory variables. The individual was treated as a random effect to account for variance due to differences in the individuals and planting microsites for each study individual. All numeric explanatory variables were standardized to have a mean of 0 and standard deviation of 1, using the "scale" function to compare effects among explanatory variables. The contribution ratio of each explanatory variable was calculated in each model as the ratio of the absolute value of the coefficient of each explanatory variable to the sum of the absolute values of the coefficients of all explanatory variables. In Japan, the seedling buyer is typically provided with the information that was added to the explanatory variables in these model analyses, but is not provided with information about the type of container, growing medium, or seed source, which were not added to the explanatory variables.

Three years after planting, approximately half of the planted seedlings had exceeded 200 cm, which is the maximum height of competing vegetation. If the height of the planted seedlings rapidly exceeds that of the competing vegetation, subsequent weeding can be unnecessary, resulting in successful and less-costly reforestation. Therefore, a decision tree analysis was performed using the packages of "rpart" [37] and "rpart.plot" [38] to evaluate what seedling characteristics at planting and MSP methods would determine the probability of seedling height exceeding 200 cm after the third GS. The analysis used a binary variable, 1 for growth to 200 cm height after the third GS, and 0 for no growth, as the objective variable, with initial height, RCD, height:RCD ratio, seedling type (bareroot or container-grown), age of planted seedlings, and MSP method (bucket or mulcher) as explanatory variables.

3. Results

3.1. *Dry Mass by Organ and Allocation for Seedlings of Each Stock Type*

Bareroot (1+2) seedlings had significantly higher stem, branch, and total thick- and fine-root DM than other stock types (Figure 2A,B,D–F). Bareroot (1+1) seedlings were insignificantly different from all stock types of container-grown seedlings in stem, branch, and root DM. Top: root and top:fine-root ratios were lower in container-grown seedlings than in bareroot seedlings (Figure 2G,H). Comparisons among container-grown stock types showed that BCC150 (0+2) had significantly lower stem, branch, and root DM than the other stock types (Figure 2A,B,D–F). Stem and total and thick-root DM were significantly higher in JFA300.C (1+2) than in JFA300.E (1+2), despite being grown in the same container for the same number of growing years (Figure 2A,D,E). Dormant Japanese larch seedlings are normally planted before flushing, but JFA300.C (1+2) and HRO200 (1+1) seedlings were already flushed, and had leaves at planting (Figure 2C).

3.2. *Survival and Growth of Each Stock Type*

Survival rate after one GS from planting ranged from 70% to 100% among seedlings of stock type with bucket and mulcher MSP (Figure 3). Bareroot (1+2) and JFA300.E (1+2) seedlings in the mulcher plot showed low survival rates. Bareroot (1+1), HRO200 (1+1), and BCC150 (0+2) in the bucket plot and bareroot (1+2) and LIECO390 (0+1) in the mulcher plot demonstrated the highest survival rate.

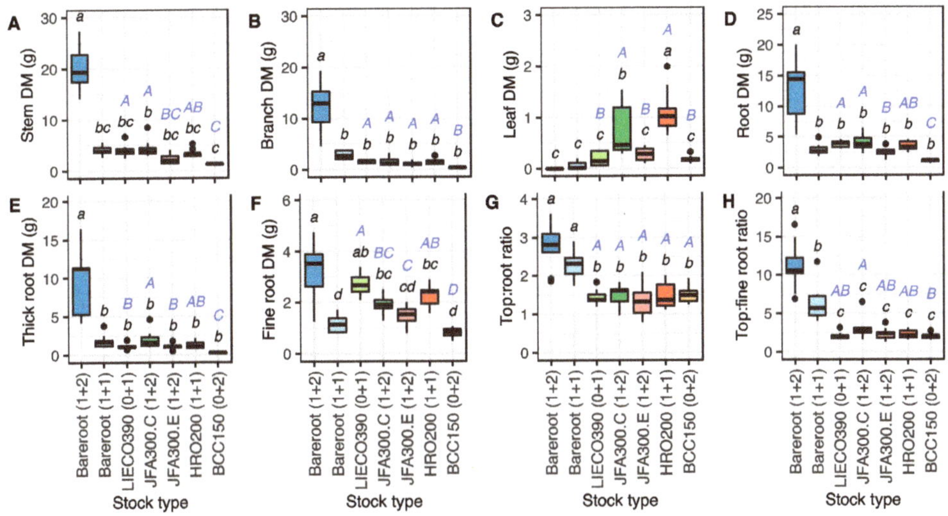

Figure 2. Dry mass by organ and allocation for each stock type. (**A**) stem, (**B**) branch, (**C**) leaf, (**D**) root, (**E**) thick root more than 2 mm in diameter, (**F**) fine root less than 2 mm in diameter, (**G**) top (stem + branch): root ratio, and (**H**) top:fine-root ratio. Different black lowercase and blue uppercase letters indicate significant differences at $p < 0.05$ among all stock types and among container-grown stock types, respectively. Details of stock types are shown in Table 1.

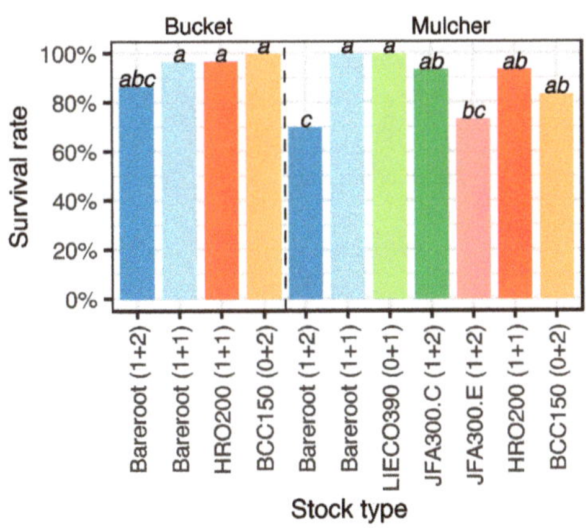

Figure 3. Survival rate after one growing season from planting in the mechanical-site-preparation (MSP) plot with bucket and mulcher. Different letters indicate significant differences at $p < 0.05$. Details of stock types are shown in Table 1.

Seedling height at planting was highest for bareroot (1+2) seedlings of 104 ± 15 cm (mean ± standard deviation), followed by LIECO390 (0+1), JFA300.C (1+2), and HRO200 (1+1), of 54 ± 7 cm, 53 ± 6 cm, and 49 ± 9 cm, respectively (Figure 4A). The lowest heights were reached by JFA300.E (1+2) and BCC150 (0+2) seedlings of 38 ± 8cm and 37 ± 3 cm, respectively. Bareroot (1+2) seedlings in the bucket plot were the tallest of the stock types in all four GSs, while JEA300.E (1+2) and BCC150 (0+2) seedlings in the mulcher plot were the lowest in all four GSs. LIECO390 (0+1) seedlings in the mulcher plot, which were the same height as many other container-grown stock-type seedlings at planting, became the tallest of the container-grown stock types after the fourth GS and were insignificantly different from bareroot (1+2) seedlings in the bucket plot.

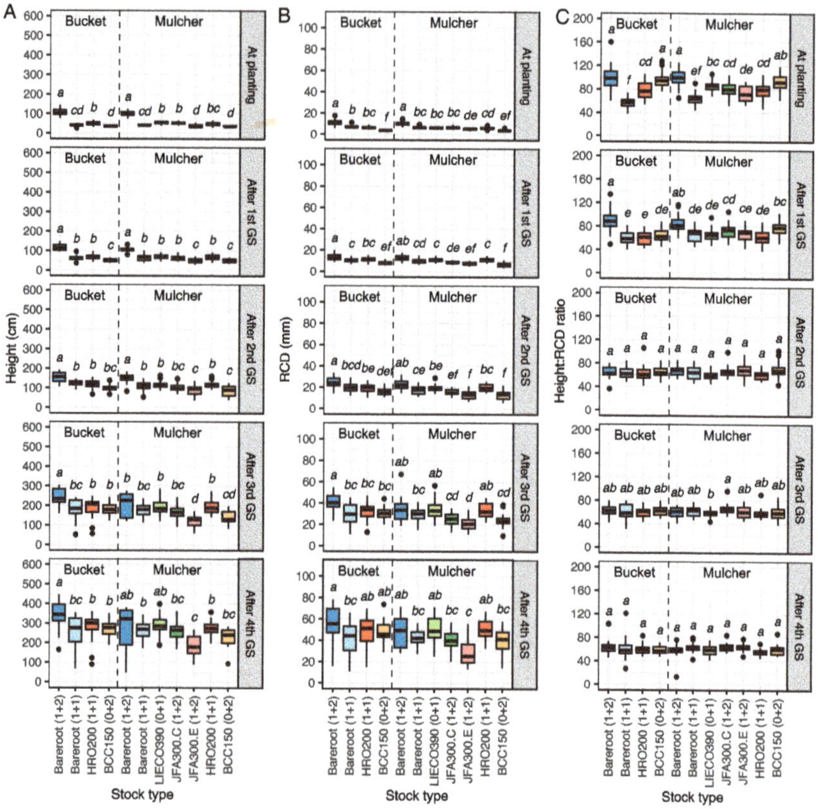

Figure 4. (**A**) Height, (**B**) root collar diameter (RCD), and (**C**) height:RCD ratio of planting seedlings of each stock type in the bucket and mulcher MSP plot at planting and at the end of the first to fourth growing seasons (GS). Different letters indicate significant differences at $p < 0.05$. Details of stock types are shown in Table 1.

At planting, bareroot (1+2) seedlings demonstrated the highest RCD of all stock types, and this trend continued throughout the four GSs, especially in the bucket plot (Figure 4B). BCC150 (0+2) had the smallest RCD of all stock types at planting. As the GS progressed, the difference in RCD of BCC150 (0+2) between the bucket and mulcher plots increased: BCC 150 (0+2) in the bucket plot had a high RCD, which was insignificantly different from that of the largest bareroot (1+2) seedlings after the fourth GS, while BCC 150 (0+2) in the mulcher plot was one of the stock types with a low RCD. JFA300.E (1+2) in the mulcher plot showed the lowest RCD among stock types after the fourth GS.

The height:RCD ratio was high, around 100 in bareroot (1+2) and BCC150 (0+2) seedlings, and low, around 60 in bareroot (1+1) seedlings, at planting (Figure 4C). After the second GS, the height:RCD ratio was around 60 for all stock types, demonstrating almost no significant differences among stock types after the second to fourth GS.

The RGRs of the height of the bareroot (1+2) seedlings in the bucket and mulcher MSP plot were the lowest of the stock types in the first growing season (Figure 5A). This trend continued in the second GS. For the other stock types, no significant differences were observed in the first and second GSs, except between bareroot (1+1) and LIECO390 (0+1) or JFA300.C (1+2) in the first growing season and between BCC150 (0+2) in the bucket plot and JFA300.C (1+2) in the mulcher plot in the second growing season. Almost no significant differences were observed during the third and fourth GS. There was no trend for JFA300.E (1+2), which had a low height until the four GS, to have a lower RGR of height. Bareroot

(1+2) seedlings had the lowest annual RGR of RCD among the stock types during the first GS (Figure 5B). Almost no significant difference in the RGR of RCD was observed among the stock types during the second to fourth GS.

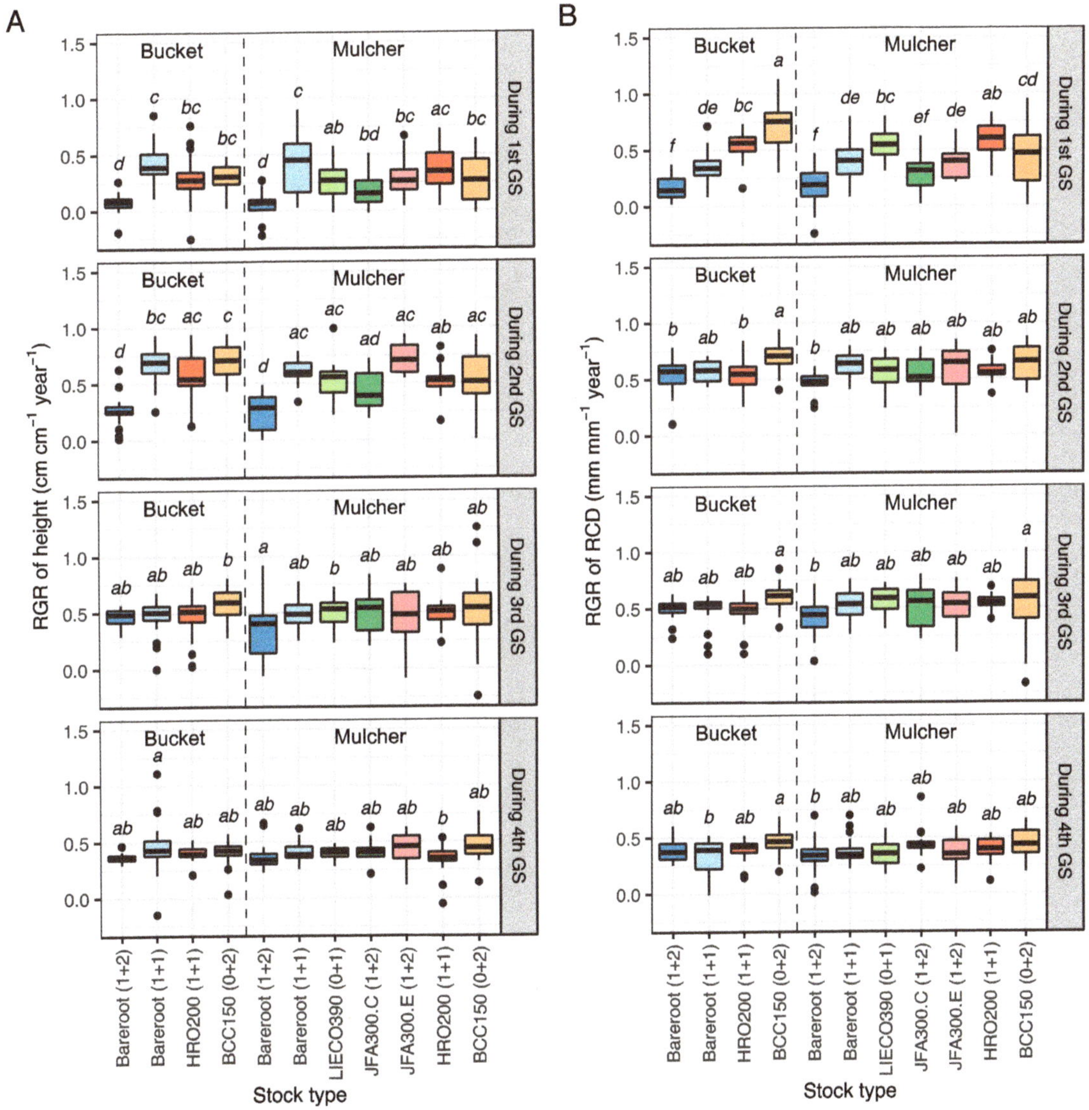

Figure 5. Relative growth rate (RGR) of (**A**) height and (**B**) root collar diameter (RCD) of planting seedlings of each stock type in the bucket and mulcher MSP plots during the first to fourth GS. Different letters indicate significant differences at $p < 0.05$. Details on stock types are shown in Table 1.

RGR for height growth during the first four years after planting was significantly lower for bareroot (1+2) than for the other stock types (Figure 6A). RGR of height tended to be higher for bareroot (1+1) in both MSP plots and BCC150 (0+2) in the bucket MSP plot, although few significant differences in RGR of height for four years were observed among the other stock types, due to the small number of individuals that could be measured for four consecutive years due to mortality from feeding damage. In the RGR of RCD for four years, bareroot (1+2) was also significantly lower than the other stock types, and BCC150 (0+2) was significantly higher (Figure 6B).

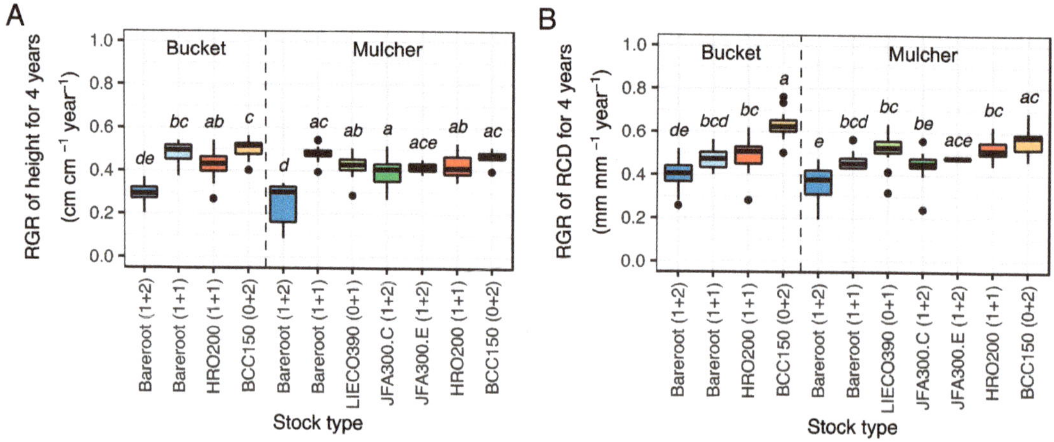

Figure 6. Relative growth rate (RGR) over the four years after planting of (**A**) height and (**B**) root collar diameter (RCD) of planting seedlings of each stock type in the MSP plot with the bucket and mulcher. Different letters indicate significant differences at $p < 0.05$. Details of stock types are shown in Table 1.

3.3. Model Analysis of the Effects of Seedling Quality and Plot on Seedling Height after Planting

The LMM analysis of seven stock types in the mulcher plot revealed a significant positive and negative effect of initial seedling height and seedling age, respectively, on seedling height after the first GS from planting (Figure 7A), and the initial tree height had the highest percentage of contribution (Figure 7B). A negative effect was also observed for seedling type (i.e., container seedlings as compared to bareroot seedlings). The significant positive effect of initial seedling height and the negative effect of seedling age continued until the fourth GS. The contribution rate of seedling age gradually increased, and became the highest contributor after the fourth GS. The significant negative effect of seedling type was not observed after the second GS.

Results of the LM analysis of four stock types planted in mulcher and bucket plots revealed significant positive effects of initial seedling height and RCD on seedling height after the first and second GS after planting (Figure 8A). The positive effects persisted until after the fourth GS, but were no longer significant after the third and fourth GS for initial height and RCD, respectively. Initial height had the highest contribution ratio to the LM among the explanatory variables after the first GS, but the contribution ratio decreased over time; meanwhile, RCD also demonstrated a high contribution rate after the second GS (Figure 8B). The plot effect was negligible after the first GS. After the second to fourth GS, the negative effect of the mulcher plot relative to the bucket plot increased with GS, and the plot effect became significant after the fourth GS. The contribution ratio was the largest among the explanatory variables. The effect of seedling type was small and insignificant throughout the four GSs.

A decision tree analysis indicated that the most important factor for a planted seedlings to exceed the 2 m height of the competing vegetation after the third GS was an initial RCD of 7.3 mm or larger, and the probability increased from 0.46 to 0.75 for this criterion (Figure 9). None of the small seedlings with an initial RCD of less than 7.3 mm and an initial height of less than 34 cm reached 2 m. Seedling age was a factor for the third bifurcation point, with seedlings less than three years old having a high probability of reaching a height of 2 m. Seedling type (bareroot or container-grown) and plot with mulcher or bucket MSP were not selected as criteria.

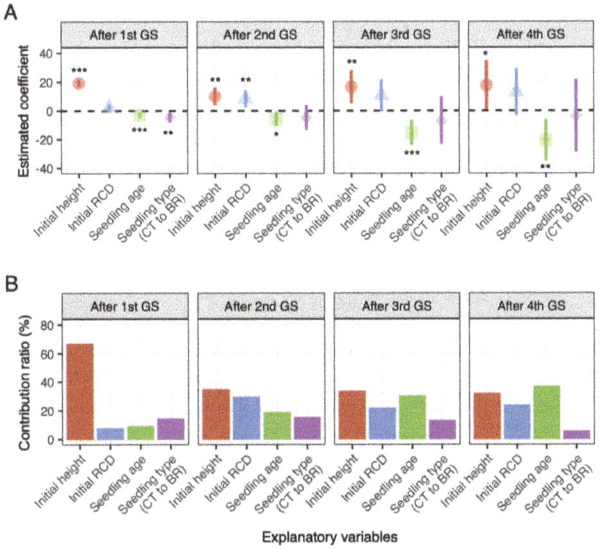

Figure 7. (**A**) Estimated coefficient and (**B**) contribution ratio of explanatory variables of the linear mixed model for each seedling height after the first to fourth GS. Numerical explanatory variables, such as initial height, RCD, and seedling age, were standardized to have a mean of 0 and standard deviation of 1. Seedling type is evaluated as the effect of container-grown (CT) seedlings relative to bareroot (BR) seedlings. Bars in panel (**A**) represent standard errors. *** $p < 0.001$, ** $p < 0.01$, * $p < 0.05$.

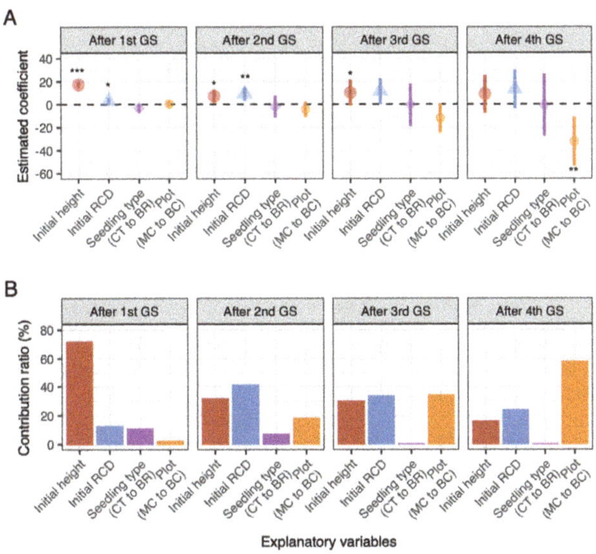

Figure 8. (**A**) Estimated coefficient and (**B**) contribution ratio of explanatory variables of linear model for each seedling height after the first to fourth GSs. Numerical explanatory variables, such as initial height and RCD, were standardized to have a mean of 0 and standard deviation of 1. Seedling type and plot are evaluated as the effect of container-grown (CT) seedlings relative to bareroot (BR) seedlings and mulcher site preparation plot (MC) relative to bucket site preparation plot (BC), respectively. Bars in panel (**A**) represent standard errors. *** $p < 0.001$, ** $p < 0.01$, * $p < 0.05$.

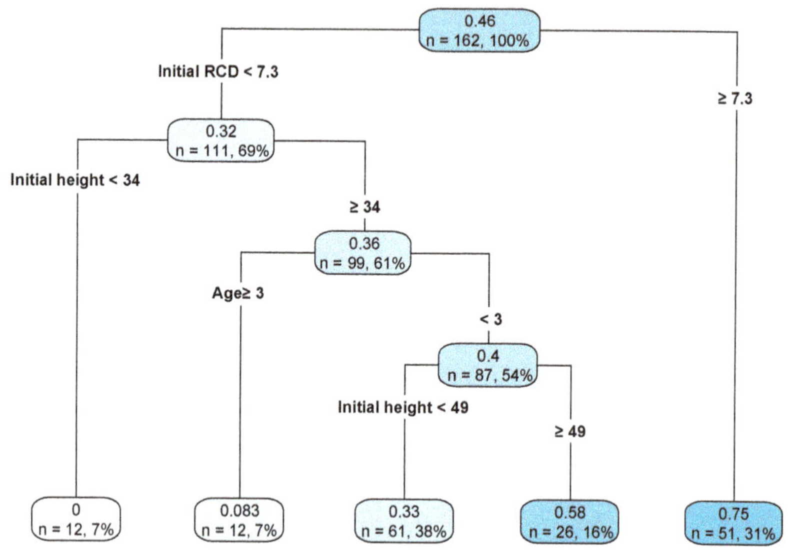

Figure 9. Results of the decision tree analysis of whether the height of planted seedlings exceeded 2 m, which is the height of competing vegetation, after the third GS. Explanatory variables include initial seedling height (cm) and RCD (mm), seedling age and type (bareroot or container-grown), and plot (mulcher or bucket site preparation). Probability of exceeding 200 cm, number of seedlings, and percentage of number of seedlings in each condition are also shown.

4. Discussion

4.1. Effects of Stock Type and MSP Method on Survival

The survival of Japanese larch seedlings after the first GS differed among stock types and plots with varying MSP methods (Figure 3). Bareroot (1+2) seedlings with a high top:root ratio of approximately three (Figure 2A) had low survival rates, especially in the mulcher plot. It is usually considered desirable to have a top:root ratio of about two for bareroot seedlings [39], and bareroot seedlings with a top:root ratio around three often have an imbalance between shoot transpiration and root water uptake, resulting in a low survival rate as the top:root ratio increases from two to three [5]. The monthly precipitation in May 2017 at this study site during planting of seedlings was 41.5 mm, which was 34% lower than in the previous years (1991–2020 average). Thus, the low survival of bareroot (1+2) seedlings in the mulcher plot is attributed to their inability to withstand the drought stress of low precipitation, due to their low drought tolerance with high top:root ratio. In the mulcher MSP, mulched woody residues covered the soil (Figure S1D), which can prevent water loss from the soil surface due to evaporation, and thus should allow for high soil moisture content [40,41]. However, the survival rate of bareroot (1+2) seedlings was lower than in the bucket MSP plot (Figure 3), wherein the soil surface was exposed, and thus was highly susceptible to desiccation. This discrepancy may have occurred due to the large roots of bareroot (1+2) seedlings (Figure 2), which required substantially large planting holes for planting, resulting in the entrance of woody debris in the planting hole and prevention of the root growth and root water uptake in the mulcher MSP plot.

JFA300.E (1+2) seedlings had low survival rates (Figure 3), despite the absence of significant differences in organ dry mass or organ allocation of the seedlings compared to other container-grown stock types (Figure 2). JFA300.E (1+2) seedlings had a nursery period of three years (Table 1) but had smaller seedling size at planting than other container-grown stock types, which had a short nursery period of one or two years (Figure 4A,B). In addition, JFA300.E (1+2) seedlings did not grow well after planting (Figures 5A and 6A); thus, these seedlings were smaller in size after the fourth GS (Figure 4A,B). These results suggest that seed lots for JFA300.E (1+2) seedlings may be of low quality, which may have resulted in poor nursery and post-planting growth and low post-planting survival [42].

4.2. Effects of Stock Type on Growth

Seedling height and RCD, which differed significantly among stock types at planting, became less different over the years after planting; after the fourth GS, no significant differences were found among most stock types (Figure 4). This phenomenon can be attributed to the gradual dilution of the effect of seedling size at planting, because the growth of each planted seedling is influenced by the environmental conditions at each seedling microsite and the intrinsic growth potential of each seedling [1]. The model analyses showed that the contribution of the negative effects of seedling age at planting on seedling height after each GS increased over time, and was the highest among the explanatory variables after four GSs (Figure 7). The negative effect of seedling age could be attributed to seedling height:RCD ratio, especially in bareroot seedlings. The RGR of height of bareroot (1+2) seedlings was low until the second GS (Figure 5A), while that of RCD was comparable to the other stock types in the second GS (Figure 5B), suggesting that bareroot (1+2) seedlings preferred RCD growth to height growth until the proper height:RCD ratio was reached. Similarly, the negative effect of high height:RCD ratio on initial height growth has also been reported in Japanese cedar [43] and Austrian pine [44]. The bareroot (1+2) seedlings in the current study were leftover bareroot (1+1) seedlings, which are commonly used in Japan. The bareroot (1+2) seedlings used in this study were grown at high density in the nursery for an additional year, without transplanting the commonly used bareroot (1+1) seedlings, resulting in a higher height: RCD ratio at planting than other stock types [45]. The high height:RCD ratio of bareroot (1+2) seedlings was eliminated after the second GS after planting (Figure 4C). Thus, the negative effect of seedling age derived from height:RCD ratio on seedling height after planting should be limited to the second GS.

On the other hand, for container-grown seedlings, the negative effect of age appears to be unrelated to height:RCD ratio. Theoretically, height:RCD and top:root ratios could increase with seedling age in container-grown seedlings, because root growth and growth density are fixed during the nursery period, resulting in a negative effect of seedling age on height growth after planting [6]. However, in this study, height:RCD ratio (Figure 4C) and top: root ratio (Figure 2G) did not increase in three-year-old container-grown seedlings, JFA300.C (1+2) and JFA300.E (1+2). This is because the three-year-old container-grown seedlings in this study took longer to reach a shipping size, due to poor growth in the container during the nursery period, and were therefore older. In fact, among the 3-year-old container-grown seedlings, JFA300E tended to have less root mass than some other 1- and 2-year-old container-grown seedlings (Figure 2F). Seed quality may be another factor contributing to the negative effect of seedling age on post-planting seedling height. The age of the stock types varied from one to three years. However, the seedling height at planting was relatively the same among the stock types, ranging from 30 cm to 50 cm, except for the bareroot (1+2) seedlings (Figure 4A). This suggests that the stock type with older seedlings at planting grew slower in the nursery. The slow growth in the nursery can be partly attributed to the use of low-quality seeds, including genetic characteristics [46], which may have influenced the subsequent slow growth after planting.

LMM analyses showed that the effect of initial seedling height and RCD contributed more to seedling height in the first to fourth GSs in Japanese larch than differences in seedling type (bareroot or container-grown seedlings) (Figures 7 and 8, respectively). The results were consistent with the general trend that once seedlings are established, field performance of bareroot and container-grown seedlings can be comparable [7] and large initial height and RCD can lead to subsequent height growth after planting [1]. The seedling height advantage is generally beneficial on sites with competing vegetation [47], and Japanese larch growth is sensitive to competing vegetation [20]; thus, the size of planted seedlings will be an important trait in Japanese larch. Among the container-grown stock types in this study, JFA300 (1+2) and BCC150 (0+2) had significantly lower initial height and RCD than LIECO390 (0+1) and HRO200 (1+1), and this significant trend continued until the third GS (Figure 4). The small initial size of BCC150 (0+2) seedlings can be attributed to the smallest cell capacity and highest growing density among the container-grown stock types [7].

4.3. Effects of Plots with Different MSP Method on Growth

The effect of plots with different MSP methods on seedling height was unclear in the comparisons among stock types (Figure 4). However, LMM analyses revealed that the effect of plots with different MSP methods on seedling height increased with rising GS after planting, and that mulcher MSP had a negative effect (significant at the fourth GS) on seedling height after planting, compared to bucket scarification MSP (Figure 8). Contrary to the seedling growth trends in this study, previous studies have shown that mulcher MSP can provide better soil nutrient availability and less growth of competing vegetation for the successful growth of planted seedlings than bucket MSP. Sikström, et al. [48] reviewed studies on the growth of seedlings planted after five major MSP techniques, including scarification. They reported that scarification removes the humus layer and thus reduces nutrient availability for seedlings planted in pure mineral soil, which may occasionally consequently weaken the positive effect of weed suppression by MSP on seedling height growth. By contrast, in mulcher MSP, soil nitrogen availability could remain the same compared to the control [49] or even increase, due to additions from woody residues generated by the mulching treatment [40]. Moreover, the amount of first summer competing vegetation after planting at this site was comparable to the bucket MSP plot up up to 2 cm thick of mulch residue in the mulcher MSP plot, and less than that of the bucket MSP plot when mulch residue thickness in the mulcher MSP plot was greater than 2 cm [30]. The discrepancy between the potentially favorable environmental conditions for rapid seedling growth in the mulcher MSP reported in these previous studies and the poor seedling growth in the current study may be partially due to the plot location at the study site. The mulcher plot was located higher upslope than the bucket plot; therefore, the mulcher plot was expected to have minimal soil moisture and strong winds, potentially resulting in reduced growth of Japanese larch [18,50,51]. In addition, seedling growth may also have been inhibited by allelopathy, caused by the fermentation of the mulch material [52,53]. The plot design of this study was inadequate to detect the effect of MSP, because there was no MSP replication and different MSP plots were placed at different slope locations. Further research with an adequately randomized study design [8] is needed to clarify whether the effect of plots on the height growth of Japanese larch seedlings, which increased in the years after planting, is due to differences in MSP methods.

4.4. Characteristics of Seedlings That Grow Faster up to the Height of Surrounding Vegetation

Seedling height after planting was affected by initial seedling height and RCD, seedling age, and plots with different MSP methods, and their effects varied with GSs after planting (Figures 7 and 8). Despite these complex effects, the decision tree analysis identified criteria traits that could grow taller faster than the height of the surrounding vegetation (Figure 9). The primary criterion was an initial RCD $\geq$ 7.3 mm (Figure 9), and the percentage of seedlings exceeding this criterion was 18%–30% for LIECO390 (0+1), JFA300.C (1+2), and HRO200 (1+1) but 0% for JFA300.E (1+2) and BCC150 (0+2) in container-grown stock types (Figure 4B, Table S1). For container-grown seedlings, the RCD of seedlings generally increases with large cell volume and low growing density of the container [54–56]; in Japanese larch, lower RCD was reported in JFA150 containers (cell volume: 150 mL, growing density: 292 cells m^{-2}) with lower cell volume and higher seedling density compared to JFA300 containers (cell volume: 300 mL, growing density: 178 cells m^{-2}) [57]. Thus, the low initial RCD of the BCC150 (0+2) can be attributed to the smallest cell volume and highest density of this container in the stock type of this study (Table 1). The small cell volume and high growing density of the BCC150 container can reduce seedling production costs [56].

However, the BCC150 container would be unsuitable for Japanese larch seedlings if the goal is to reduce weeding costs, because they did not grow taller after planting.

JFA300.E (1+2) was another stock type that failed to meet the primary criterion for height after three GSs (initial RCD ≥ 7.5 mm) (Figure 4B, Table S1). As mentioned above, this stock type had a low initial RCD and a low height, despite a long nursery period of three years, indicating poor growth in containers during the nursery. This suggests that the quality of the seed, including genetic quality, used for this stock type may be poor [8], although the quality of seed collected from ordinal plantations had never been evaluated in this study. Seedling age was selected as the third determinant in the decision tree (Figure 9), and contribution of the seedling age to the seedling height after planting increased over time (Figures 7 and 8). These results implied that in order to produce high-growth seedlings of Japanese larch after planting, it is important that the seedlings are planted in the appropriate morphology, as well as that the seedlings are not old, reflecting high growth during the nursery period, which can be partly attributed to good seed quality. In general, tree growth is genetically controlled [58,59]. Japanese larch breeding in Japan began in the 1950s, and more than 530 trees have been selected to produce improved seeds [13]. However, the establishment of seed orchards has not developed, and most of the seeds used in nurseries are not from the selected trees, but from ordinal plantations without genetic assessment. Early establishment of seed orchards composed of the selected trees is important for the production of high-growth seedlings. In Japanese larch, seed weight had a positive effect on the RCD of seedlings grown in containers for one year, and the Seed Quality Index (SQI) calculated from three near-infrared wavelengths showed a good separation of seed weight and germination potential [57]. Therefore, the SQI may be useful for selecting high-quality seeds from those collected from ordinal plantations, and consequently for producing high-growth seedlings. High growth in the nursery, associated with the use of high-quality seed, will also help reduce costs by reducing the number of production years and will increase the size of containerized seedlings, which are currently more expensive but smaller than bareroot seedlings.

5. Conclusions

A four-year consecutive measurement of seven Japanese larch stock types planted on two different MSPs showed that the combination of large bareroot seedlings and mulcher MSP can reduce post-planting survival. Model analysis showed that initial height, RCD, seedling age, and plot influenced post-planting seedling height, but seedling type (bareroot or container-grown seedlings) did not, and the effect of initial size decreased with the number of years since planting. The determinant tree analysis indicated that the most important factor for fast-growing tall seedlings after planting, which is suitable for reducing weeding costs, was a large diameter, regardless of seedling type (bareroot or container-grown). Older seedlings had lower survival and height growth after planting, reflecting high top: root and height:RCD ratios in bareroot seedlings and slow nursery growth in container seedlings.

Supplementary Materials: The following supporting information can be downloaded at https://www.mdpi.com/article/10.3390/f14040784/s1, Figure S1: Photographs of the mechanical site preparation (A,C) and the forest floor after preparation (B,D); Table S1: Percentage of number of planted seedlings that had an initial root collar diameter (RCD) of 7.3 mm or larger and number of observations (n) by stock type.

Author Contributions: Conceptualization, H.H. and T.Y.; methodology, H.H., I.T. and T.Y.; formal analysis, H.H. and I.T.; investigation, H.H., I.T, M.K., T.Y., N.F., H.U. and S.S.; data curation, H.H. and T.Y.; writing—original draft preparation, H.H.; writing—review and editing, H.H., I.T, M.K., T.Y., N.F., H.U. and S.S.; visualization, H.H. and S.S.; supervision, H.U. and S.S.; project administration, H.U. and S.S.; funding acquisition, H.U. and S.S. All authors have read and agreed to the published version of the manuscript.

Funding: This research was funded by the Ministry of Agriculture, Forestry, and Fisheries of Japan (grant number 18064868) and the New Energy and Industrial Technology Development Organization ("Decarbonized electric robot for forest management in rural area").

Data Availability Statement: The data presented in this study are available on request from the corresponding author.

Acknowledgments: We would like to thank Shimokawa Town, especially Tomohiro Saitou, for providing the planting site for this study, and Hokkaido Tree Nursery Cooperative for providing seed information. We are also grateful to Kazuhito Kita, Hirokazu Kon, Akira Uemura, Kenichi Yazaki, Mitsutoshi Umemura, Tetsuto Sugai, Tatsuya Sasaki, and Haruka Yamamoto for their assistance in the field and laboratory work.

Conflicts of Interest: The authors declare no conflict of interest. The funders had no role in the design of the study; in the collection, analyses, or interpretation of data; in the writing of the manuscript, or in the decision to publish the results.

References

1. Grossnickle, S.C.; MacDonald, J.E. Why seedlings grow: Influence of plant attributes. *New For.* **2018**, *49*, 1–34. [CrossRef]
2. Grossnickle, S.; MacDonald, J. Seedling Quality: History, Application, and Plant Attributes. *Forests* **2018**, *9*, 283. [CrossRef]
3. Mattsson, A. Predicting field performance using seedling quality assessment. *New For.* **1997**, *13*, 227–252. [CrossRef]
4. Duryea, M.L. Nursery cultural practices: Impacts on seedling quality. In *Forestry Nursery Manual: Production of Bareroot Seedlings*; Duryea, M.L., Landis, T.D., Perry, C.R., Eds.; Springer: Dordrecht, The Netherlands, 1984; Volume 11, pp. 143–164.
5. Grossnickle, S.C. Why seedlings survive: Influence of plant attributes. *New For.* **2012**, *43*, 711–738. [CrossRef]
6. Landis, T.D.; Dumroese, R.K.; Haase, D.L. Seedling processing, storage and outplanting. In *The Container Tree Nursery Manual*; USA Department of Agriculture, Forest Service: Washington, DC, USA, 2010; Volume 7, p. 192.
7. Grossnickle, S.C.; El-Kassaby, Y.A. Bareroot versus container stocktypes: A performance comparison. *New For.* **2016**, *47*, 1–51. [CrossRef]
8. Pinto, J.R.; Dumroese, R.K.; Davis, A.S.; Landis, T.D. Conducting seedling stock type trials: A new approach to an old question. *J. For.* **2011**, *109*, 293–299. [CrossRef]
9. Nagamitsu, T.; Nagasaka, K.; Yoshimaru, H.; Tsumura, Y. Provenance tests for survival and growth of 50-year-old Japanese larch (*Larix kaempferi*) trees related to climatic conditions in central Japan. *Tree Genet. Genomes* **2013**, *10*, 87–99. [CrossRef]
10. Qu, L.; Wang, Y.; Masyagina, O.; Kitaoka, S.; Fujita, S.; Kita, K.; Prokushkin, A.; Koike, T. Larch: A promising deciduous conifer as an eco-environmental resource. In *Conifers—Recent Advances*; Gonçalves, A.C., Fonseca, T.F., Eds.; IntechOpen: London, UK, 2022; pp. 1–37.
11. Da Ronch, F.; Caudullo, G.; Tinner, W.; de Rigo, D. Larix decidua and other larches in Europe: Distribution, habitat, usage and threats. In *European Atlas of Forest Tree Species*; San-Miguel-Ayanz, J., de Rigo, D., Caudullo, G., Houston Durrant, T., Mauri, A., Eds.; Publication Office of the European Union: Luxembourg, 2016.
12. Greenwood, M.S.; Roth, B.E.; Maass, D.D.; Irland, L.C. Near rotation-length performance of selected hybrid larch in Central Maine, U.S.A. *Silvae Genet.* **2015**, *64*, 73. [CrossRef]
13. Kurinobu, S. Forest tree breeding for Japanese larch. *Eurasian J. For. Res.* **2005**, *8*, 127–134.
14. Harayama, H.; Tsuyama, I.; Kuramoto, S.; Uemura, A.; Kitao, M.; Han, Q.; Yamada, T.; Sasaki, S. Effects of weed competition on the survival and initial growth of planted seedlings of Japanese larch (*Larix kaempferi*). *J. Jpn. For. Soc.* **2018**, *100*, 158–164, (In Japanese with English summary) [CrossRef]
15. Tsuyama, I.; Harayama, H.; Kita, K. Evaluation on the effectiveness of containerized seedlings in Hokkaido. *Boreal For. Res.* **2018**, *66*, 69–72. (In Japanese) [CrossRef]
16. Agathokleous, E.; Kitao, M.; Komatsu, M.; Tamai, Y.; Saito, H.; Harayama, H.; Uemura, A.; Tobita, H.; Koike, T. Effects of soil nutrient availability and ozone on container-grown Japanese larch seedlings and role of soil microbes. *J. For. Res.* **2020**, *31*, 2295–2311. [CrossRef]
17. Harayama, H.; Tobita, H.; Kitao, M.; Kon, H.; Ishizuka, W.; Kuromaru, M.; Kita, K. Enhanced summer planting survival of Japanese larch container-grown seedlings. *Forests* **2021**, *12*, 1115. [CrossRef]
18. Kitao, M.; Agathokleous, E.; Harayama, H.; Kitaoka, S.; Uemura, A.; Yazaki, K.; Tobita, H. Tolerance of Japanese larch to drought is modified by nitrogen and water regimes during cultivation of container seedlings. *Eur. J. For. Res.* **2022**, *141*, 699–712. [CrossRef]
19. Agathokleous, E.; Kitao, M.; Komatsu, M.; Tamai, Y.; Harayama, H.; Koike, T. Single and combined effects of fertilization, ectomycorrhizal inoculation, and drought on container-grown Japanese larch seedlings. *J. For. Res.* **2022**, 1–18. [CrossRef]
20. Harayama, H.; Tsuyama, I.; Uemura, A.; Kitao, M.; Han, Q.; Kuramoto, S.; Utsugi, H. Growth and survival of hybrid larch F1 (*Larix gmelinii* var. *japonica* × *L. kaempferi*) and Japanese larch under various intensities of competition. *New For.* **2022**, 1–17. [CrossRef]

21. Kitao, M.; Harayama, H.; Yazaki, K.; Tobita, H.; Agathokleous, E.; Furuya, N.; Hashimoto, T. Photosynthetic and growth responses in a pioneer tree (Japanese white birch) and competitive perennial weeds (*Eupatorium* sp.) grown under different regimes with limited water supply to waterlogging. *Front. Plant Sci.* **2022**, *13*, 835068. [CrossRef]
22. Löf, M.; Dey, D.C.; Navarro, R.M.; Jacobs, D.F. Mechanical site preparation for forest restoration. *New For.* **2012**, *43*, 825–848. [CrossRef]
23. Dumas, N.; Dassot, M.; Pitaud, J.; Piat, J.; Arnaudet, L.; Richter, C.; Collet, C. Four-year-performance of oak and pine seedlings following mechanical site preparation with lightweight excavators. *Silva Fenn.* **2021**, *55*, 10409. [CrossRef]
24. Nakagawa, M.; Kanno, M.; Yasaka, M. A weeding-duration model for *Larix kaempferi* plantations in Hokkaido, northern Japan. *J. For. Res.* **2017**, *16*, 319–324. [CrossRef]
25. Masaki, T.; Oguro, M.; Yamashita, N.; Otani, T.; Utsugi, H. Reforestation following harvesting of conifer plantations in Japan: Current issues from silvicultural and ecological perspectives. *Reforesta* **2017**, *3*, 125–141. [CrossRef]
26. Oya, S.; Kuramoto, S.; Koyama, Y.; Nakazawa, M.; Taki, S.; Utsugi, H. Effects of mechanical site preparation on controlling competing vegetation and weeding reduction. *J. Jpn. For. Eng. Soc.* **2021**, *36*, 99–110, (In Japanese with English summary) [CrossRef]
27. Yamada, T.; Sasaki, S.; Kuramoto, S.; Uemura, A.; Harayama, H.; Utsugi, H.; Saito, T. Operational efficiency and effects on silvicultural works of the crusher for site preparation. *J. Jpn. For. Eng. Soc.* **2018**, *33*, 67–71. (In Japanese) [CrossRef]
28. Harayama, H.; Uemura, A.; Tsuyama, I.; Sasaki, S.; Yamada, T.; Utsugi, H.; Kuramoto, S. Weed suppression effect of site preparation by a crusher. *Boreal For. Res.* **2016**, *64*, 61–62. [CrossRef]
29. Jain, T.; Sikkink, P.; Keefe, R.; Byrne, J. *To Masticate or not: Useful Tips for Treating Forest, Woodland, and Shrubland Vegetation*; Rocky Mountain Research Station: Logan, UT, USA, 2018; p. 55.
30. Harayama, H.; Uemura, A.; Tsuyama, I.; Sasaki, S.; Yamada, T.; Watanabe, I.; Utsugi, H. Effect of crashed branch on weed outbreak by a crusher for site preparation. *Boreal For. Res.* **2018**, *66*, 73–76. (In Japanese) [CrossRef]
31. Hunt, R. Plant growth curves. In *The Functional Approach to Plant Growth Analysis*; Edward Arnold Ltd.: London, UK, 1982.
32. R Core Team. *R: A Language and Environment for Statistical Computing*; R Foundation for Statistical Computing: Vienna, Austria, 2021.
33. Hothorn, T.; Bretz, F.; Westfall, P. Simultaneous inference in general parametric models. *Biom. J.* **2008**, *50*, 346–363. [CrossRef]
34. Kosmidis, I.; Firth, D. Jeffreys-prior penalty, finiteness and shrinkage in binomial-response generalized linear models. *Biometrika* **2021**, *108*, 71–82. [CrossRef]
35. Lenth, R.V.; Bolker, B.; Buerkner, P.; Giné-Vázquez, I.; Herve, M.; Jung, M.; Love, J.; Miguez, F.; Riebl, H. Singmann HEmmeans: Estimated Marginal Means, aka Least-Squares Means. Available online: https://cran.r-project.org/web/packages/emmeans/emmeans.pdf (accessed on 3 February 2023).
36. Bates, D.; Maechler, M.; Bolker, B.; Walker, S. Fitting linear mixed-effects models using lme4. *J. Stat. Softw.* **2015**, *76*, 1–48. [CrossRef]
37. Therneau, T.; Atkinson, B. Rpart: Recursive Partitioning and Regression Trees. Available online: https://cran.r-project.org/web/packages/rpart/rpart.pdf (accessed on 3 February 2023).
38. Milborrow, S. Rpart.plot: Plot 'rpart' Models: An Enhanced Version of 'plot.rpart'. Available online: https://cran.r-project.org/web/packages/rpart.plot/rpart.plot.pdf (accessed on 3 February 2023).
39. Bernier, P.Y.; Lamhamedi, M.S.; Simpson, D.G. Shoot:root ratio is of limited use in evaluating the quality of container conifer stock. *Tree Plant. Notes* **1995**, *46*, 102–106.
40. Rhoades, C.C.; Battaglia, M.A.; Rocca, M.E.; Ryan, M.G. Short- and medium-term effects of fuel reduction mulch treatments on soil nitrogen availability in Colorado conifer forests. *For. Ecol. Manag.* **2012**, *276*, 231–238. [CrossRef]
41. Massman, W.J.; Frank, J.M.; Jimenez Esquilin, A.E.; Stromberger, M.E.; Shepperd, W.D. Long term consequences of a controlled slash burn and slash mastication to soil moisture and CO_2 at a southern Colorado site. In Proceedings of the 27th Conference on Agricultural and Forest Meteorology, Boston, MA, USA, 22 May 2006.
42. Carrasquinho, I.; Gonçalves, E. Genetic variability among Pinus pinea L. provenances for survival and growth traits in Portugal. *Tree Genet. Genomes* **2013**, *9*, 855–866. [CrossRef]
43. Yagihashi, T.; Nakaya, T.; Nakahara, K.; Nasuno, S.; Hitsuma, G.; Noguchi, M.; Yagi, T.; Saitoh, T.; Matsumoto, K.; Yamada, T.; et al. Correlation between height: Diameter ratio and shoot growth in containerised and bare-root seedlings of *Cryptomeria japonica*. *J. Jpn. For. Soc.* **2016**, *98*, 139–145, (In Japanese with English summary). [CrossRef]
44. Ivetic, V.; Grossnickle, S.; Skoric, M. Forecasting the field performance of Austrian pine seedlings using morphological attributes. *IForest—Biogeosci. For.* **2017**, *10*, 99–107. [CrossRef]
45. Schultz, R.C.; Thompson, J.R. Effect of density control and undercutting on root morphology of 1 + 0 bareroot hardwood seedlings: Five-year field performance of root-graded stock in the central USA. *New For.* **1997**, *13*, 301–314. [CrossRef]
46. Landis, T.; Tinus, R.; Barnett, J. Seedling Propagation. In *The Container Tree Nursery Manual*; USA Forest Service: Washington, DC, USA, 1999; Volume 6, p. 167.
47. Thiffault, N.; Jobidon, R.; Munson, A.D. Comparing large containerized and bareroot conifer stock on sites of contrasting vegetation composition in a non-herbicide scenario. *New For.* **2014**, *45*, 875–891. [CrossRef]
48. Sikström, U.; Hjelm, K.; Holt Hanssen, K.; Saksa, T.; Wallertz, K. Influence of mechanical site preparation on regeneration success of planted conifers in clearcuts in Fennoscandia—A review. *Silva Fenn.* **2020**, *54*, 10172. [CrossRef]

49. Moghaddas, E.E.Y.; Stephens, S.L. Thinning, burning, and thin-burn fuel treatment effects on soil properties in a Sierra Nevada mixed-conifer forest. *For. Ecol. Manag.* **2007**, *250*, 156–166. [CrossRef]
50. Harayama, H.; Kita, K.; Kon, H.; Ishizuka, W.; Tobita, H.; Utsugi, H. Effect of planting season on survival rate, growth and ecophysiological properties of container seedlings of Japanese larch (*Larix kaempferi*). *J. Jpn. For. Soc.* **2016**, *98*, 158–166, (In Japanese with English summary) [CrossRef]
51. Mitsuda, Y.; Yoshida, S.; Imada, M. Use of GIS-derived environmental factors in predicting site indices in Japanese larch plantations in Hokkaido. *J. For. Res.* **2001**, *6*, 87–93. [CrossRef]
52. Bachheti, A.; Sharma, A.; Bachheti, R.K.; Husen, A.; Pandey, D.P. Plant allelochemicals and their various applications. In *Co-Evolution of Secondary Metabolites*; Springer: Berlin/Heidelberg, Germany, 2019; pp. 1–25.
53. Kato-Noguchi, H.; Kurniadie, D. Allelopathy and Allelochemicals of Leucaenaleucocephala as an Invasive Plant Species. *Plants* **2022**, *11*, 1672. [CrossRef]
54. Dominguez-Lerena, S.; Herrero Sierra, N.; Carrasco Manzano, I.; Ocaña Bueno, L.; Peñuelas Rubira, J.L.; Mexal, J.G. Container characteristics influence *Pinus pinea* seedling development in the nursery and field. *For. Ecol. Manag.* **2006**, *221*, 63–71. [CrossRef]
55. Aphalo, P.; Rikala, R. Field performance of silver-birch planting-stock grown at different spacing and in containers of different volume. *New For.* **2003**, *25*, 93–108. [CrossRef]
56. Landis, T.D. Containers: Types and functions. In *The Container Tree Nursery Manual*; USA Forest Service: Washington, DC, USA, 1990; Volume 2, p. 87.
57. Kita, K.; Kon, H.; Ishizuka, W.; Matsuda, O. Characterization of *Larix kaempferi* seeds selected by near-infrared spectroscopy for germination and post-germination growth in nursery containers. *J. For. Res.* **2022**, *27*, 158–167. [CrossRef]
58. Jansson, G.; Hansen, J.K.; Haapanen, M.; Kvaalen, H.; Steffenrem, A. The genetic and economic gains from forest tree breeding programmes in Scandinavia and Finland. *Scand. J. For. Res.* **2016**, *32*, 273–286. [CrossRef]
59. Jansen, S.; Geburek, T. Historic translocations of European larch (Larix decidua Mill.) genetic resources across Europe—A review from the 17th until the mid-20th century. *For. Ecol. Manag.* **2016**, *379*, 114–123. [CrossRef]

forests

MDPI

Article

The Effect of Biochar Amendment, Microbiome Inoculation, Crop Mixture and Planting Density on Post-Mining Restoration

Degi Harja Asmara [1,*,†], Suzanne Allaire [2], Meine van Noordwijk [3,4] and Damase P. Khasa [1]

1 Centre for Forest Research and Institute for Integrative Systems Biology, Université Laval, Quebec, QC G1V 0A6, Canada
2 GECA Environnement, 4018 Ave des Pyrénées, Quebec, QC G1P 3L9, Canada
3 World Agroforestry (ICRAF), Bogor 16115, Indonesia
4 Plant Production Systems, Wageningen University, 6700 AK Wageningen, The Netherlands
* Correspondence: degiharja@gmail.com
† This manuscript is part of a Ph.D. thesis by the first author, available online at https://corpus.ulaval.ca/entities/publication/6b074276-e0b9-4f45-bfec-37fc262a5599 (accessed on 17 April 2023).

Abstract: Ecological restoration with a multispecies and multifunctional approach can accelerate the re-establishment of numerous ecosystem services. The challenges with land that is degraded, damaged, or destroyed post-mining are the low productivity of soil and the high potential for contaminants. Herein, we evaluated the multispecies and multifunctional approach to restoration strategy through a mixture of woody and herbaceous species, microsymbiont and biochar amendments, and plant spacing. The experiments were conducted using greenhouse and field trials located in Quebec, Canada. We used a mixture of tree species (*Alnus viridis* (Chaix) DC. ssp. crispa (Aiton) Turrill, *Picea glauca* (Moench) Voss, *Populus tremuloides* Michx. and *Salix arbusculoides* Andersson) and herbaceous species (*Avena sativa* L., *Festuca rubra* L. and *Trifolium repens* L.) on two types of gold-mine waste materials (fine tailing and waste rock). The biochar amendment and microbial inoculation were applied on both greenhouse and field trials. We found both positive and negative effects of plant spacing, biochar amendment and inoculation depending on their interactions. The net positive effect was shown by combining high plantation density, biochar, and inoculation factors on *Alnus viridis* ssp. crispa. Overall, plantation density was shown to be the most important factor in generating the net positive effect. We suggest that the mechanism was correlated with the improvement in microclimate through soil plant water conservation and microbial activity enhancement over soil temperature modification. Hence, we propose to put emphasis on microclimate improvement for accelerating the restoration processes, along with other combined factors, including microbial inoculation and biochar amendment.

Keywords: restoration; facilitation; post-mining; plantation spacing; biochar; inoculation

Citation: Asmara, D.H.; Allaire, S.; van Noordwijk, M.; Khasa, D.P. The Effect of Biochar Amendment, Microbiome Inoculation, Crop Mixture and Planting Density on Post-Mining Restoration. *Forests* **2023**, *14*, 856. https://doi.org/10.3390/f14040856

Academic Editors: Timothy A. Martin and Guntis Brūmelis

Received: 16 February 2023
Revised: 31 March 2023
Accepted: 18 April 2023
Published: 21 April 2023

1. Introduction

Since the industrial era, land degradation, damage and destruction (3D) (hereafter, collectively referred to as degradation), have become some of the most significant environmental problems globally. These problems have diminished the biodiversity, functioning and resilience of ecosystems, which in turn negatively affects the resilience and sustainability of social–ecological systems [1]. The tradeoffs between economic benefits and the loss of ecosystem functions seem to be unsustainable in terms of natural resources. The ecological restoration paradigm, which focuses on enhancing ecosystem services and increasing resilience, is believed to be the best environmentally sustainable practice [2]. Accelerating land restoration with similar principles could be beneficial considering the expansion in demand for land and food [3].

Multi-species mixtures of trees and crops in agroforestry bring about a unique set of ecological interactions that can be positive, neutral or negative among different species [4–6].

If the agroforestry system formed by different multi-species mixtures of trees and crops makes it more resilient, eases recovery from disturbances and accelerates successional processes [2,7,8], then this system may in total accrue ecosystem goods and services greater than the output of those species if they were grown separately on an equal area of land [5]. The multi-species approach has been suggested in restoration practices [9–12]. However, the mechanism by which the species diversity and the ecosystem function is sometimes confusing [13], and most of the restoration experiments have used the trial-and-error approach [10], which makes it difficult to predict the optimum method in various cases. Therefore, understanding both the biophysical processes and the mechanisms involved in the allocation of resources is essential for the development of ecologically sound agroforestry systems that are sustainable, economically viable, and socially acceptable [4].

The diversity of the soil ecosystem has the same role as the diversity of aboveground species in providing ecosystem services. A healthy soil ecosystem is formed by various species of microorganism with specific roles and functions [14]. While some plants are hosts for other symbiotic microorganisms, their existence and diversity are highly correlated [15]. The introduction of inoculation with mixed microorganism species is expected to restore the soil ecosystem and ameliorate the soil conditions. At the same time, some microorganisms are also able to extract contaminants from soil, which are often used in phytobial remediation technologies [16,17]. Degraded post-mining soil is not an ideal material for plant growth. The physico-chemical and microbiological characteristics of the materials are too poor for optimal plant growth [18,19]. Fertilization may help the plant to grow better, but it can be costly and not sustainable in ecological restoration [20]. The introduction of soil amendments could be a sustainable option supporting plant growth, as it has been shown to improve soil properties and functions relevant to agronomic and environmental performance [21,22]. Hypothesized mechanisms for such a potential improvement are mainly enhanced water and nutrient retention (as well as improved soil structure and drainage). Furthermore, there is experimental evidence that soil microbial communities and their activity, which have key roles in sustaining soil health and functioning, are directly affected by the addition of biochar to soils [23–26].

Here, we evaluated the mixture of woody and herbaceous plant species with the introduction of microsymbionts through inoculation and the application of biochar amendments for accelerating the restoration processes. The spacing effect was also tested to find out the interaction mechanism between the plant species and their micro-environment. The aim of the research is to find the best method for the restoration of post-mining sites.

2. Materials and Methods

2.1. Biochar and Hydrogel Amendments

The biochar used in this experiment was the commercial Award-Maple-700 made through pyrolysis of maple bark at 700 °C for 20 min from the company Award Rubber (Windsor, QC, Canada). The physicochemical properties of the biochar are pH 8.39 ± 0.75, cation exchange capacity (CEC) 26.3 ± 2.2 $cmol^+ kg^{-1}$, C total 65.4%, ash content 14.2%, N 0.58 ± 0.01%, P 805 ± 22 mg kg^{-1}, $K_{exchangeable}$ 8.09 ± 0.4 $cmol^+ kg^{-1}$, $Na_{exchangeable}$ 2.03 ± 0.05 $cmol^+ kg^{-1}$, $Mg_{exchangeable}$ 2.23 ± 0.22 $cmol^+ kg^{-1}$, $Ca_{exchangeable}$ 13.9 ± 1.6 $cmol^+ kg^{-1}$, Electrical Conductivity (EC) 0.48 dS m^{-1}, total porosity 0.8 $m^3 m^{-3}$, and true specific density (ρs)1.76 ± 0.03 g cm^{-3}.

The commercial hydrogel from Solid Rain Corp. (San Diego, CA 92101, USA) was used in this experiment. This hydrogel is a soil amendment with the main component $(C_3H_3KO_2)_n$ (potassium polyacrylate). This polymeric material, with its hydrophilic structure, can hold a large amount of water in its three-dimensional networks. We used a dosage of 15 kg per hectare as recommended by the usage instructions of the product.

2.2. Plant-Microbial Organisms

The selected plant species were *A. viridis* subsp. *crispa*, *P. glauca*, *P. tremuloides* and *S. arbusculoides*. These four species are native to North America region and are commonly found growing in the Abitibi-Témiscamingue region. Each of those species can be dominant or codominant depending on the type of habitat [27]. The composition of the habitat may change over time with the dynamics of the ecosystem and successional processes. In this experiment, we combined all those four species at the initial stage and expected to obtain the potential benefit of different species' composition for accelerating restoration processes. *P. tremuloides* and *S. arbusculoides* are the fastest-growing species, followed by *A. viridis* subsp. *crispa*, which is also relatively fast-growing. The three species are light-demanding species and are considered pioneer species, while *P. glauca* is slow-growing species and is considered a mid- to late-successional species with shade-tolerant characteristics [28]. All the listed species are tolerant and adapted to poor soil and disturbed sites and are often used for restoration and rehabilitation projects [19,28–31], especially *A. viridis* subsp. *crispa*, which can grow well on poor soils because of its association with the nitrogen-fixing actinobacterium *Frankia* spp. and mycorrhizal fungi [32]. Herbaceous species associations were also included as one of the treatment factors. The herbaceous species included oat (*A. sativa*), red fescue (*F. rubra*), and white clover (*T. repens*). *A. sativa* is a grass species that is grown for its seeds for human consumption and as a livestock feed. The other grass species *F. rubra* is known for its tolerance of heavy metal contamination and is often used for phytoremediation in post-mining restoration [33]. *T. repens* is also quite tolerant of heavy metal contamination and also has the ability to fix nitrogen with Rhizobia [34]. As a legume species, *T. repens* can fix nitrogen up to 80 g N ha^{-1} h^{-1} in contaminated soil [34]. Apart from those benefits, herbaceous species have a faster turnover rate, which contributes to soil organic accumulation, which can be advantageous for associated woody species.

Fine tailings and waste rocks have poor soil nutrients and organic matter, which may limit microbial activity in these challenging materials. Therefore, we included microorganism inoculation as part of the experimental factors. The microorganisms applied as inoculants were *Cadophora finlandia*, *Tricholoma scalptiratum*, *Azobacter chroococcum*, *Pseudomonas putida*, *Frankia alni*, and the commercial inoculum of arbuscular mycorrhizal fungus (AMF) *Rhizophagus irregularis* produced by the Company Premier Tech Biotechnologies (Rivière-du-Loup, QC, CA). The microorganisms were inoculated on tree seedlings based on their known symbiotic associations as follows: *C. finlandia*, *T. scalptiratum*, *A. chroococcum* and *P. putida* on *P. glauca* and *S. arbusculoides*; *C. finlandia*, *P. putida*, *F. alni* and *R. irregularis* on *A. virdis*; *T. scalptiratum*, *A. chroococcum*, *P. putida* and *R. irregularis* on *P. tremuloides*. For the herbaceous species, the inoculant was added to the hydroseeding mixture.

2.3. Greenhouse Mesocosm Experiment

The rectangular plastic containers of 34 cm × 54 cm × 18 cm were used as a mesocosm experiment unit. They were filled with two types of waste gold mine materials (fine tailing and waste rock). Biochar and Hydrogel amendments, a micro-symbiont inoculation, and combination of tree and herbaceous crop species (HerbMix) were used. The experimental design was a split–split plot with 3 blocks and 36 treatment combinations, resulting in a total number of 108 experimental units. A mixture of woody species was randomly planted in a Latin square arrangement, as shown in Figure 1, with 8 cm spacing. The purpose of this arrangement was to give a balanced interaction for all 4 plants randomly allocated amongst 16 plots, such that each species appears once in each of four column blocks and once in each of four row blocks. For each species, we had 4 individual plants, where 3 were planted on the border and another one inside the square.

The woody plants were first propagated for one week in small 10 × 20 mm pellets Jiffy-7 Forestry (Stuewe & Sons, Inc., Tangent, ON 97389, USA). Half of the seedlings were inoculated with specific microorganisms and planted in the designed containers. The crops were planted using the hydro-seeding medium in which the specific symbiotic microorganisms were mixed for the inoculation factor. Biochar was applied at a rate of 0.0075 m^3/m^2. The hydrogel was applied at 20 g/L of water mixed with soil.

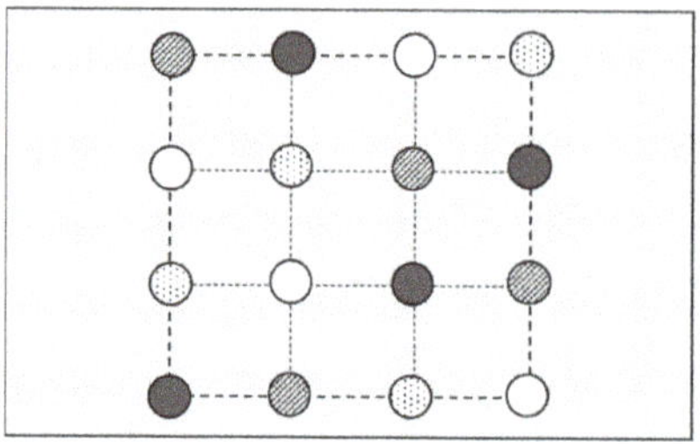

Figure 1. Tree seedling plantation arrangements. The seedling positions are shown using colored circles (gray shading), and the color legend shows different species.

Fertilization 20-8-20 (50 ppm) was applied once at the beginning of the experiment. The temperature of the greenhouse was maintained at 23 °C (daytime) and 16 °C (nighttime), with an average humidity of 50%. The experiment was set up in June 2016 and lasted for three months. After three months, soil respiration was measured using the LICOR LI-6400XT Portable Photosynthesis, Fluorescence, Respiration System (Lincoln, NE 68504, USA). The 6400-09 Soil CO_2 Flux Chamber was installed on the LI-6400XT system for measuring the CO_2 flux from the soils. The soil core at a 6 cm depth was used as the interface of the soil surface and the flux chamber. The plant shoot and root biomass were harvested at the end of experiment, and the dry weight was measured.

2.4. Field Trials

The field trials were established on two mining sites, Sigma-Lamaque (now called Eldorado Gold Lamaque, EGL) and Metanor Resources (now called BonTerra Resources, BTR) in Abitibi-Témiscamingue region, Quebec, Canada. The location of these two mining sites is shown in Figure 2. Two types of waste materials were selected: fine tailing and waste rock. The field coordinates are 48°06′38.4″ N and 077°44′44.74″4 W (fine tailing—EGL), 48°06′20.7″ N and 077°45′43.1″ W (waste rock—EGL), 49°29′40.1″ N and 076°08′49.9″ W (fine tailing—BTR), 48°59′03.8″ N and 075°46′18.4″ W (waste rock—BTR). The average daily temperature at BTR site is 1 °C, with the temperature range between −23° and 23 °C. The average precipitation is 702.3 mm and the snowfall 226.2 cm. The weather information is based on data obtained between 1981 and 2010 from the Canada Environment and Natural Resources website (http://climate.weather.gc.ca/climate_normals/index_e.html, accessed on 1 March 2020). Our weather station installed in June 2016 at EGL site showed the precipitation of 703.8 mm, with temperature range between −24.1°C and 25.4 °C and a mean annual temperature of 4.3 °C.

Gold mining operations produce waste materials such as soil, rock, and fine tailing during gold extraction. Waste rock is often stored in heaps or dumps on the mine site. Tailings are finely ground and can contain leftover processing chemicals such as arsenic (As). These tailings are usually deposited in the form of a water-based slurry in tailings ponds that are left to evaporate over time [35,36]. The tailings are often stored underwater to reduce contact with the atmosphere and prevents oxidation [36]. In a dry climate, evaporation from ponded tailings water and wet tailings can lead to a salinity concentration. The tailings in our field sites are mainly composed of biotite and Fe (Taner et al., 1986). The chemical analyses [37] are as follows: sulfur (0.48 to 0.51%), Al (5500 to 6100 mg kg^{-1}), Ca (21,000 to 23,000 mg kg^{-1}), Fe (14,000 to 16,000 mg kg^{-1}), Mg (4000 to 4500 mg kg^{-1}), P (0 to 560 mg kg^{-1}), and K (86 to 100 mg kg^{-1}). Zn, Mn, Cu, Mo, and Na were found in low concentrations and there was no N in the tailings. The pH of tailings was between 8.55 and 8.68. The arsenic (As) and cyanide concentrations were quite high (8 to 9 mg/kg and 3.7 to 6.3 mg/kg, respectively). Fine tailings have very low hydraulic conductivity in the range of 10^{-4} to 10^{-5} cm/sec, with a grain size <74 µm [35], while waste rock has very high hydraulic conductivity ranging from 10^{-1} to 10^{2} cm/sec, with particles' grain size ranging from sand (625 µm–2 mm) to gravel-free particles (<2 mm) [36]. Soil material with large particle size and high hydraulic conductivity such as waste rock is not suitable for plant growth. This material has a low water holding capacity and may lack water in dry periods. Silty clay soil such as fine tailing with very low hydraulic conductivity is also not good for plant growth. This silty clay can be sticky and plastic when wet and prone to drainage problems but hard when dry [38]. Both materials have very extreme physico-chemical properties that are not suitable for plant growth, and the ideal soil texture is between loam and silt, with a pH between 5.8–6.5 [38].

Figure 2. The site location at the gold mine in Quebec, Canada.

Set up of Experimental Design Field Trials

The field trial was set up as a split–split plot design arrangement. The trial had 12 combinations of factors and 4 replication blocks in each waste rock and fine tailing sites on two mining sites, with a total of 192 plots. The treatment factors were biochar amendment, micro-symbiont inoculation, and a combination of tree and herbaceous crop species. The plot dimensions were 5 m × 8 m, and the tree plantation arrangement is shown in Figure 3. The plant seedling position was arranged in a patch with different inner spacings. Each patch contained 4 seedlings of different species: *P. tremuloides*, *A. viridis* subsp. *crispa*, *S. arbusculoides* and *P. glauca*. The inner spacings were 20 cm × 20 cm, 40 cm × 40 cm and 60 cm × 60 cm with 4 replicates in each plot. Each block consisted of 8 plots with tree seedlings and 4 plots with herbaceous crops. The block was arranged from the North (block 1) to the South (block 4) direction, and the distance between blocks was 4 m.

The seeds were germinated on 20 × 32 mm pellets Jiffy-7 Forestry in the greenhouse. The pellet was made from peat and coir. After 3 months, half of the seedlings were inoculated with specific microorganisms. The seedlings were then grown in the greenhouse for 4 months. The seedlings were moved outside the greenhouse for one week before being transported to the planting site. The planting took place in June 2016 and was monitored for two growing sessions in September each year.

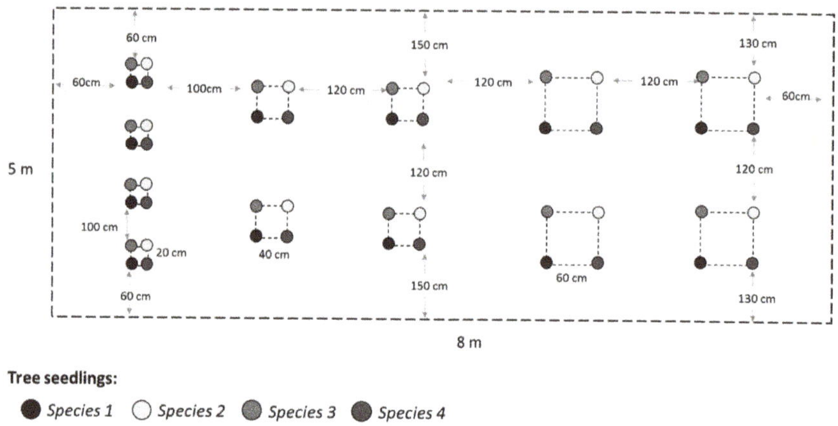

Figure 3. A plot showing tree seedling plantation arrangements. The seedling position is shown by the circle and the grayscale legend shows a distinct species.

The herbaceous crops were planted using hydro-seeding method with commercial mulch Beno-Vert made from recycled paper (Soprema-Quebec, Quebec, QC, Canada). The seedling rate ratio was based on the common seedling rate of oats (*A. sativa*) as a companion crop, which is 50 to 75 kg/ha [39]. However, here, we increased the seedling rate ratio for 10% (84 kg/ha) to compensate for the higher mortality rate on degraded soil. For the other herbaceous species, the weight ratio was adjusted for the same number of seeds as oats, 8 kg/ha for red fescue (*F. rubra*) and 4 kg/ha for white clover (*T. repens*). The application used Beno-Vert and additional Beno-Tack, a vegetal adhesive for the complement binder. The Beno-Vert application was 1500 kg/ha and Beno-Tack 60 kg/ha mixed with 250 g/ha of 15-30-15 fertilizer in water (40,000 L/ha). Therefore, the crop seed composition was 8 kg/ha of red fescue seed, 4 kg/ha of white clover, and 84 kg/ha of oat seeds, mixed with Beno-Vert solution. Biochar application was 75 m^3/ha or about 20 ton/ha, mixed with the soil on the surface down to 5 cm depth in fine tailing. The tilling machine was used for mixing the biochar. On the waste rock site, the biochar was spread on the surface without mixing.

Soil moisture sensors were installed in two blocks of fine-tailing material at the EGL site. The probes were installed at 15 cm and 30 cm depths on each experiment plot within the two blocks. Lysimeters were also installed on the same blocks at a 15 and 30 cm depth, and water was sampled weekly. A weather station was also installed on this site to measure daily precipitation rate, air temperature, relative humidity, and soil moisture (at 15 cm depth).

2.5. Statistical and Data Analyses

The data collected in the field trials were stem diameter and height increments for tree species. The herbaceous plants were not measured because they only grew during the first growing season. The plant diameter was measured at the ground surface level. The measurement was performed at the beginning of the experiment (June 2016), in September 2016 for the first growing season, and in September 2017 for the second growing season. Since data were collected in a shorter time range (4 months) during the first growing season and could have been biased due to the adaptation factors, we decided to use only growth data from the second growing season (one-year growth). Apart from that, we also excluded the height increment data because we found bias on some plants that were broken and/or re-sprouting, resulting in negative increments and high variabilities.

For all statistical analyses, we used the general linear mixed-effects regression model for a split–split plot design [40,41]. The fitting uses the restricted maximum likelihood (REML) method from the lme4 package [42] in R Statistic software (version 4.2.1, Vienna,

Austria) [43] for both greenhouse experiments and field trials with unbalanced data. A general statistical term with the assumption of fixed-effect (α, β) and split–split plot factors ($\varnothing$, ρ, δ) is as follows [40]:

$$\begin{aligned} y_{dhijqrt} &= \mu + \theta_h + \alpha_i + (\varnothing 2)_{q_2} + (\rho 2)_{r_2} + \epsilon^{W}_{d(h)} + \beta_j + \alpha\beta_{ij} + (\varnothing 3)_{q_3} + (\varnothing_2 \varnothing_3)_{q_2 q_3} \\ &\quad + (\rho 3)_{r_3} + (\rho_2 \rho_3)_{r_2 r_3} + \epsilon^{SS}_{ijq_3 r_3(dhq_2 r_2)} + \delta_t + (\beta\delta)_{jt} + \epsilon^{SS}_{t(dhijqr)}, \\ \theta_h &\sim N(0,\, \sigma^2_\theta),\ \epsilon^{W}_{d(h)} \sim N(0,\, \sigma^2_W),\ \epsilon^{S}_{ijq_3 r_3(dhq_2 r_2)} \sim N(0,\, \sigma^2_S),\ \epsilon^{SS}_{t(dhijqr)} \sim N(0,\, \sigma^2_{SS}) \end{aligned} \tag{1}$$

θ_h's, $\epsilon^{W}_{d(h)}$'s, $\epsilon^{SS}_{ijq_3 r_3(dhq_2 r_2)}$'s, $\epsilon^{SS}_{t(dhijqr)}$'s are mutually independent.

d, h, i, j, q, r, t are the observation levels for each corresponding factor.

The dependent variable for greenhouse experiments are dry plant biomass and root:shoot ratio. The fixed effects are tailings, soil supplement, inoculation, and herbaceous mixture (HerbMix). The random effects on a split–split plot design are inoculation nested under supplements, under tailings, and under blocks. The statistical term for lmer method on R software is as follows [40,41]:

$$Y = Tailings * Supplement * Inoculation * HerbMix + (1 \mid Block/Tailings/Supplement/Inoculation) + \epsilon \tag{2}$$

where Y is a dependent variable for plant biomass and the shoot:root ratio.

The field trial data analyses excluded the spacing factor on the waste rock site because the planted seedlings had a mortality of up to 80%, which made the spacing arrangement no longer consistent. The analysis was then split into two regression models, with the spacing factor (fine tailing only) and without the spacing factor (fine tailing and waste rock).

The dependent variable for field trials was the plant diameter growth. The fixed effects are species, tailings, initial diameter (InitDiameter), biochar, inoculation, and herbaceous mixture (HerbMix). The random effects on the split–split plot design were herbaceous mixture nested under inoculation under biochar, under blocks, under tailings, and under site location. The statistical term for the lmer method on R software is as follows:

$$Growth = Species * Tailings * InitDiameter * Biochar * Inoculation * HerbMix + (1 \mid Location/Tailings/Block/Biochar/Inoculation/HerbMix) + \epsilon \tag{3}$$

The statistical term for field trials with the inclusion of spacing effect on fine tailing only (removing the tailings factor) is as follows:

$$Growth = Species * InitDiameter * Biochar * Inoculation * HerbMix * Spacing + (1 \mid Location/Block/Biochar/Inoculation/HerbMix) + \epsilon \tag{4}$$

The plot of marginal effects interaction terms is displayed with error bars showing the 95% confidence intervals, unless mentioned otherwise in the captions.

3. Results

3.1. Greenhouse Experiment

The woody species were not grown in the greenhouse experiment, and the average survival rate was only about 10% in the mesocosms. Thus, species-specific aboveground biomass data analysis was only applied to herbaceous crop species, but the total above- and below-ground biomass was for all species, including the woody species.

The total biomass was dominated by *A. sativa* and seemed to be higher in fine tailing than in waste rock material, as shown in Figure 4. Waste rock material has low water retention and high hydraulic conductivity, which may lead to nutrient leaching with daily watering. On the other hand, fine tailing has high water retention, which allows he conservation of nutrients, but the retention was too high to permit penetration of the water deeper in the soil. This fact seems only beneficial for herbaceous crop species with shallow fibrous root characteristics.

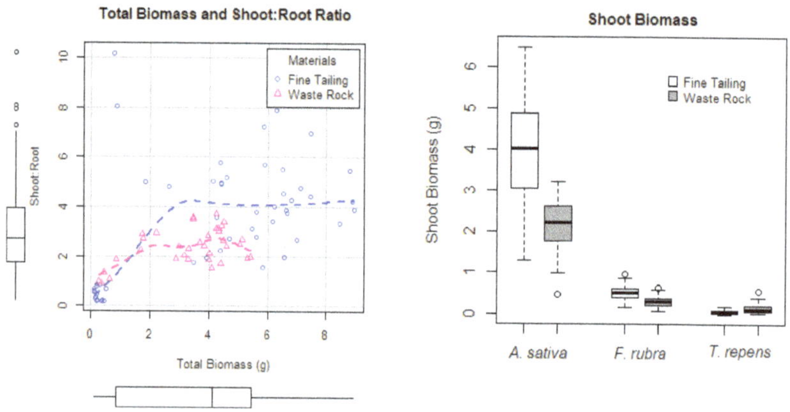

(a) (b)

Figure 4. (**a**) Dry weight biomass and shoot:root ratio of herbaceous and woody species, compared between fine tailing and waste rock growth materials after 3 months of growth. The loess regression curve shows a tendency of constant shoot:root ratio on bigger plants. (**b**) The biomass yield of herbaceous species on fine tailing material tends to be higher compared to that on waste rock material.

The statistical analysis is shown for fixed effects in Table S1 and for random effects in Table S2. The total biomass has a significant interaction between herbaceous mixing with tailings as well as herbaceous mixing and amendments. The shoot:root ratio was shown to be lower on waste rock material (Figure 4), which could be an indicator of higher limitations of nutrients compared to fine tailing material. However, statistically, we found no correlation between the shoot:root ratio with all treatment factors in our experiment (Table S1), although in Figure 3, we can see some differences between tailing materials. The stunted seedlings of the tree treatment (perennial species) with a low survival rate, caused a high variability in the shoot:root ratio, and thus the statistical results for the tree treatment should be interpreted cautiously.

Figure 5 shows that some interaction effects of the main factors on the total biomass were significant, but we could not find other slightly significant effects on the interactions between some main effects at $p < 0.05$ (see Table S1). Herein, the biochar amendment had a negative effect on total biomass, while hydrogel showed a slightly positive effect but as not significant at $p < 0.05$.

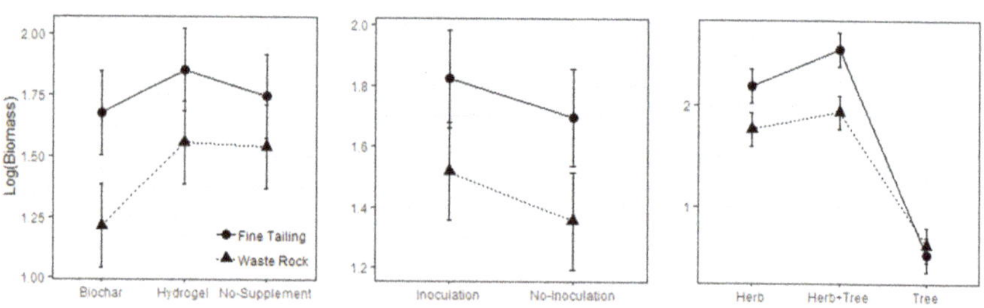

(a) (b) (c)

Figure 5. (**a**) The interaction effect of substrates materials and soil amendment on total biomass. (**b**) The interaction effect of substrates materials and microbial inoculation on total biomass. (**c**) The interaction effect of substrates materials and plantation method on total biomass. The error bars represent the 95% confidence interval (CI) of means.

The measured soil temperature using the LICOR temperature sensor showed that the biochar treatment has a slightly higher temperature. Our assumption is that biochar might increase the soil surface temperature (by lowering the soil albedo), which may accelerate the evaporation and limit the available water for the plant.

The inoculation showed a positive effect on biomass yield ($p < 0.01$). The consistent effect between fine tailing and waste rock materials showed that microbial inoculation helped accelerate plant growth in both tailing materials. The positive effect of tree species inclusion on total biomass could be more due to the additional biomass from tree species. However, the analysis of the aboveground biomass of herbaceous species also showed the same positive effect on the inclusion of tree species, suggesting that there could be an indication of facilitative effects from tree species. In a separate analysis, we found that the benefits of the addition of tree species are higher with biochar treatment, thus suggesting that the inclusion of tree species might help in reducing the negative effect of biochar on soil surface temperature. Since we did not observe this specific effect of biochar on this aspect, this hypothesis might need to be verified with another experiment.

3.2. The Field Trials

The plant mortality was high at the waste rock site (60% at Sigma and 80% at Metanor), while fine tiling showed a better survival rate (2% and 10% mortality on the Sigma and Metanor sites, respectively) during the first month after initial planting. Surprisingly, the rest of the plants on the waste rock were able to survive until the second growing session with a mortality of less than 10%. Thus, we assumed that the cause of high mortality (up to 80%) was only during the initial adaptation to the transplant shock, which could be due to the harsh climate and worse soil conditions on the waste rock material. The spacing analysis was then excluded for the waste rock materials as it no longer had the proper spacing arrangement.

We analyzed plant diameter increment as dependent variable, and the factors of species, initial diameter, time of measurement, soil material, biochar, inoculation, and species mixing. The analysis of the variance table is shown in Tables S3 and S5, while the standard deviation for random effect is shown in Tables S4 and S6. Table S3 is the analysis of variance for all data on waste rock and fine tailing but without the spacing factor, while Table S5 is the analysis of variance for fine tailing only with the inclusion of the spacing factor.

We found interactions between all factors ($p < 0.01$) and that it is difficult to interpret all the interactions at once. Thus, we used the marginal interaction effect from the model analysis. The interaction was mostly consistent between fine tailing and waste rock tailings. Figure 6 shows the marginal interaction effect for each plant species and tailings with biochar treatment on plant diameter increments.

The marginal interaction effect in Figure 6 shows that *A. viridis* subsp. *crispa* has the biggest diameter increment compared to the other tree species (*P. glauca, P. tremuloides* and *S. arbusculoides*). The inoculation treatment had a positive effect on *A. viridis* subsp. *crispa* but showed no significant effect on *P. glauca* or *S. arbusculoides*. The inoculation effect was the opposite between fine tailing and waste rock on *P. tremuloides*.

Biochar showed positive interactions with inoculation treatment on fine tailing for *A. viridis* subsp. *crispa* and *P. tremuloides* (Figure 6). However, the effect was the opposite on waste rock for *A. viridis* subsp. *crispa*. The interaction of biochar and inoculation seems to have no significant effect on the rest of the species.

The effect of biochar was found to be negative in a greenhouse experiment on herbaceous biomass in both fine tailing and waste rock tailings (Figure 5). Thus, the biochar effect varies depending on the plant species and environment. Figure 6 showed that the biochar effect also differed for various plant spacings. The various effect also applies to inoculation treatment on different plant species, the environment, and plant spacings. Hence, the interaction cannot be interpreted easily because of the ecophysiological complexity of plant responses to the different amendments and plant spacings.

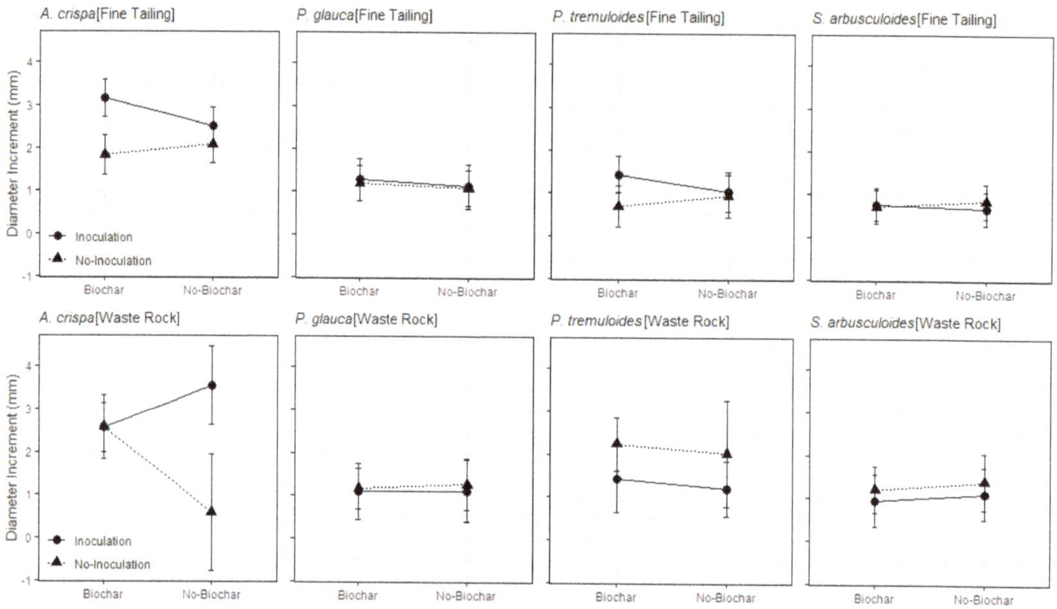

Figure 6. The marginal interaction effect of plant species, tailing, biochar, and inoculation treatment on plant diameter increments. The error bars represent the 95% confidence interval (CI) of means.

The spacing treatment showed unexpected results, as shown in Figure 7. The plant mixing with the highest density or the smallest inner spacing did not necessarily show the lowest growth rate because of higher competition. Only *S. arbusculoides* without biochar and *P. tremuloides* without biochar and inoculation showed the positive linear trend with the spacing. For most other interactions, the growth rate was shown to be decreasing from the largest spacing (60 × 60 cm) to the middle spacing (40 × 40 cm) and increasing again with the smallest spacing (20 × 20 cm), except for *S. arbusculoides*.

Different responses were also shown by *A. viridis* subsp. *crispa* and *P. tremuloides* with biochar and inoculation treatments, with an increasing growth rate from the largest spacing to the intermediate spacing and a decreasing effect for the smallest spacing. The greatest soil water loss was observed with the smallest spacing, where competition dominated, while facilitation was observed at the intermediate spacing and little to no interactions for the largest plant spacing.

The positive balance between competitive and facilitative effects was also amplified by the addition of biochar and inoculation treatments (Figure 7). The positive response of *A. viridis* subsp. *crispa* and *P. tremuloides* with the biochar and inoculation treatments at 40 × 40 cm spacing showed the impact of density or spacing configuration on the interaction of the treatments (Figure 7). The 40 × 40 cm spacing was the optimum spacing for *A. viridis* subsp. *crispa* and *P. tremuloides*, where a balance is achieved between competition and facilitation.

The addition of herbaceous crop plants did not show a significant effect on woody plant growth (Figure 8), except one noted interaction shown on herbaceous crops and inoculation treatment on *A. viridis* subsp. *crispa*. Without inoculation, the herbaceous plants tend to improve the growth rate of *A. viridis* subsp. *crispa*. The herbaceous plants may perform as a cover crop in this case, which maintains the evaporation and soil surface temperature around the woody plants. However, the plant response changed when the inoculation was added, thereby improving the herbaceous growth and the competitive effect on the perennial woody plant.

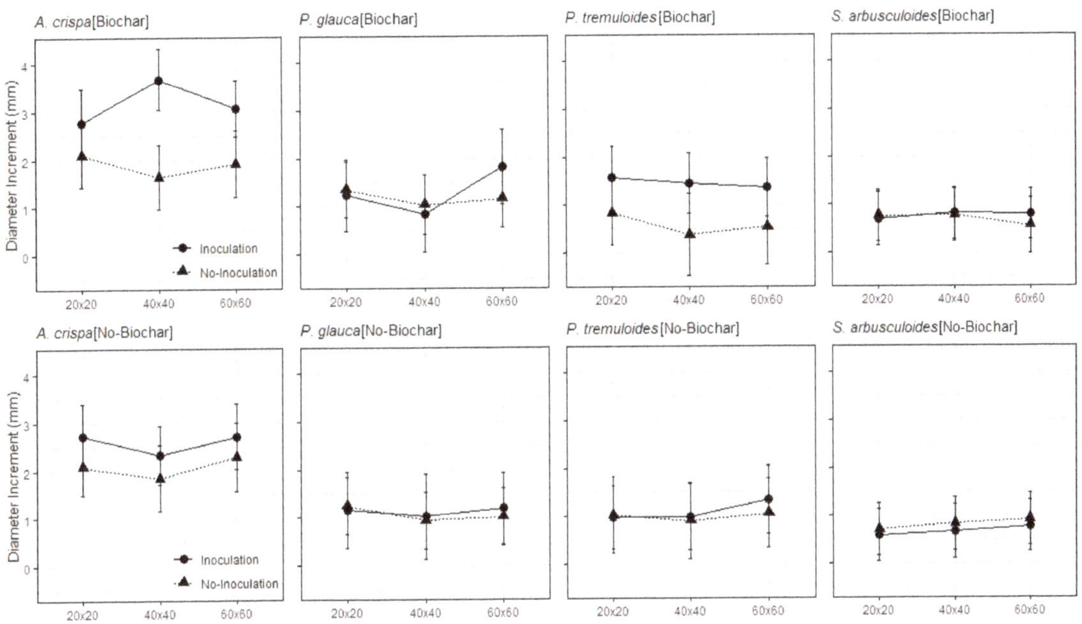

Figure 7. The interaction effect of biochar, inoculation, and spacing treatments in the fine tailing site. The error bars represent the 95% confidence interval (CI) of means.

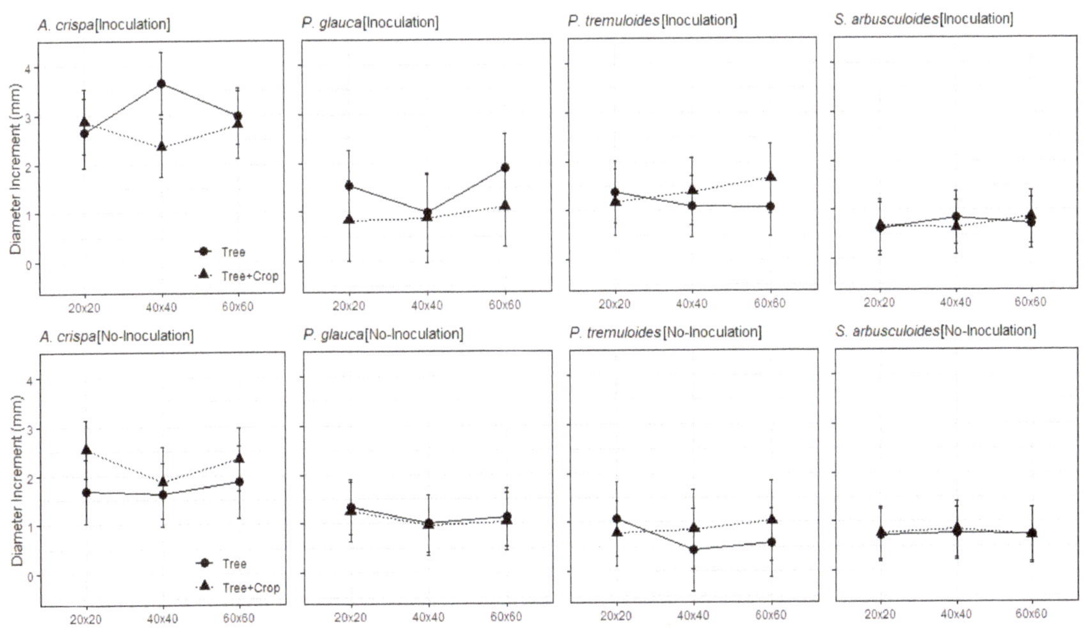

Figure 8. The interaction effect of inoculation, tree-crop, and spacing treatments in the fine tailing site. The error bars represent the 95% confidence interval (CI) of means.

4. Discussion

The physical characteristics of the fine tailing with very high moisture retention [44] and very low hydraulic conductivity [35] were not ideal for plant growth. The particle size was between that of clay and silt, which has a high water content at permanent wilting point (±27%) [45]. The soil moisture content at 15 cm below the surface was always below 30% for the whole year based on our measurements. Although the site had a mean rainfall of 700 mm, the soil moisture never reached the field capacity level (±40%), suggesting that the water was unable to infiltrate the soil.

Fine tailing and waste rock soil tailing had inadequate characteristics for plant growth in both greenhouse and field trials. The tree species exhibited better survival than herbaceous species in the field trial after two growing sessions, and better survival in the field trial than in the greenhouse trial. This can be explained by the fact that the woody perennial species used in the field trials are native and ecologically well adapted to the boreal Forest of Abitibi-Témiscamingue region, while the herbaceous species were species that were allochthonous to the region. These results corroborate those of [19,46], who showed the importance of using ecologically well-adapted native mycorrhizal fungi with their host plants for a successful long-term revegetation program. On the other hand, the herbaceous species showed better growth in the greenhouse experiment. Indeed, because of the overgrowth of the herbaceous species, the perennial woody plant species were suppressed in our greenhouse mesocosm experiment since the first month of plantation, although these annual herbaceous crop species started dying after 3 months of experiment time. Plant seedling size was shown to be important for the survivability and adaptability of tree species. While the biochar and inoculation treatments did not influence plant survivability, they did have an effect on the biomass yield productivity.

4.1. Biochar Amendment

Biochar is a highly stable and rich source of carbon and is a potential carbon sink in relation to climate change mitigation [47]. Biochar is also known as a soil amendment for improving the soil's properties and functions in agronomic applications [47–49]. It has the capacity to enhance the water and nutrient retention and improves soil structure and drainage [47]. There is evidence of an effect of biochar on microbial activity and plant symbiosis, which improve crop productivity [47,49]. The possible mechanisms involved include the immobilization of plant-available N, and the mineralization of labile, high-C-to-N fractions of biochar into microbial biomass [48]. Other possible mechanisms are the alteration of soil physico-chemical properties, mycorrhiza helper bacteria, plant–fungus signaling interference and detoxification of allelochemicals on biochar, and provision of refugia from fungal grazers [26,50].

In opposite to those reports above, our greenhouse experiment with biochar amendments showed a negative effect on total biomass yield (Figure 4). In fact, the plants did not benefit from the improvement in soil water retention physical characteristics by the biochar, while the control plants grew better with only daily watering treatment. The field trials also showed a negative effect of biochar without the inoculation treatment. However, when combined with inoculation in the field trial, the biochar showed a positive effect, especially on *A. viridis* subsp. *Crispa* and *P. tremuloides* (Figure 6). However, surprisingly, the effect became negative with bigger spacing (60 × 60 cm), when the plant competition was lower. This could mean that the biochar and inoculation effect was less strong than the effect of density. At the same time, we also noted the positive effect of density in our trials, which could be due to an improvement in the microclimate condition due to higher density [51–53].

There was another effect of biochar amendment on soil, which was the reduction in soil albedo [54]. The lower albedo of biochar can make the soil warmer, which can result in more soil water evaporation compared to the higher albedo, resulting in a negative effect on plant growth. Some interesting results on the effect of biochar on soil temperature in the temperate zone have shown that it can increase the average soil temperature, but it has a lower temperature on the hottest day of the year [55]. This means that biochar was also able to stabilize the soil temperature in an extreme zone. The stable soil temperature was favorable for the plant and also for the soil microbial activity [56]. The improvement in microbial activity in warmer soil could be another interaction mechanism between biochar and microbial inoculation in our field trial experiments.

4.2. Microbial Inoculation

A healthy soil ecosystem is formed by various microbial species with specific roles and functions [14]. We believe that soil ecosystem biodiversity is as important as above-ground biodiversity. We also believe that above- and belowground biodiversity is highly correlated, as some microorganisms may require specific host plants [15]. Thus, providing the mixture of root inoculants for the plants is expected to introduce and increase below-ground biodiversity. The mixture of microbial processes is expected to regulate the nutrient mineralization, biological nitrogen fixation, and other functions that can improve soil physico-chemical and microbiological properties [57–59].

The inoculation treatment showed a positive effect on plant growth in the greenhouse experiment. It had a slightly positive interaction with biochar, although it was not significant. The inoculation was beneficial for plant growth in both fine tailings and waste rock tailings. Since the mining waste tailings are mostly deprived of beneficial microorganisms [18], the inoculation with beneficial microbes was shown to be a good way of introducing beneficial symbiotic microorganisms into these challenging reclamation tailings, as reported in other studies [19].

The field trials showed a similar positive effect of microbial inoculation on *A. viridis* subsp. *crispa* and *P. tremuloides*, but not for *P. glauca* and *S. arbusculoides*. The different responses could be due to the specificity and efficiency of different microsymbionts [19]. The symbiotic relationship can range from parasitism to true mutualism depending on the microsymbiont, the plant host, and the soil fertility [59–61]. *Alnus viridis* subsp. *crispa* as an ectomycorrhizal plant benefits symbiotic mutualism due to both nitrogen-fixing *Frankia* actinomycete and mycorrhizal fungi [32]. *P. tremuloides* is not a nitrogen-fixing plant but can form both ectomycorrhizal and arbuscular mycorrhizal [62] and benefit from the co-occurrence of the mycorrhizal network with *A. viridis* subsp. *crispa* [63–65]. This hypothesis can be explored by another specific experiment involving *P. tremuloides*, *A. viridis* subsp. *crispa* and their interaction with mycorrhiza.

The benefit of inoculation on *A. viridis* subsp. *crispa* and *P. tremuloides* was shown to be enhanced by the addition of biochar. Biochar is known as a soil amendment that provides a good environment for mycorrhiza colonization [49,66,67]. However, the direct mechanism of how it affects mycorrhizal dynamics is still unclear [49]. The known mechanism in correlation with mycorrhiza is through soil chemical and physical alteration [48,49]. Another possible mechanism is through the soil temperature stabilization by the reduction in soil albedo, as discussed above. In fact, the microorganisms are known to be sensitive to soil temperature and microclimate conditions [56,68–71].

4.3. Mixed System Interactions

The field trials showed a positive effect of the herbaceous crop on tree species without inoculation treatment. One of the reasons for this finding is the well-known role of cover crops in reducing the evapotranspiration [72]. The herbaceous crops became disadvantageous and competitive to the woody perennial species when the inoculation treatment was applied (Figure 8). The literature on the use of bioinoculants in agroforestry systems is very scanty [73]. We conducted the first test on the use of bioinoculants in boreal agroforestry in the context of ecological restoration of post-mining areas with the aim of improving nutrient availability for plants while reducing the use of inorganic or organic fertilizers, pesticides, and water. More research is needed in that area to develop a broad conceptual framework and methodology that is supported by robust scientific data for the large-scale use of bioinoculants in the ecological restoration industry.

In general, we also found the positive effect of density which is supported by the principles of "Allee" effect in ecological theory [52,74,75]. The positive effect on higher density could be explained by the below-ground facilitative mechanism or aboveground microclimate improvement. The improved microclimate could be an important factor on spacing configuration, which correlates with the sensitivity of microorganisms on soil temperature and the effect of biochar on soil albedo [55,56].

We found that the interactions between components in mixed systems were not straightforward and required a comprehensive scenario for the intervention of the system. Figure 9 shows the ecophysiological complexity of plant responses to the different interactions between the factors that we studied in our experiment. The addition of biochar, inoculation treatment, and herbaceous crops could improve plant growth, and at the same time, it can have negative effect. The planting density treatment not only affects the plant competition, but also alters the microclimate around the plants. These conditions may change as the plants grow and yield different outcomes in the long term. The modeling effort can help estimate the growth dynamics in this restoration processes at the later stages. Some aspects that need to be considered in the long-term stages are the nutrient cycle and phytoremediation processes. The species mixing and planting configuration are other elements that need to be considered in the modeling scenario.

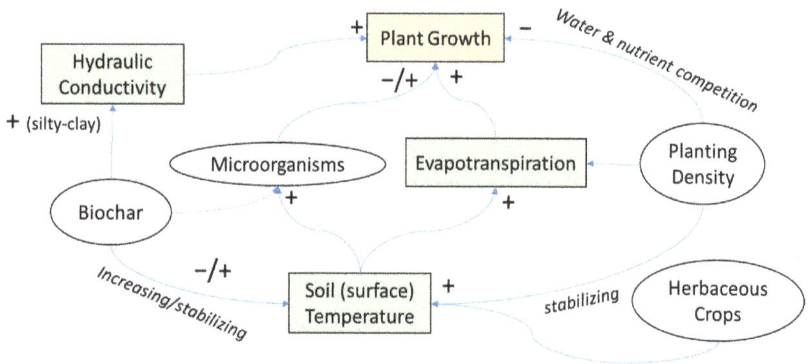

Figure 9. Hypothetical interactions between component factors in a restoration trial experiment. The interactions are complex, and all the factors can contribute to plant growth dynamics.

Soil temperature, soil cover, evaporation, and evapotranspiration affect soil water availability. Therefore, the comparison of volumetric water content between biochar-amended and control soils in field experiments may be confounded by indirect effects, that is, on plant growth and soil thermal properties. In addition to the chemical stabilization of nutrients, the modification of the physical structure of the bulk soil may result in biochar not simply increasing the capacity of soil to retain water, but also nutrients in the soil solution. The multispecies approach has shown some advantages in our restoration experiment. Mixing species at the early stage of planting proved to have no significant effect on spacing competition, although the outcome can differ in the later stages, when the plant is growing and both aboveground and below-ground competition occurs. At the early stage, we found a positive balance between plant competition and microclimate improvement on high-density plantation. Another facilitative effect was shown on the inoculation treatment between *A. viridis* subsp. *crispa* and *P. tremuloides*. *Alnus viridis* subsp. *crispa* may have served as a nurse species for *P. tremuloides* through mycorrhizal network associations. The addition of herbaceous crop species showed a positive effect on the perennial plant growth rate through their function as cover crops. In the long term, we expect to find more interactions between the species through their functions on nutrient cycling and successional dynamics.

The positive effect of plant density confirms the "Allee" effect, which showed the benefit of living in groups for inducing the facilitation within individuals [52]. The practical implementation of high-density planting can be very costly. The proposed "Nucleation" method introduced by [51] could be a low-cost alternative. The seedlings are planted in patches or "islands" to facilitate forest recovery that is less expensive than planting large areas. However, a study in tropical forest restoration in Costa Rica highlighted the importance of broad spatial replicated studies to account for high variability and make generalizable restoration recommendations [76]. The improvement in microclimate on high-density cluster planting is known to increase the survivability of the seedlings [51,77,78]. The combination with other enabling biotechniques such as microbial inoculation and biochar amendment may improve the whole successional process. Reinstalling the biological life through microbial inoculation of seedlings planted in reconstructed anthroposols after mining operations has shown successful plant growth and health and improved soil quality [19,32].

Biochar has the capacity to increase the hydraulic conductivity on the soil or tailings with very fine grain size, such fine tailing waste tailings, and at the same time, it is also able to reduce the hydraulic conductivity on the tailings with large particle size such as waste rock [48,49,79]. Biochar also reduces the soil albedo, which may increase the average soil surface temperature [54,55]. Warmer soil temperature can have negative effects on some ecosystems, but it seems to be advantageous for colder climate zones, as warmer soil temperature may increase the soil microbial activity and accelerate the plant recovery processes [56].

To improve the microclimate, it is suggested focus be placed on the initial planting, along with planting configuration and multispecies approach. In the long term, the nutrient cycle, phytoremediation processes and successional dynamics are some other aspects that are also important. The combination of annual and perennial plants in agroforestry systems can be applied to accelerate the nutrient cycle and successional processes for forest recovery of severely disturbed ecosystems [3]. The herbaceous crops serve as cover crops for supporting other woody plants [72,80]. We believe that this multispecies and multifunctional approach can be advantageous for ecological restoration projects of mining sites.

5. Conclusions

There were both positive and negative effects on plant spacing, biochar amendment and inoculation, depending on their interactions. The net positive effect was shown by combining high plantation density, biochar and inoculation factors on *Alnus viridis* ssp. crispa. Overall, planting density was shown to be the most important factor in generating a net positive effect. We suggest that the mechanism was correlated with microclimate improvement through soil plant water conservation and microbial activity enhancement over soil temperature modification. Hence, we propose emphasizing microclimate improvement for accelerating the restoration processes, along with other combined factors, including microbial inoculation and biochar amendment.

Supplementary Materials: The following supporting information can be downloaded at: https://www.mdpi.com/article/10.3390/f14040856/s1, Table S1: The ANOVA P-value for total biomass, shoot biomass of *A. sativa*, *F. rubra*, *T. repens*, and shoot:root ratio; Table S2: The standard deviation of random effect factors on split-split plot design for each dependent variable data group of total biomass, shoot biomass of *A. sativa*, *F. rubra*, *T. repens*, and shoot:root ratio; Table S3: Type III Analysis of Variance Table with Satterthwaite's method on diameter growth of tree species on waste rock and fine tailing without spacing factor; Table S4: The standard deviation of random effect factors on split-split plot design for regression model on data group of fine tailing and waste rock without spacing effect. The total observation data is 2795; Table S5: Type III Analysis of Variance Table with Satterthwaite's method on diameter growth of tree species on fine tailing only with the inclusion of spacing factor; Table S6: The standard deviation of random effect factors on split-split plot design for regression model on data group of fine tailing with spacing effect. The total observation data is 2132.

Author Contributions: Conceptualization and methodology, D.H.A., S.A. and D.P.K.; formal analysis, D.H.A. and M.v.N.; writing—original draft preparation, D.H.A.; writing—review and editing, all authors. All authors have read and agreed to the published version of the manuscript.

Funding: This research was funded by Fonds de recherche du Québec—Nature et technologies (FRQNT). Project number: MI-193447.

Institutional Review Board Statement: Not applicable.

Informed Consent Statement: Not applicable.

Data Availability Statement: Not applicable.

Acknowledgments: We thank Marie-Eve Beaulieu, Marie R. Coyea, Francois Larochelle and Marie-Andrée Paré for their lab and greenhouse work. We thank Martin Beaudoin Nadeau, Sébastien Lange, Marc Guimont, Roudy Jean, Stephen Fredrick Omondi, Chantal Giroux, Léonie Côté, Catherine Ruest, Gabriel Grenier, André Gagné and Ira Ratna Sari for their fieldwork. This research was supported by a joint research program on sustainable development in the mining sector developed by Fonds de recherche du Québec—Nature et technologies (FRQNT) in collaboration with Quebec's Ministry of Natural Resources and Forests. We thank the mining companies Bonterra (formerly Ressources Métanor), Eldorado Gold (formerly Integra Gold), and Fournier et Fils for their logistical support. NSERC Discovery Grant (DPK) and GECA Environnement (SA) supports are also acknowledged.

Conflicts of Interest: The authors declare no conflict of interest.

References

1. Gann, G.D.; McDonald, T.; Walder, B.; Aronson, J.; Nelson, C.R.; Jonson, J.; Hallett, J.G.; Eisenberg, C.; Guariguata, M.R.; Liu, J.; et al. International Principles and Standards for the Practice of Ecological Restoration. Second Edition. *Restor. Ecol.* **2019**, *27*, S1–S46. [CrossRef]
2. Choi, Y.D.; Temperton, V.M.; Allen, E.B.; Grootjans, A.P.; Halassy, M.; Hobbs, R.J.; Naeth, M.A.; Torok, K. Ecological Restoration for Future Sustainability in a Changing Environment. *Ecoscience* **2008**, *15*, 53–64. [CrossRef]
3. Vieira, D.L.M.; Holl, K.D.; Peneireiro, F.M. Agro-Successional Restoration as a Strategy to Facilitate Tropical Forest Recovery. *Restor. Ecol.* **2009**, *17*, 451–459. [CrossRef]
4. Jose, S.; Holzmueller, E.J.; Gillespie, A.R. Tree–Crop Interactions in Temperate Agroforestry. In *North American Agroforestry: An Integrated Science and Practice*; American Society of Agronomy, Inc.: Madison, WI, USA, 2009; pp. 57–74.
5. Atangana, A.; Khasa, D.; Chang, S.; Degrande, A. *Tropical Agroforestry*; Springer: Dordrecht, The Netherlands, 2014; ISBN 978-94-007-7722-4.
6. Ong, C.K.; Black, C.R.; Wilson, J. (Eds.) *Tree-Crop Interactions: Agroforestry in a Changing Climate*, 2nd ed.; CAB International: Wallingford, UK; Boston, MA, USA, 2015; ISBN 978-1-78064-511-7.
7. Solbrig, O.T. Plant Traits and Adaptive Strategies: Their Role in Ecosystem Function. In *Biodiversity and Ecosystem Function*; Schulze, E.-D., Mooney, H.A., Eds.; Ecological Studies; Springer: Berlin/Heidelberg, Germany; New York, NY, USA, 1994; ISBN 978-3-540-55804-0.
8. Leakey, R. Environmental Resilience and Agroforestry. In *Agroforestry-The Future of Global Land Use*; Springer: Berlin/Heidelberg, Germany, 2012; pp. 11–12.
9. Palmer, M.A.; Ambrose, R.F.; Poff, N.L. Ecological Theory and Community Restoration Ecology. *Restor. Ecol.* **1997**, *5*, 291–300. [CrossRef]
10. Palmer, M.A.; Zedler, J.B.; Falk, D.A. Ecological Theory and Restoration Ecology. In *Foundations of Restoration Ecology*; Springer: Berlin/Heidelberg, Germany, 2006; pp. 3–26.
11. Martínez-Garza, C.; Howe, H.F. Restoring Tropical Diversity: Beating the Time Tax on Species Loss. *J. Appl. Ecol.* **2003**, *40*, 423–429. [CrossRef]
12. Lamb, D. *Regreening the Bare Hills*; World Forests; Springer: Dordrecht, The Netherlands, 2011; Volume 8, ISBN 978-90-481-9869-6.
13. Swift, M.J.; Izac, A.-M.N.; van Noordwijk, M. Biodiversity and Ecosystem Services in Agricultural Landscapes—Are We Asking the Right Questions? *Agric. Ecosyst. Environ.* **2004**, *104*, 113–134. [CrossRef]
14. Paul, E.A. *Soil Microbiology, Ecology and Biochemistry*; Academic Press: Cambridge, MA, USA, 2014; ISBN 0-12-391411-6.
15. van der Heijden, M.G.A.; Klironomos, J.N.; Ursic, M.; Moutoglis, P.; Streitwolf-Engel, R.; Boller, T.; Wiemken, A.; Sanders, I.R. Mycorrhizal Fungal Diversity Determines Plant Biodiversity, Ecosystem Variability and Productivity. *Nature* **1998**, *396*, 69–72. [CrossRef]
16. Weyens, N.; van der Lelie, D.; Taghavi, S.; Vangronsveld, J. Phytoremediation: Plant–Endophyte Partnerships Take the Challenge. *Curr. Opin. Biotechnol.* **2009**, *20*, 248–254. [CrossRef]
17. Ali, H.; Khan, E.; Sajad, M.A. Phytoremediation of Heavy Metals—Concepts and Applications. *Chemosphere* **2013**, *91*, 869–881. [CrossRef]
18. Bois, G.; Piché, Y.; Fung, M.Y.P.; Khasa, D.P. Mycorrhizal Inoculum Potentials of Pure Reclamation Materials and Revegetated Tailing Sands from the Canadian Oil Sand Industry. *Mycorrhiza* **2005**, *15*, 149–158. [CrossRef]
19. Nadeau, M.B.; Laur, J.; Khasa, D.P. Mycorrhizae and Rhizobacteria on Precambrian Rocky Gold Mine Tailings: I. Mine-Adapted Symbionts Promote White Spruce Health and Growth. *Front. Plant Sci.* **2018**, *9*, 1267. [CrossRef]

20. Nadeau, M.B.; Quoreshi, A.; Khasa, D.P. Ecological Restoration and Bioremediation of Canadian Mining Boreal Ecosystems. In *Microbes for Restoration of Degraded Ecosystems*; Bagyaraj, D.J., Jamaluddin, Eds.; New India Publishing Agency: Delhi, India, 2016; pp. 259–284. ISBN 978-93-85516-66-5.
21. Joseph, S.; Lehmann, J. *Biochar for Environmental Management: Science and Technology*; Earthscan: London, UK, 2009; ISBN 1-84407-658-X.
22. Woolf, D.; Amonette, J.E.; Street-Perrott, F.A.; Lehmann, J.; Joseph, S. Sustainable Biochar to Mitigate Global Climate Change. *Nat. Commun.* **2010**, *1*, 56. [CrossRef]
23. Ogawa, M. Symbiosis of People and Nature in the Tropics. III. *Trop. Agric. Using Charcoal. Farming Jpn.* **1994**, *28*, 21–35.
24. Rondon, M.A.; Lehmann, J.; Ramírez, J.; Hurtado, M. Biological Nitrogen Fixation by Common Beans (*Phaseolus vulgaris* L.) Increases with Bio-Char Additions. *Biol. Fertil. Soils* **2007**, *43*, 699–708. [CrossRef]
25. Steiner, C. *Biochar Carbon Sequestration*; University of Georgia, Biorefining and Carbon Cycling Program: Athens, Greece, 2008; p. 30602.
26. Warnock, D.D.; Lehmann, J.; Kuyper, T.W.; Rillig, M.C. Mycorrhizal Responses to Biochar in Soil–Concepts and Mechanisms. *Plant Soil* **2007**, *300*, 9–20. [CrossRef]
27. Barbour, M.G.; Billings, W.D. (Eds.) *North American Terrestrial Vegetation*, 2nd ed.; Cambridge University Press: Cambridge, UK, 2000; ISBN 978-0-521-55027-7.
28. Abrahamson, I. Picea Glauca, White Spruce. In *Fire Effects Information System*; USDA, Forest Service, Rocky Mountain Research Station, Fire Sciences Laboratory: Ft. Collins, CO, USA, 2015.
29. Esser, L.L. Salix Arbusculoides. In *Fire Effects Information System*; USDA Forest Service, Rocky Mountain Research Station, Fire Sciences Laboratory: Ft. Collins, CO, USA, 1992.
30. Howard, J.L. Populus Tremuloides. In *Fire Effects Information System*; USDA Forest Service, Rocky Mountain Research Station, Fire Sciences Laboratory: Ft. Collins, CO, USA, 1996.
31. Matthews, R.F. Alnus Viridis Subsp. Crispa. In *Fire Effects Information System*; USDA Forest Service, Rocky Mountain Research Station, Fire Sciences Laboratory: Ft. Collins, CO, USA, 1992.
32. Roy, S.; Khasa, D.P.; Greer, C.W. Combining Alders, Frankiae, and Mycorrhizae for the Revegetation and Remediation of Contaminated Ecosystems. *Can. J. Bot.* **2007**, *85*, 237–251. [CrossRef]
33. Wong, M.H. Metal Cotolerance to Cooper, Lead, and Zinc in Festuca Rubra. *Environ. Res.* **1982**, *29*, 42–47. [CrossRef] [PubMed]
34. Rother, J.A.; Millbank, J.W.; Thornton, I. Nitrogen Fixation by White Clover (*Trifolium repens*) in Grasslands on Soils Contaminated with Cadmium, Lead and Zinc. *J. Soil Sci.* **1983**, *34*, 127–136. [CrossRef]
35. Aubertin, M.; Bussiere, B.; Chapuis, R.P. Hydraulic Conductivity of Homogenized Tailings from Hard Rock Mines. *Can. Geotech. J.* **1996**, *33*, 470–482. [CrossRef]
36. Kossoff, D.; Dubbin, W.E.; Alfredsson, M.; Edwards, S.J.; Macklin, M.G.; Hudson-Edwards, K.A. Mine Tailings Dams: Characteristics, Failure, Environmental Impacts, and Remediation. *Appl. Geochem.* **2014**, *51*, 229–245. [CrossRef]
37. Nadeau, M.B. *Development of a New Green Technology for the Revegetation of Abandoned Gold Mine Tailings Using Specific Symbionts Associated with Picea Glauca*; Mémoire de Maîtrise, Département des Sciences du Bois et de la Forêt, Université Laval: Québec, QC, Canada, 2015.
38. Gardner, C.M.; Laryea, K.B.; Unger, P.W. *Soil Physical Constraints to Plant Growth and Crop Production*; Land and Water Development Division, Food and Agriculture Organization: Rome, Italy, 1999.
39. Lanini, W.T.; Orloff, S.B.; Vargas, R.N.; Orr, J.P.; Marble, V.L.; Grattan, S.R. Oat Companion Crop Seeding Rate Effect on Alfalfa Establishment, Yield, and Weed Control. *Agron. J.* **1991**, *83*, 330. [CrossRef]
40. Dean, A.; Voss, D.; Draguljić, D. Split-Plot Designs. In *Design and Analysis of Experiments*; Springer International Publishing: Cham, Switzerland, 2017; pp. 703–764. ISBN 978-3-319-52250-0.
41. Pardo, S. Mixed Models and Variance Components. In *Statistical Analysis of Empirical Data: Methods for Applied Sciences*; Springer International Publishing: Cham, Switzerland, 2020; pp. 107–119. ISBN 978-3-030-43328-4.
42. Bates, D.; Mächler, M.; Bolker, B.; Walker, S. Fitting Linear Mixed-Effects Models Using Lme4. *arXiv* **2014**, arXiv:1406.5823.
43. R Core Team. *R: A Language and Environment for Statistical Computing*; R Core Team: Vienna, Austria, 2022.
44. Blowes, D.W. The Environmental Effects of Mine Wastes. *Proc. Explor.* **1997**, *97*, 887–892.
45. Saxton, K.E.; Rawls, W.J. Soil Water Characteristic Estimates by Texture and Organic Matter for Hydrologic Solutions. *Soil Sci. Soc. Am. J.* **2006**, *70*, 1569. [CrossRef]
46. Nadeau, M.B.; Laur, J.; Khasa, D.P. Mycorrhizae and Rhizobacteria on Precambrian Rocky Gold Mine Tailings: II. Mine-Adapted Symbionts Alleviate Soil Element Imbalance for a Better Nutritional Status of White Spruce Seedlings. *Front. Plant Sci.* **2018**, *9*, 1268. [CrossRef]
47. Jeffery, S.; Verheijen, F.G.A.; van der Velde, M.; Bastos, A.C. A Quantitative Review of the Effects of Biochar Application to Soils on Crop Productivity Using Meta-Analysis. *Agric. Ecosyst. Environ.* **2011**, *144*, 175–187. [CrossRef]
48. Sohi, S.P.; Krull, E.; Lopez-Capel, E.; Bol, R. A Review of Biochar and Its Use and Function in Soil. In *Advances in Agronomy*; Elsevier: Amsterdam, The Netherlands, 2010; Volume 105, pp. 47–82. ISBN 978-0-12-381023-6.
49. Lehmann, J.; Rillig, M.C.; Thies, J.; Masiello, C.A.; Hockaday, W.C.; Crowley, D. Biochar Effects on Soil Biota–A Review. *Soil Biol. Biochem.* **2011**, *43*, 1812–1836. [CrossRef]
50. Tarkka, M.T.; Frey-Klett, P. Mycorrhiza Helper Bacteria. In *Mycorrhiza*; Springer: Berlin/Heidelberg, Germany, 2008; pp. 113–132.
51. Corbin, J.D.; Holl, K.D. Applied Nucleation as a Forest Restoration Strategy. *For. Ecol. Manag.* **2012**, *265*, 37–46. [CrossRef]

52. Courchamp, F.; Berec, L.; Gascoigne, J. *Allee Effects in Ecology and Conservation*; Oxford Biology, Oxford University Press: Oxford, UK, 2008; ISBN 978-0-19-857030-1.
53. Asmara, D.H.; Allaire, S.; van Noordwijk, M.; Khasa, D.P. Tree Establishment on Post-Mining Waste Soils: Species, Density, and Mixture Effects. *Can. J. For. Res.* **2021**, *52*, 79–89. [CrossRef]
54. Verheijen, F.G.A.; Jeffery, S.; van der Velde, M.; Penížek, V.; Beland, M.; Bastos, A.C.; Keizer, J.J. Reductions in Soil Surface Albedo as a Function of Biochar Application Rate: Implications for Global Radiative Forcing. *Environ. Res. Lett.* **2013**, *8*, 044008. [CrossRef]
55. Zhang, Q.; Wang, Y.; Wu, Y.; Wang, X.; Du, Z.; Liu, X.; Song, J. Effects of Biochar Amendment on Soil Thermal Conductivity, Reflectance, and Temperature. *Soil Sci. Soc. Am. J.* **2013**, *77*, 1478. [CrossRef]
56. Curiel Yuste, J.; Baldocchi, D.D.; Gershenson, A.; Goldstein, A.; Misson, L.; Wong, S. Microbial Soil Respiration and Its Dependency on Carbon Inputs, Soil Temperature and Moisture. *Glob. Chang. Biol.* **2007**, *13*, 2018–2035. [CrossRef]
57. Swift, M.J.; Anderson, J.M. Biodiversity and Ecosystem Function in Agricultural Systems. In *Biodiversity and Ecosystem Function*; Schulze, E.-D., Mooney, H.A., Eds.; Springer: Berlin/Heidelberg, Germany, 1994; pp. 15–41. ISBN 978-3-642-58001-7.
58. Fortin, J.A.; Plenchette, C.; Piché, Y. Les Mycorhizes. In *La Nouvelle Révolution Verte*; MultiMonde, Q., Ed.; Université Laval: Quebec City, QC, Canada, 2008.
59. Smith, S.E.; Read, D.J. *Mycorrhizal Symbiosis*, 3rd ed.; Repr.; Elsevier: Amsterdam, The Netherlands; Academic Press: Cambridge, MA, USA, 2009; ISBN 978-0-12-370526-6.
60. Smith, F.A.; Grace, E.J.; Smith, S.E. More than a Carbon Economy: Nutrient Trade and Ecological Sustainability in Facultative Arbuscular Mycorrhizal Symbioses. *Res. Review. New Phytol.* **2009**, *182*, 347–358. [CrossRef]
61. Franklin, O.; Näsholm, T.; Högberg, P.; Högberg, M.N. Forests Trapped in Nitrogen Limitation-an Ecological Market Perspective on Ectomycorrhizal Symbiosis. *New Phytol.* **2014**, *203*, 657–666. [CrossRef] [PubMed]
62. Khasa, P.D.; Chakravarty, P.; Robertson, A.; Thomas, B.R.; Dancik, B.P. The Mycorrhizal Status of Selected Poplar Clones Introduced in Alberta. *Biomass Bioenergy* **2002**, *22*, 99–104. [CrossRef]
63. Walder, F.; Niemann, H.; Natarajan, M.; Lehmann, M.F.; Boller, T.; Wiemken, A. Mycorrhizal Networks: Common Goods of Plants Shared under Unequal Terms of Trade. *Plant Physiol.* **2012**, *159*, 789–797. [CrossRef] [PubMed]
64. Fellbaum, C.R.; Mensah, J.A.; Cloos, A.J.; Strahan, G.E.; Pfeffer, P.E.; Kiers, E.T.; Bücking, H. Fungal Nutrient Allocation in Common Mycorrhizal Networks Is Regulated by the Carbon Source Strength of Individual Host Plants. *New Phytol.* **2014**, *203*, 646–656. [CrossRef] [PubMed]
65. Bücking, H.; Mensah, J.A.; Fellbaum, C.R. Common Mycorrhizal Networks and Their Effect on the Bargaining Power of the Fungal Partner in the Arbuscular Mycorrhizal Symbiosis. *Commun. Integr. Biol.* **2016**, *9*, e1107684. [CrossRef] [PubMed]
66. Anderson, C.R.; Condron, L.M.; Clough, T.J.; Fiers, M.; Stewart, A.; Hill, R.A.; Sherlock, R.R. Biochar Induced Soil Microbial Community Change: Implications for Biogeochemical Cycling of Carbon, Nitrogen and Phosphorus. *Pedobiologia* **2011**, *54*, 309–320. [CrossRef]
67. Robertson, S.J.; Rutherford, P.M.; López-Gutiérrez, J.C.; Massicotte, H.B. Biochar Enhances Seedling Growth and Alters Root Symbioses and Properties of Sub-Boreal Forest Soils. *Can. J. Soil Sci.* **2012**, *92*, 329–340. [CrossRef]
68. Graham, J.H.; Leonard, R.T.; Menge, J.A. Interaction of Light Intensity and Soil Temperature with Phosphorus Inhibition of Vesicular-Arbuscular Mycorrhiza Formation. *New Phytol.* **1982**, *91*, 683–690. [CrossRef]
69. Hawkes, C.V.; Hartley, I.P.; Ineson, P.; Fitter, A.H. Soil Temperature Affects Carbon Allocation within Arbuscular Mycorrhizal Networks and Carbon Transport from Plant to Fungus. *Glob. Chang. Biol.* **2008**, *14*, 1181–1190. [CrossRef]
70. Shukla, A.; Kumar, A.; Jha, A.; Chaturvedi, O.P.; Prasad, R. Ajit Gupta Effects of Shade on Arbuscular Mycorrhizal Colonization and Growth of Crops and Tree Seedlings in Central India. *Agrofor. Syst.* **2009**, *76*, 95–109. [CrossRef]
71. Hasselquist, N.J.; Metcalfe, D.B.; Inselsbacher, E.; Stangl, Z.; Oren, R.; Näsholm, T.; Högberg, P. Greater Carbon Allocation to Mycorrhizal Fungi Reduces Tree Nitrogen Uptake in a Boreal Forest. *Ecology* **2016**, *97*, 1012–1022. [CrossRef]
72. Snapp, S.S.; Swinton, S.M.; Labarta, R.; Mutch, D.; Black, J.R.; Leep, R.; Nyiraneza, J.; O'neil, K. Evaluating Cover Crops for Benefits, Costs and Performance within Cropping System Niches. *Agron. J.* **2005**, *97*, 322–332. [CrossRef]
73. Araujo, A.S.F.; Leite, L.F.C.; Iwata, B.d.F.; Lira, M.d.A.; Xavier, G.R.; Figueiredo, M.d.V.B. Microbiological Process in Agroforestry Systems. A Review. *Agron. Sustain. Dev.* **2012**, *32*, 215–226. [CrossRef]
74. Holmgren, M.; Scheffer, M.; Huston, M.A. The Interplay of Facilitation and Competition in Plant Communities. *Ecology* **1997**, *78*, 1966–1975. [CrossRef]
75. Brooker, R.W.; Maestre, F.T.; Callaway, R.M.; Lortie, C.L.; Cavieres, L.A.; Kunstler, G.; Liancourt, P.; TielböRger, K.; Travis, J.M.J.; Anthelme, F.; et al. Facilitation in Plant Communities: The Past, the Present, and the Future. *J. Ecol.* **2007**, *96*, 18–34. [CrossRef]
76. Holl, K.D.; Zahawi, R.A.; Cole, R.J.; Ostertag, R.; Cordell, S. Planting Seedlings in Tree Islands versus Plantations as a Large-scale Tropical Forest Restoration Strategy. *Restor. Ecol.* **2011**, *19*, 470–479. [CrossRef]
77. Corbin, J.D.; Robinson, G.R.; Hafkemeyer, L.M.; Handel, S.N. A Long-Term Evaluation of Applied Nucleation as a Strategy to Facilitate Forest Restoration. *Ecol. Appl.* **2016**, *26*, 104–114. [CrossRef] [PubMed]
78. Bertoncello, R.; Oliveira, A.A.; Holl, K.D.; Pansonato, M.P.; Martini, A.M.Z. Cluster Planting Facilitates Survival but Not Growth in Early Development of Restored Tropical Forest. *Basic Appl. Ecol.* **2016**, *17*, 489–496. [CrossRef]

79. Kammann, C.; Graber, E.R. *Biochar Effects on Plant Ecophysiology. Biochar for Environmental Management: Science, Technology and Implementation*; Routledge: London, UK, 2015; pp. 391–420.
80. Bodner, G.; Loiskandl, W.; Kaul, H.-P. Cover Crop Evapotranspiration under Semi-Arid Conditions Using FAO Dual Crop Coefficient Method with Water Stress Compensation. *Agric. Water Manag.* **2007**, *93*, 85–98. [CrossRef]

forests

MDPI

Review

Why Healthy Pine Seedlings Die after They Leave the Nursery

David B. South [1,*], Tom E. Starkey [2] and Al Lyons [3]

[1] College of Forestry, Wildlife Sciences and the Environment, Auburn University, Duncan Drive, Auburn, AL 36849, USA
[2] Forest Nursery & Regeneration Consultant, 220 Northridge Dr, Anderson, SC 29621, USA; tomstarkey3@gmail.com
[3] Manulife Investment Management, 3891 Klein Road, Harpersville, AL 35078, USA; alyons@manulife.com
* Correspondence: southdb@auburn.edu

Abstract: Artificial regeneration is successful when high-performing seedlings are transported with care to the planting site, stored for a short period in an environment without desiccation or fungal growth, and planted in a deep hole, so roots are in contact with moist soil. One of the requirements for success is the ability to avoid common planting mistakes. Due, in part, to the use of container stock plus an increase in rainfall, the average first-year survival of pine seedlings (89%) in the southern United States is about 15% greater now than 45 years ago. However, when survival is less than 50% six months after planting, some landowners seek reimbursement for their loss. Some assume poor seedling quality was the cause without realizing that anaerobic soils or sudden freeze events, shallow planting holes, pruning roots, a lack of rain or underground insects can kill pines. With a focus on pines planted in the southern United States, we list non-nursery factors that have killed seedlings in North America, Africa and Europe.

Keywords: planting depth; drought; freezing injury; herbivory; mortality; survival; insects

Citation: South, D.B.; Starkey, T.E.; Lyons, A. Why Healthy Pine Seedlings Die after They Leave the Nursery. *Forests* **2023**, *14*, 645. https://doi.org/10.3390/f14030645

Academic Editors: Timothy A. Martin, Mohammed S. Lamhamedi, Damase P. Khasa and Steeve Pépin

Received: 2 January 2023
Revised: 6 March 2023
Accepted: 18 March 2023
Published: 21 March 2023

1. Introduction

When most factors are optimum, the 11-month survival of bareroot pine seedlings can exceed 90% [1,2], but in some years, survival is less than 75%. It seems the initial survival of pine seedlings in 13 southern United States (SUS) has gradually increased over time (Figure 1). The average survival of loblolly pine (*Pinus taeda* L.) for 58 sites during the 1980s was 74% [2,3], and now it averages about 89%. This increase is likely due to improvements in the seedling quality of bareroot stock combined with planting more container stock. In the SUS, container stock now represents one-fourth of the pine seedling production.

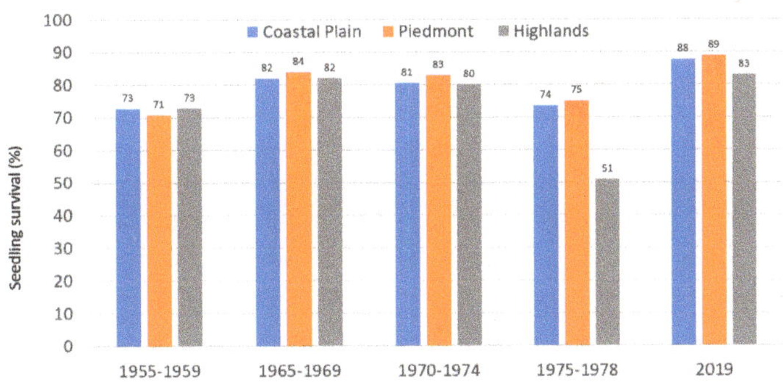

Figure 1. Early survival of pine seedlings reported for various periods in the southern United States. Adapted from [4–8]. Coastal Plain, Piedmont and Highland provinces were defined by Schultz [9]. The least significant difference (α = 0.05) for a one-tail test is −9.3% and ±11.4% for a two-tailed test.

Most agree that rainfall plays a major role in the survival of planted seedlings. If average rainfall has increased, this might help explain some of the increase in the rate of survival. In fact, for one region in the SUS, the average rainfall for 2019 to 2021 was about 19% greater than for the period from 1955 to 1959 and 15% greater than the period from 1975 to 1978 (Figure 2). Although annual rainfall was near normal in 2019, reported survival for the SUS was 88%, and about one out of 20 planted areas had survival low enough to be replanted. Average mortality can exceed 15% in some years (Figure 3).

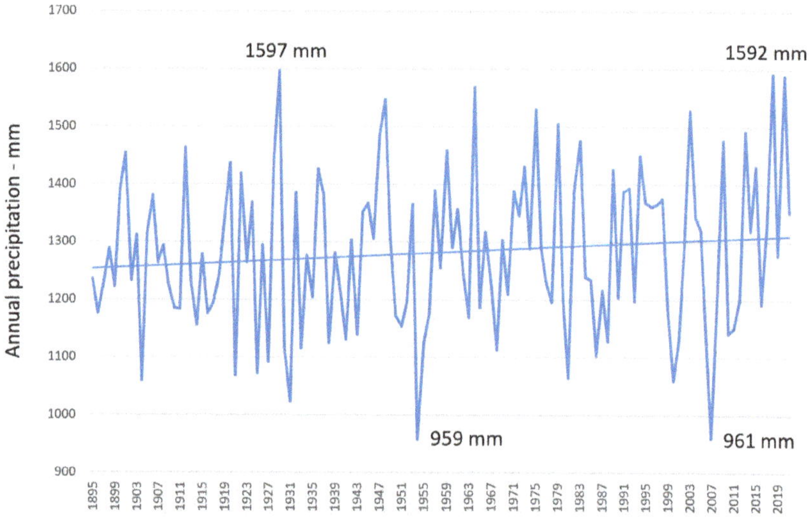

Figure 2. Average precipitation (1297 mm for years 2000 to 2021) for the southeast region of the USA (AL, FL, GA, SC, NC, VA) is about 45 mm greater now than 100 years ago. Since 1895, there have been 8 years with rainfall greater than 1500 mm, and 6 of those occurred after 1960.

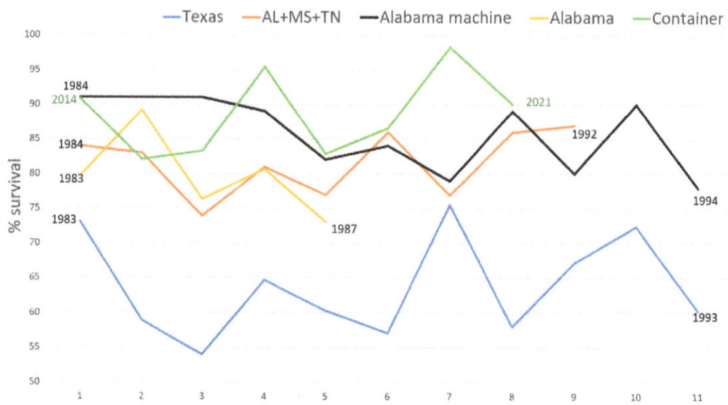

Figure 3. Survival of loblolly pine varies with year, region, stock type and planting method. For this graph, survival values before 1995 represent bareroot stock. The green line represents hand-planted container-grown stock, and the black line represents machine-planted bareroot stock. The average rainfall for East Texas (1176 mm) is about 17% less than the average rainfall for North Alabama (1418 mm). Data were provided by Brad Barber (Texas), Steve McKeand (container), Jerry Frame (Alabama machine; AL + MS + TN) and Harry Vanderveer (Alabama).

The objective of this review is to list non-nursery factors that contribute to the initial mortality of pine seedlings. Nursery managers, foresters and landowners may find this list of over 50 factors useful when customers ask what can be done to increase survival. Too often, landowners want reimbursement for mistakes that were made after seedlings left the nursery. The outline of the paper is as follows. Section 2 presents the literature sources, and Section 3 discusses the importance of keeping good records. Factors affecting survival before planting and during transportation are presented in Sections 4 and 5. Section 6 covers planting, and Section 7 covers beneficial treatments applied after planting. Weather factors are listed in Section 8, and Section 9 involves replanting. Section 10 covers costs, and Section 11 involves use of statistics by researchers.

2. Materials and Methods

Since 1975, we have collected a large number of papers published after 1900. The papers were sourced from books, proceedings and repository sites, such as Treesearch, TreeCD, Agricola and Google Scholar. We excluded papers about angiosperms and pseudoreplication and retained papers that contained survival data for pine seedlings grown in nurseries in North America, Canada, Europe and Africa. A few references to the survival of Sitka spruce (*Picea sitchensis* (Bong.) Carr.) and Douglas-fir (*Pseudotsuga menziesii* (Mirbel) Franco) were included. Many papers involved seedling quality, but these were excluded as these effects were adequately discussed in other unsystematic reviews. Papers were retained if they involved factors that killed pine seedlings after they passed the nursery gate. To keep citations to a minimum, a small portion of the total available papers ended up as citations. This review does not address a specific research question, and it does not cite all of the published literature on pine survival. As a result, it is not a systematic review of all literature pertaining to a single question.

Powerpoint software was used to construct graphs to illustrate treatment effects on the survival of planted stock. Among the 23 graphs, about half present survival data for both the *Y*-axis (treatment) and *X*-axis (control). This graphical technique was utilized since it is relatively easy for landowners to understand. It also illustrates that treatment response is often greater when mean survival is less than 80%.

3. Record Keeping

When seedling survival is unexpectedly low, landowners need to know why so that mistakes are not repeated in the future. Sometimes the reason for low survival is obvious, but occasionally, consultants are hired to determine the cause. When experts arrive soon after the onset of mortality, dying seedlings can be examined, and the cause may be determined. However, when expert arrival is delayed, important clues begin to dissipate and are gone after seedlings have dried and decomposed. When this happens, accurate records provide needed clues (see planting log in Supplementary Materials).

Unfortunately, key records are often missing. Important information includes lifting date, storage length, method of transportation to the site, seed origin, seedling morphology, planting date, temperature in storage, planting depth, soil moisture when planting, fertilizer placement and rate. We recommend recording the standardized precipitation index at planting (see Supplementary Materials). If an herbicide was applied, the total amount purchased and the total area treated should be recorded along with the application date. Weather records are very important. In 1977, weather stations in the Cumberland Plateau in Tennessee recorded the coldest January since 1895. This likely explains why pine survival averaged 51% in the highland region of the SUS (Figure 1).

Pine seedlings can die within 30 days of planting [10] or death can linger for 6 months to 5 years or more. When mortality is acute (1 to 3 months after planting), the cause can generally be traced to one factor. Heat exposure due to improper handling/storage or debarking weevils are examples of quick mortality. However, more often, chronic mortality occurs within the first five years after planting. The root cause can then be difficult to ascertain as a multitude of factors generally contribute to mortality. For example, machine-planted seedlings receiving cold damage on soils with high available water hold capacity with adequate rainfall might not die. However, cold damaged seedlings can contribute to chronic mortality when combined with factors such as root pruning, planting seedlings with the root-collar near the surface, nematodes and low soil moisture combining to eventually kill the seedling due to too much transpiration. Mortality analysis needs to be conducted by foresters with extensive reforestation experience to avoid repeating mistakes of the past. Often the blame is placed on the nursery when the actual cause is experienced after the seedlings leave the nursery.

4. Preparing the Site

Prior to 1980, site preparation in the SUS was dominated by mechanical methods such as shear rake and pile, bedding and disking. These treatments helped to facilitate machine planting and reduced the growth of hardwood sprouts. Site preparation and road maintenance contractors utilized the smaller crawler-tractors in the winter to pull planting machines, and during the spring and summer, they used the same machines to cultivate the soil and conduct road work. Keeping crews and machines working year-long is a significant economic advantage. Integrated forest products foresters favored mechanical site prep, and many experienced foresters of that day commented that woodland organizations were reluctant to change from mechanical soil cultivation to chemical site preparation for fear of losing a "mechanical empire", for lack of a better description. Since machine planting often results in deeper planted seedlings, average survival from machine planting can be greater than with contract hand-planting crews. After effective chemical weed control methods were adopted, companies realized there was little additional benefit from disturbing the soil surface. As a result, the frequency of soil cultivation declined, and the use of hand-planting increased.

4.1. Burning

Each year, prescribed burning occurs in the SUS on about 9% of sites before planting pine seedlings. The goal is to control small hardwoods that compete with newly planted pines and to make hand-planting easier [11]. At a flatwood site in Florida, burning before planting increased survival by 7% [12]. Coosa County, Alabama, was in an extreme drought from July 2007 to March 2008, and seedlings planted in February on burned sites averaged 66% survival, while survival on not burned areas averaged 54% (Figure 4). When seedlings were hand planted in February 2008, burning increased survival by 21%. Possible reasons for this effect include less hardwood competition for moisture, duff in planting holes, fewer live insects in the topsoil and more rainfall reaching the soil. At some locations, burning increases soil moisture [12,13].

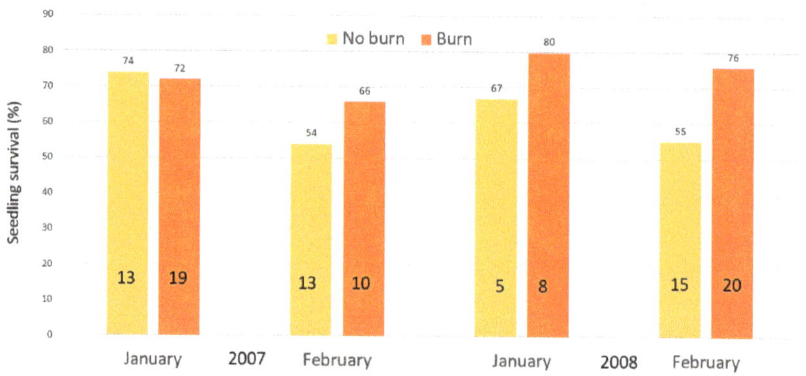

Figure 4. Prescribed burning (before planting in January or February) increased the survival of bareroot loblolly pine seedlings during a severe drought in central Alabama (data from Al Lyons; number on bar represents number of sites per mean). The increase in survival was significant except for January 2007. The least significant difference ($\alpha = 0.05$) for a one-tail test is −11% and ±14% for a two-tailed test.

Site preparation burning in the SUS has declined over the last four decades for various reasons. In 1980, 98% of the area planted to pines in the lower Coastal Plain was burned prior to planting. During the era of integrated forest product companies, woodland organizations had a significant workforce and enough equipment to conduct road construction, maintenance and site preparation (mechanical, burning, etc.). After these companies monetized timberlands, this workforce capacity was lost, which lowered the amount of burning. Only a few contractors had the capacity to conduct burning. There were also environmental (smoke management), liability (smoke, wildfire) and soil nutrient concerns that limited wide-scale burning [14]. Currently, site preparation burning is limited to sites with enough slash and debris to slow hand planting and reduce planting quality or where there is a high density of wildling pines. (Appendix A Figure A1).

4.2. Bedding

In the SUS, bedding has been used on sites where high-water tables reduce soil oxygen and reduce pine survival [12,15]. In southeast Georgia, burning and bedding increased survival by 7% (Figure 5). However, at some locations, bedding increased seedling mortality by more than 10% [16]. The additional mortality was attributed to rough, unsettled beds full of air pockets. Therefore, planting too soon after bedding can increase seedling mortality. Bedding quality affects survival as seedlings planted in low dips generally have higher mortality. This is common in single-pass beds, with bed quality increasing with double-pass bedding. Double bedding might increase the survival of slash pine (*Pinus elliottii* Engelm.) by 5 percentage points. In the lower Coastal Plain, about 31% of planted areas are single-bedded while 9% are double-bedded [8].

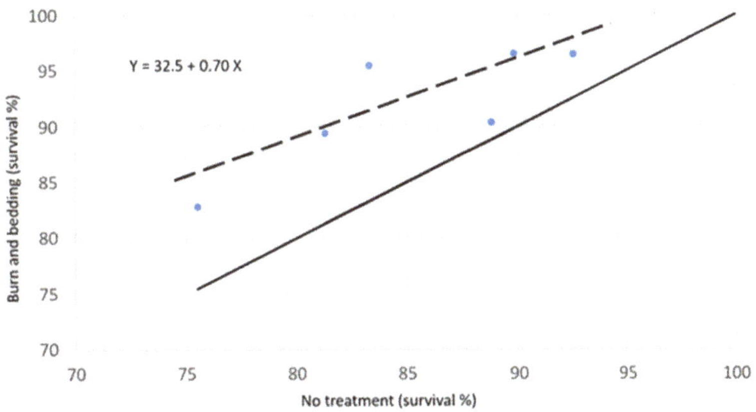

Figure 5. Burning followed by bedding increased fourth-year survival of slash pine at six sites in southeastern Georgia [17]. The least significant difference ($\alpha = 0.01$: two-tailed test is 4.5%). Points above the solid line represent sites where burning and bedding increased survival. $p > |t| = 0.19$ for intercept and 0.0464 for slope; $R^2 = 0.67$.

4.3. Disking

On some sites, flat disking may increase early volume growth due to a 9% increase in survival, but there may be no increase in volume where disking has no effect on survival [18–20]. Disking is not a common site preparation method today and was most common during integrated forest product management.

4.4. Subsoiling

Subsoiling (or ripping) compacted soil before hand-planting can increase survival since seedlings are planted deeper and less foliage is exposed to drying winds. On some sites, ripping soil before planting increased pine survival by 7% or more ([21–24]; Figure 6).

When compared to digging a 20 cm deep hole, ripping to a depth of 40 cm increased the survival of patula pine (*Pinus patula* Schiede ex Schltdl. & Cham.) by 16% [24]. Seedlings should be planted four months after ripping to allow the rain to settle the soil. When the soil in the rip is not yet settled, planting seedlings on the off-set berm will likely improve survival [25]. If the soil has not settled and seedlings are planted in the ripped zone, mortality may result due to reduced root-soil contact or soil erosion burying seedlings [26].

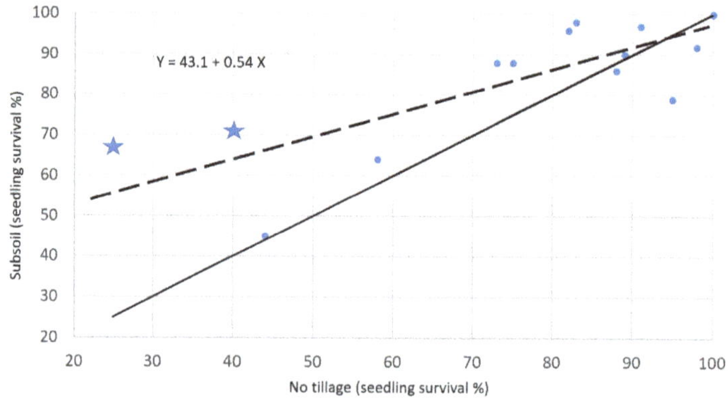

Figure 6. Subsoiling increased the survival of loblolly pine seedlings (n = 15) by an average of 8 percent [2], and the increase was 13% on sites with less than 90% survival (n = 10). Sites with less than 50% survival were planted in early 1995, a year with below-average rainfall. At two sites in Saluda County, South Carolina (Stars), March to May (1995), rainfall was 160 mm below normal. Points above the solid line represent sites where subsoiling increased survival. $p > |t| = 0.001$ for intercept and 0.0025 for slope; $R^2 = 0.58$.

4.5. Scalping

Scalping is a recommended practice before planting on former cropland sites. In some cases, scalping increased survival by more than 30% [27–29]. This increase is due to reducing weed competition and reducing the populations of insects and diseases. On sandy soils, scalping might not reduce the rate of water infiltration. However, on fine-textured soils, scalping reduces infiltration and can hold water like a pond. "If the soils are very wet, or the soils are very heavy (high clay content), scalped rows may hold water and drown the seedling" [30].

4.6. Liming

Often lime is not needed in bareroot pine nurseries unless the soil pH is below 4.5, and lime is also not operationally applied in North American plantations. In research trials, applying 4480 to 6700 kg ha^{-1} of dolomite did not appear to increase the mortality of loblolly pine [31,32]. However, incorporating a higher rate of 11,200 kg of agricultural-grade lime before planting western white pine (Pinus monticola Dougl. ex D. Don) reduced survival on two sites in 1992 by more than 9% [33].

5. From Nursery Gate to Planting Site

Boxes, bags and bales of seedlings are transported to the planting site in open trucks, insulated vans and refrigerated vans [34]. Seedling mortality increases when open trucks-trailers transport seedlings for long distances during −4 °C days. Therefore, transportation should be in refrigerated vans or covered vehicles that are preferably insulated [35,36].

5.1. Improper Storage

When soil is moist, it is generally best for landowners to plant seedlings soon after seedlings arrive at the planting site. When weather delays planting, then southern pine seedlings should be stored in a cool environment to reduce heat, which encourages fungal growth on roots and foliage. In a trial in Tennessee, survival was greater when seedlings were planted soon after delivery than when seedlings were planted after 9 weeks of storage in ambient conditions (Figure 7).

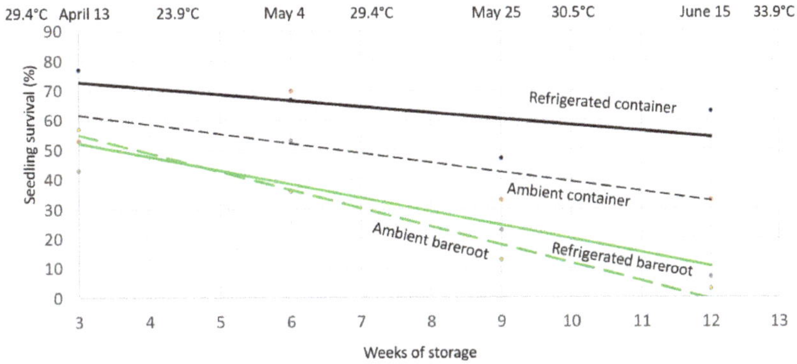

Figure 7. Effect of date of planting, storage length and storage method on survival of loblolly pine planted in Scott County, Tennessee [37]. The data presented assume 3 weeks of storage prior to planting on April 13. Storing seedlings reduced survival, as did storing seedlings without refrigeration. The highest air temperature recorded for each 3-week period is indicated at the top of the graph. There were 30 seedlings planted per treatment-date combination. Bareroot seedlings, likely produced at Pinson, Tennessee, and container seedlings (likely Spencer-Lamaire root-trainers) were grown in a greenhouse at the University of Tennessee. The least significant difference is −13.7% (α = 0.05: one-tailed test). For ambient bareroot seedlings, $p > |t|$ = 0.0057 for intercept and 0.012 for slope; R^2 = 0.97. The slopes of the remaining lines were not significantly different from zero (p = 0.13 for refrigerated bareroot, 0.30 for ambient container and p = 0.36 for refrigerated container).

A few landowners have access to coolers and store seedlings for a few weeks before planting. Some coolers are equipped to keep the humidity high while others act as a dehumidifier and can dry seedlings over time. Spraying water into the cooler on a daily basis will help maintain the humidity within the cooler. If seedlings are not stored properly, they can dry during storage and mortality is increased. Small quantities of seedlings may be stored by heeling trees in sawdust or soil.

In some countries, containers with seedlings are shipped to the planting site, and pines are extracted just prior to planting. When freezing weather is imminent before all seedlings are planted, the remaining containers with seedlings can be placed on the ground and covered with an insulated tarp. This reduces the risk of root injury to unplanted seedlings.

5.2. Water in Bags

Landowners should be careful not to puddle water in bags since "excess water can drown root tips or promote mold on the seedlings" [38]. In one trial, adding 2 L of water to bags increased the mortality of bareroot spruce seedlings that were stored for 2 weeks at 20 °C [39]. In another trial with bareroot loblolly pine, seedlings were lifted in January, and 86 mL of water was added per 1000 g of seedlings. Seedlings were stored in kraft-polyethylene (KP) bags for 2 weeks at 2 °C, and this reduced root growth potential by more than 75% [40]. This occurred even though no mold or fungal growth was observed on the roots or foliage.

5.3. Improper Transportation

The preferred way to transport large quantities of seedlings is in a refrigerated van [41], and small quantities may be transported in insulated trailers [34]. The temperature should be recorded to ensure seedlings do not freeze or overheat. Seedlings have been killed when transported in the back of a truck during freezing temperatures.

Sometimes seedling packages are dropped when unloading seedlings at the planting site. Dropping container-grown lodgepole pine 30 times (1 m height) did not reduce seedling survival [42]. However, dropping other conifers may reduce root growth potential and decrease height growth [43–45].

5.4. Transportation Too Far North

Genetics plays a role in seedling survival, especially in terms of response to a hard freeze. For example, some southern Coastal Plain sources of loblolly pine do not survive well when planted in southern Illinois, where temperatures might reach −20 °C [46]. Likewise, lower survival is expected when planting Florida sources of pine in Virginia. For example, the survival of a local Virginia source of longleaf pine at age 10 years had 80% survival, while a seed source from Florida had 51% survival [47].

Due to first-hand experience with de-acclimation freezes, several forest companies in the SUS reject flawed computer-based recommendations to plant pines 700 km north of their native location (i.e., location of the mother tree). Most loblolly pine seedlings are planted within 5° latitude of the native origin (≈550 km). For example, in a survey of members of the North Carolina State University Tree Improvement Program, only one member experienced freeze damage when a southern coastal family was planted too far north [48]. No member reported outright failure due to establishing plantations using a single family. Only about 8% of progeny-test plots established with container-grown loblolly pine were abandoned due to survival rates lower than 60%.

6. Planting

6.1. Root Pruning

Most tree planting guides place too much emphasis on avoiding bent roots and not enough emphasis on keeping roots. The "bent-root-kills" myth has lowered seedling survival because of publications that recommend pruning up to 50% of roots to avoid bent lateral roots and L-shaped taproots. The unjustified fear of a 3-cm bend of the taproot at the bottom of a 20-cm hole encourages (1) removing a portion of the taproot and (2) making a 13-cm deep hole to match a 13-cm pruned taproot. When soil is dry, both practices contribute to an increase in mortality.

In order to make tree planting quicker, one planting guide states to "prune roots to a uniform length by aligning root collars in bunches before pruning" and to prune roots no shorter than 13 cm. In fact, pruning to a 7.6 cm length can reduce pine survival in a dry year by 13 to 28% [49]. These recommendations suggest it would be OK to prune a 20-cm taproot (with six lateral roots) so that after pruning, the taproot would be 13 cm long with four lateral roots remaining. Of course, cutting the tap root will make hand-planting easier and will reduce the frequency of taproots with a 3-cm bend at the end. However, removing fibrous roots lowers root-growth potential [50,51] and increases mortality. For example, pruning 20 cm taproots to a 13 or 16 cm length reduced loblolly pine survival by 4% [52,53] and at one location, pruning to a 7.6 cm length reduced survival by 18%. For this reason, a few tree planting guides say "do not allow planters to prune roots." A contributing reason why machine planting often results in better survival is large roots with long taproots do not slow the rate of planting. As a result, machine planters do not prune roots.

6.2. Root Stripping

Some hand planters will strip roots just before they insert the seedling into the hole. This practice (typically performed by moving the root through a closed fist) removes some of the small fibrous roots and ectomycorrhiza. As a result, root mass might be reduced by 2%. However, the ability of the seedling to produce new roots can decrease by more than 40% [51], which can increase mortality by 12% or more [54].

6.3. Root Exposure

Drying roots before planting can reduce root growth, which leads to mortality [55]. In one trial, exposing pine roots outdoors for 15 min reduced new root growth by 70% [56] and reduced survival by 100% [57]. Exposures of 5 to 10 min on a sunny day can reduce pine survival by 5% to 18% [58–61], and 30 min reduced survival by 10% to 35% (Figure 8). To increase the probability of survival, planting stock should be protected from direct sun and kept cool. Planter handling and transport should minimize seedling exposure to prevent air drying [35,56].

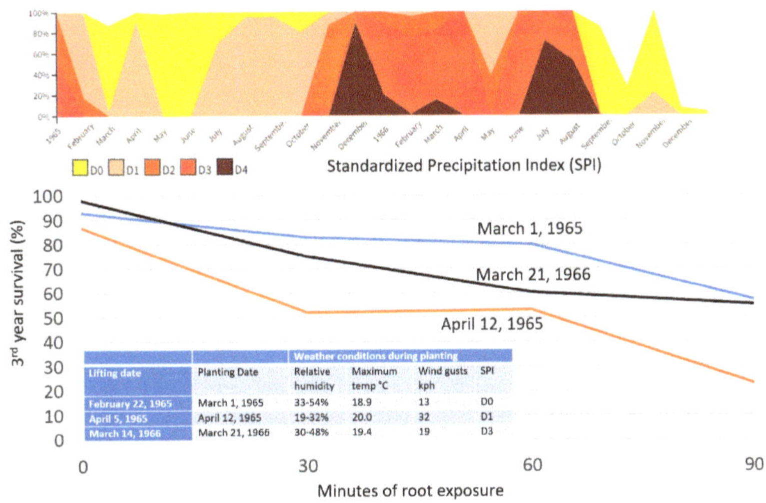

Figure 8. Exposing roots to 30 min of sun in Virginia [62] and then planting in moist soil (Standardized Precipitation Index; SPI = D0; February rain in Appomattox County = 94 mm) reduced survival by 10%. In contrast, when planted in drier soil (SPI = D1; April 1965 rain = 53 mm), the same exposure reduced survival by 35%. Likewise, planting 30-min exposed seedlings in dry soil the following year (SPI = D3; March 1966 rain = 52 mm) reduced survival by 23%. Data from [62]. The least significant difference ($\alpha = 0.05$) for a one-tail test is -10% and $\pm12\%$ for a two-tailed test.

6.4. Machine Planting

Pine seedlings are planted either by machine or by hand. From a 2020 survey of 376,000 ha in the SUS, 40% were planted by machine (Figure 9). Hand planting is often used on rough sites in the Piedmont and, when available, machine planting is preferred on flat lands in the Coastal Plain. When availability and the costs are reasonable, most regeneration foresters in the SUS prescribe machine planting since the probability of good survival is greater than that for hand planting bareroot stock. In Florida, 30 out of 64 sites were planted with machines [63].

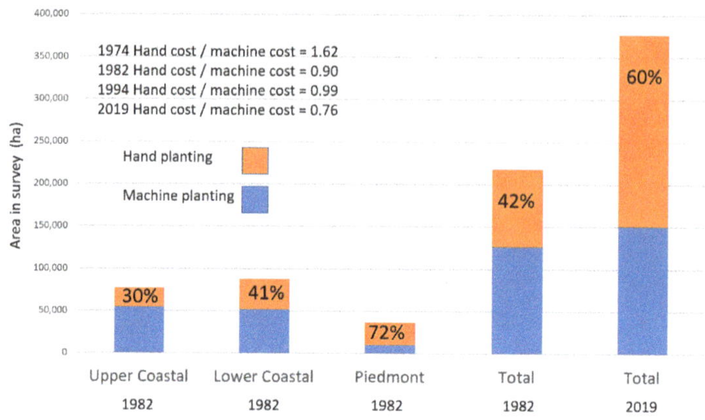

Figure 9. As the relative cost of machine planting increases, the proportion of hectares planted by hand increases. Data from [8,64].

In Louisiana (December 2000 to March 2001), machine planting on 11,220 ha averaged 87% survival, while hand planting (34,564 ha) averaged 80% survival. For one company in Alabama, machine planting bareroot pine averaged 86% survival, while hand planting averaged 75% survival (Figure 10). The increase in survival was mostly due to placing roots deeper in the soil [65] because machines do not get tired of making deep holes at the end of the day and because hand planters were allowed to prune roots. In years when rainfall was below average, machine planting resulted in 16% greater survival than hand planting. In contrast, when (January-June) rainfall exceeded 800 mm, survival with machine planting was only 5% better than hand planting. Some machines do not plant container seedlings as well as hand planters [66]. In fact, some are reluctant to plant container-grown longleaf pine with machines since the risk of burying the plug is great [67]. In one trial, machine planting of container-grown stock resulted in 64% and 70% survival in scalped and non-scalped areas, respectively [68].

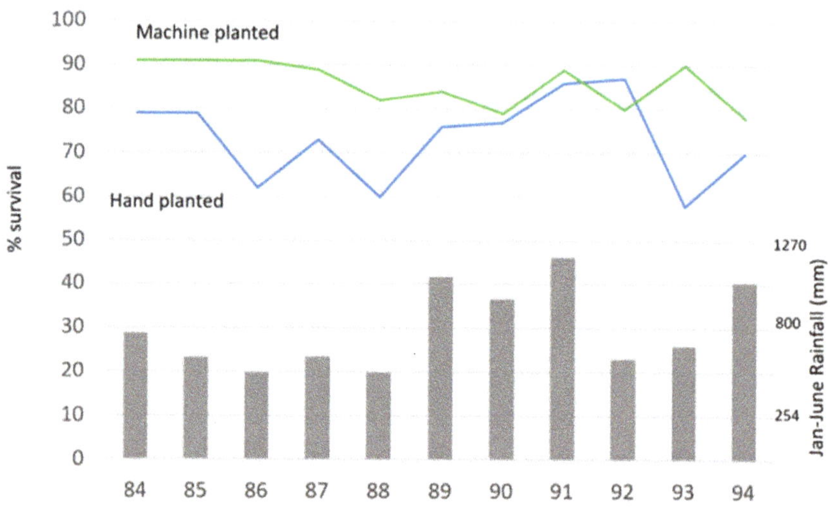

Figure 10. Survival of bareroot loblolly pine in northeast Alabama varied with rainfall and planting method. Over the 11-year period, early survival averaged 86% and 75% for machine and hand planting, respectively. The least significant difference ($\alpha = 0.05$: one-tail test is -6.4%). Data from [69].

Various brands of machines are used to plant seedlings [70,71], and some machines can plant 400 to 800 seedlings an hour, regardless of stock type. In 1982, the time required to plant 10 ha of upper coastal plain soil was 11 h for a machine-planting crew or 15 h for a hand-planter [64]. In contrast, some container-only machines plant less than 400 seedlings per hour [72,73]. The time required to machine plant 1000 pine seedlings depends on the spacing between seedling rows. In theory, a spacing of 5 m between rows would require half the time as a typical 2.5 m row spacing, which would lower the cost of machine planting.

The cost of planting 1000 seedlings could be $120 with hand planters or $170 with machines. If average survival is greater with machine planting, the $50 greater cost might increase survival by 0% for wet years and 20% during dry years (Figure 10). When planting 1000 seedlings per ha, a 10% increase in survival equates to an additional 100 living seedlings. When a seedling costs 20 cents and planting costs 12 cents, 100 dead seedlings would equal a loss of $32. Many landowners choose hand-planting and are unwilling to spend $50 in order to save $32 in dry years. However, at a cost of 50 cents per seedling, a $50 investment in machine planting might save $62.

6.5. Hand Planting Tools

Many types of tools are available for hand planting but generally, the size of the root system is a factor when selecting tools for operational use. A metal pipe used to plant a 30 cm^3 container plug is not suitable for planting a seedling that was grown in a 1000 cm^3 container. A planting spade is the preferred tool in New Zealand, while a shovel is popular in Oregon [74]. In the SUS, bareroot seedlings are typically planted with either a planting bar or hoedad [75]. All four tools can be used to make a 25 cm deep hole. When the hole is deep enough, soil moisture is adequate, and the planting method is correct, good survival can be achieved. Tools designed to make a 10 cm deep hole are not recommended for planting bareroot seedlings.

6.6. Family Block Planting

Planting advanced, half-sib and full-sibling families of loblolly pine in family blocks is a common practice in the SUS. In 2019, over 45% of loblolly pine stands were established using second-cycle seedlings [8]. In theory, the method of planting (family-block vs. mixed genotypes) does not affect overall seedling survival during the first year. However, some foresters notice lower than expected survival when improved genotypes are planted in single-family blocks. Due to record keeping, one organization noted a certain family had 10% lower survival than others [48]. Large landowners can use this knowledge to their advantage when planting half-sibling blocks the following year.

In contrast, some landowners own small forests, and they might plant seedlings once every 25 years. If they decide to plant a mixture of 30 genotypes, they might achieve 96% survival in a good rainfall year or 56% survival in a droughty year (Figure 11). However, if they are unlucky and plant a fast-growing family with the lowest survival, they may get only 29% survival during the bad year.

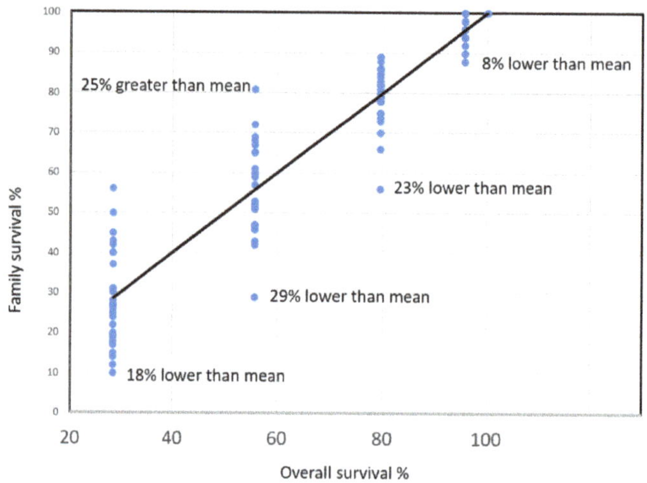

Figure 11. When planted in a good environment, the range in seedling survival among 30 loblolly pine families (100% to 88%) is relatively low [76]. As site conditions worsen, it becomes easier to identify families with low potential for survival when records are maintained and reviewed. Data from [76].

6.7. Seed Source

There is significant genetic variation within the natural range of loblolly pine, extending from southern New Jersey to southeast Texas. Limited to the north by cold temperatures and to the west by rainfall, native loblolly pine races have developed that are adapted to local climates. The adaption to local climatic conditions is referred to as geographic variation. Seeds harvested from these geographic areas vary in their potential for adaptation and survival depending on where they are planted. Seeds sourced from warmer climates within the loblolly range grow faster than those from more northerly climates. However, the principal factor influencing survival within natural ranges is the average yearly minimum temperature at the seed source's native location [77]. Moving warmer climate loblolly pine seeds far north can result in poor survival [46] and long-term adaptability issues due to cold, snow and ice damage. Eastern loblolly pine seed sources, such as Atlantic coastal material, grow faster than western sources. The slower-growing seed sources of southeast Texas are more tolerant of drought, as they have adapted to lower rainfall and drought conditions. These western sources of loblolly pine cease growth immediately at the onset of drought, whereas seed sources from eastern geographic regions tend to continue to grow. Movement of Atlantic coastal material into the western Gulf coast region is common, which could reduce survival during drought conditions.

6.8. Root Soaking

Ideally, roots should not be allowed to dry in storage or during transportation. Since this does occur [78–80], soaking roots in water might increase the probability of seedling survival ([44,79,81]; Figure 12). In Pennsylvania [80], loblolly pine roots were soaked for 24 h before planting, and nine more seedlings survived when compared to nonsoaked seedlings (40 seedlings planted per treatment). However, soaking roots for 4 h before planting reduced survival in Mississippi for some unknown reason [81].

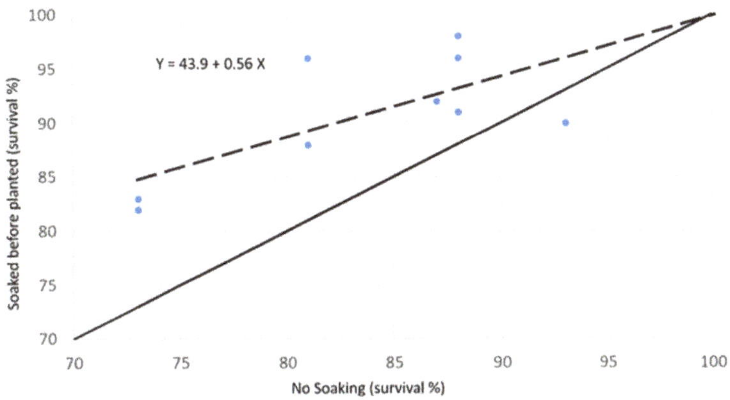

Figure 12. Loblolly pine seedlings were lifted and packed into bundles in January at the New Kent Nursery in Virginia (1979–1981). Seedlings were stored in an unheated garage for periods of 4 to 9 weeks. Just before planting, selected bundles were soaked in water for 1 h and then outplanted. Seedlings removed from other bundles were planted without any soaking. On average, soaking seedlings prior to planting increased survival by 7% [82]. Points above the solid line represent sites where soaking roots increased survival. Data from [82]. $p > |t| = 0.046$ for intercept and 0.036 for slope; $R^2 = 0.48$.

6.9. Washing Roots

Soil often adheres to the roots of bareroot seedlings, so some landowners may be tempted to wash pine roots before planting. Although soaking roots for a few minutes can be beneficial when roots have been desiccated, washing roots is not recommended since it can reduce survival. Six decades ago, Thomas Swofford said, "When you are washing your seedling roots with water you are also destroying some of your mycorrhiza, as well as some of your rootlets" [83]. In one trial, washing loblolly pine roots reduced survival by 13% [84].

6.10. Root Coatings

Several moisture-holding materials have been applied to seedling roots to protect them from desiccation prior to outplanting. These include sphagnum moss, kaolin clay and hydrophilic polymers, also known as polyacrylamide gels. Polyacrylamide gels began to appear in the late 1960's [85]. Today, pine roots in the SUS are treated with polyacrylamide gels to (1) protect seedlings from desiccation or (2) to improve water relations after planting. The amount of water mixed with 1 kg of polyacrylamide can vary from 300 to 1100 kg. The amount of gel used per 1000 seedlings can range from 2 to 6 kg. Using the right type of gel can increase survival (Figure 13), but the wrong type can decrease survival after planting [86–88]. A clay dip was popular in the past, but a gel treatment provides better protection, costs less, is less messy and requires less transportation and storage space. This explains why only three nurseries treated roots with a clay dip in 2022. The preference for a clay dip is not based upon research but instead is based on tradition, and because the clay is readily visible on the roots. When nurseries first switched from clay to gels, landowners complained they could not see the gel on the roots.

When environmental conditions are favorable, and roots have been protected, there are no expected gains from applying gel to roots before outplanting. However, when conditions are not favorable, gel-treated roots can reduce mortality [87]. When seedlings treated with hydrogels were exposed for 2 h before planting, survival was 40% higher than roots treated with roots dipped in water. Similarly, in 8 of 20 studies, applying a gel might increase survival by 40% [89].

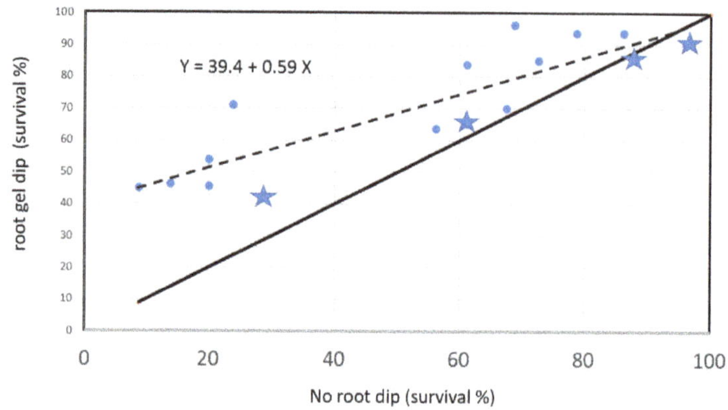

Figure 13. Effect of applying a gel root-dip on survival of bareroot red pine seedlings (*Pinus resinosa* Ait.) and jack pine (*Pinus banksiana* Lamb.) [88]. Seedlings not exposed to the sun are represented by stars, while the remaining dots were exposed to the sun for 5, 10 or 20 min prior to planting. Points above the solid line represent sites where the root gel increased survival. $p > |t| = 0.0001$ for intercept and 0.0001 for slope; $R^2 = 0.77$.

6.11. Clipping Foliage

Tall bareroot pine seedlings with too much foliage have an increased risk of mortality when planted on dry sites or in years with limited rainfall [90–94]. How much foliage is too much for a pine seedling depends on the amount of root mass, species, stock type and date of lifting from the nursery. A landowner has three options when delivered seedlings do not meet the target seedling's shoot/root mass ratio. Ideally, the nursery will exchange seedlings for stock that is closer to the desired target seedling. When this is not possible, pine seedlings with tall shoots can be top-pruned to the desired height or can be planted with the root-collar 10 to 15 cm below the surface. Planting seedlings deeper on well-drained sites reduces the amount of transpiring foliage.

For bareroot longleaf pine, clipping needles (e.g., length about 15 cm after clipping) just before planting reduces transpiration and typically increases survival (Figure 14). In one study, clipping needles reduced transpiration by 30% and increased survival by 10% or more [92]. However, clipping needles down to 2.5 cm is too short, and this will reduce survival [93,94].

6.12. Planting Hole Depth

There are two schools of thought regarding planting hole depth. One school says to make a hole the same depth as the taproot, while the other school believes in making a deeper hole using a machine or shovel. In well-drained soils, a 20 cm deep planting hole will increase survival (of stock with a 30 cm shoot) when compared to a 12 cm deep hole. This is because placing roots closer to a moist soil zone will increase survival. When sufficient rainfall results in high survival, planting 30 cm tall pines deep with 15 cm above ground will increase survival. In some cases, the height of pine seedlings is only 10 cm long, and a deep hole is not necessary for such small seedlings. Planting small container-grown pine seedlings with the root-collar 5 to 6 cm deep is recommended in Finland [95], and an 8 cm depth can reduce damage from insects and drought [96].

6.13. Container Plug Exposed

There are two schools of thought regarding planting depth for container-grown longleaf pine (*Pinus palustris* Mill.) Some say the top of the plug should be completely covered with soil to prevent "wicking" moisture out of the plug [38,97–100]. This idea might have originated at the Wind River Nursery, where containers (diameter 6.4 cm; height 25 cm) were fabricated using paper towels. At the planting site, the paper containers were planted in holes about 28 cm deep. Betts [100] said, "If the toweling sticks above the ground, it will act as a wick and pull the moisture from the area around the tree roots to the soil surface where it is lost by evaporation. A handful of dirt, duff or other debris scattered over the top of the tube can help seal off the planted tube from loss of moisture to the atmosphere." This extra work would not have been needed if the recommendation at that time had been to place the bottom of the container 30 cm below the surface instead of placing the root-collar at the level of the soil surface. A few years later, regeneration foresters were told to plant container plugs deeper and cover the plug with soil to prevent drying [98,99]. Planting container-grown pine seedlings with root-collar 10 cm below the surface is a valid practice for pines other than longleaf pine. For 30 cm tall seedlings, this practice reduces the transpiration rate and, in dry periods, will increase seedling survival in well-drained soils [101,102].

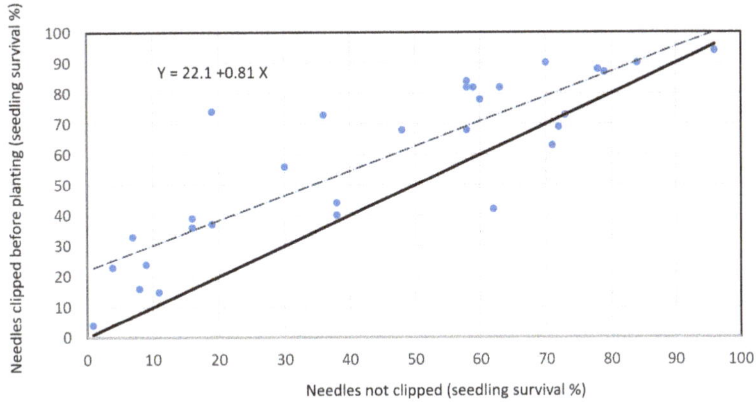

Figure 14. Comparison in survival between clipped and nonclipped longleaf pine seedlings [94]. The dashed line represents the regression equation (n = 30), and the solid line represents equal survival of the two treatments. Points above the solid line represent cases where clipping needles increased survival. Data from [94]. $p > |t| = 0.0001$ for intercept and 0.0001 for slope; $R^2 = 0.74$.

In contrast, the other school recommends bareroot [103] and container-grown longleaf pine seedlings should be planted with the root-collar 1 to 3 cm above the soil surface. Studies indicate that exposing the top 2.5 cm of the plug did not reduce initial survival [104,105]. In fact, covering the plug with 1 cm of soil increased mortality in Alabama [104]. Although numerous trials with other species show a survival benefit for deeper planting of bareroot stock [102,106,107], there are no data to support the belief that planting longleaf pine seedlings with 1 cm of soil over the plug will reduce mortality. In fact, many bareroot and container-grown longleaf seedlings have died because ponded water covered the terminal bud or because erosion covered the terminal bud with soil.

6.14. *Planting in Wrong Month*

There are two schools of thought regarding the best time to plant container-grown seedlings in the SUS. One group recommends planting in moist soil from October to December 1, while the other school says fully developed seedlings can be planted in any month of the year. Although researchers can sow seeds in any month, operational nurseries typically sow seeds in April or May, and seedlings are ready to plant in October [108]. Planting seedlings before December allows for new root growth, and established seedlings are typically tolerant of root-inhibiting herbicides applied the following May. Many researchers plant longleaf pine between 1 November and 1 February (Figure 15).

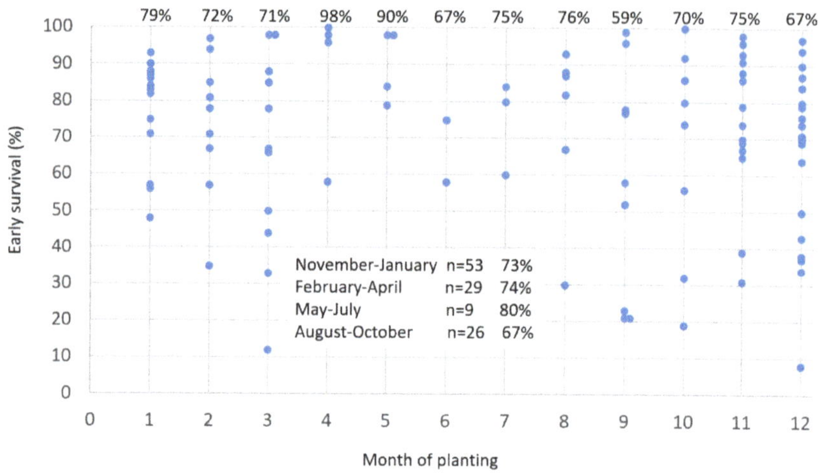

Figure 15. Some researchers prefer to plant container-grown longleaf pine seedlings in moist soil during the fall before December 1, but some wait till the end of December when students are not in school. Sometimes low survival in December is related to planting seedlings with an insufficient amount of freeze tolerance [109]. Each dot (n = 117) represents one value from a published research trial. For this graph, median survival is 79%, and 24% of the dots are below the 60% line.

Technically, there is no wrong month to plant pine seedlings, only a wrong time to plant. For bareroot pines, the planting season in the deep SUS can extend from October to February, but a dry February can be the wrong time to plant pines (Figure 15). An estimate of the amount planted by month might be 1% in October, 5% in November, 15% in December, 40% in January, 25% in February, 14% in March and <1% in April. In Virginia, bareroot seedlings had 81% to 94% survival when planted before 21 December (Table 1). In Louisiana, March and April are the worst months for achieving high survival with bareroot stock [110], while these months are acceptable in regions north of Tennessee and North Carolina [111,112].

Table 1. Survival of bareroot and container-grown loblolly pine seedlings after exposure to dark chilling in a cooler or outdoor chilling (natural chilling) in seedbeds at the Garland Gray Nursery in Virginia. Data from [112]. Rainfall and freeze data from Appomattox, VA. Dark chilling does not increase the freeze tolerance of loblolly pine seedlings [109], and dark freezer storage does not increase the freeze tolerance of ponderosa pine (*Pinus ponderosa* Douglas) seedlings [113]. The chilling hours listed below are hours between 0 and 7.2 °C.

Stock Type	Packing Date 2018	Planting Date 2018–2019	Rain during Month mm	Dark Chilling Hours	Natural Chilling Hours	Survival %	Freeze Event 2018–2019
Bareroot	15 October	15 October	99	<8	0	81	11 November −5.5 °C
Bareroot	15 November	15 November	179	<8	158	86	29 November −5.5 °C
Bareroot	18 December	18 December	173	<8	549	94	27 December −3.9 °C
==	January	Frozen soil	86	==	==	==	21 January −12.2 °C
Container	15 October	15 October	99	<8	0	100	11 November −5.5 °C
Container	15 October	15 November	179	744	0	86	23 November −3.3 °C
Container	15 October	18 December	173	1464	0	43	27 December −3.9 °C
Container	15 October	15–16 April	133	4368	0	96	None

March and April plantings should be avoided in the SUS since seedlings do not have enough time to establish roots before the hot-dry season. For example, planting loblolly pine seedlings on 1st March in Louisiana had 64% survival [114], while April plantings had 38% and 43% survival in Tennessee and Kentucky, respectively [115,116]. Landowners who plant in moist soil in October or November are planting prior to the cool-wet season while those who plant in March (below 37° N) are planting just prior to the hot (possibly dry) season.

Container-grown seedlings lifted on 13 November 2001, stored for two months, and planted on 22 January (after freezing temperatures) survived well. In contrast, seedlings stored for 10 weeks and then planted on 22 December with little natural chilling had 38% survival after a −7.2 °C freeze (5 January) (Figure 16). In contrast, a −1.7 °C event on January 26 likely did not injure seedlings planted four days earlier. Mortality due to freezes in December is referred to as the "December dip" [117].

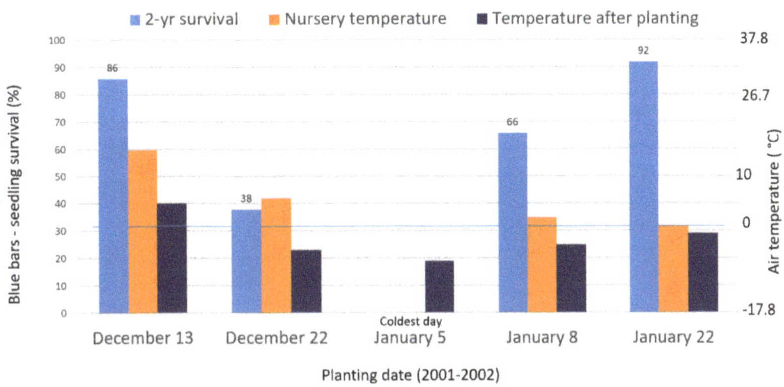

Figure 16. The effect of planting date on survival of container-grown longleaf pine seedlings stored for 10 weeks in a nursery cooler at +1.7 °C. Data from [118]. Seedlings were packed in boxes and placed into storage on 2, 16, 30 October and 13 November 2001. The lowest temperature at the nursery (for the lifting date) is represented by orange bars while black bars represent low temperatures recorded at Lumberton, North Carolina, a few days after planting. Accumulated natural chilling hours (0 to 7.2 °C) for Goldsboro, NC was 125 h for 13 November 2019. A temperature of −7.2 °C on 5 January likely contributed to the mortality of seedlings planted two weeks earlier.

6.15. Adding Water

When the soil is dry, some landowners add water at the time of planting. In South Africa, container-grown pine seedlings are planted during the summer rainfall season, and 1 or 2 L of water is sometimes added to the planting hole (Figure 17). Some mechanized tree planters are equipped with water tanks to water seedlings as they are planted [119]. Although uncommon, large-scale irrigation systems are sometimes used to establish trees in areas with limited rain. In one trial, the survival of irrigated and non-irrigated pines was 100% and 30%, respectively [120].

6.16. Loose Planting

Except for incorrect planting depth, loose planting may be the most important cause of seedling mortality [10,121]. The test for loose planting involves holding two or three fascicled needles between the thumb and finger and pulling upwards. If the seedling is pulled out of the hole, the seedling was too loose, and there was insufficient contact between the soil and roots [121].

6.17. Pulling Seedlings Up

Some outdated planting guides say that a "curled root will kill the seedling." Using this myth, a 1989 hand-planting planting guide says to push the roots "deep into the planting hole. Pull the seedling back to the correct planting depth." This method is not used by machine planters who plant the root-collar about 14 to 17 cm below ground [65,107]. If hand planters take the extra time to use the "pull-up" technique, the root-collar will be near the ground line, and they will get paid less due to planting fewer trees. However, there are no studies to show that when planted deeply, a curled root will kill the seedlings. Therefore, the "pull-up" method may move the roots away from moist soil, expose more foliage, and in dry months, this can reduce survival. Higher survival from machine planting (which typically plants roots in an L-shape) may be, in part, due to not using the "pull up" method. In Canada, planting the root-collar 7.6 cm below the surface increased survival when compared to conventional planting with the root-collar near the soil surface [122].

Figure 17. In some areas in KwaZulu-Natal, container-grown seedlings are planted during the rainy season (summer). Sometimes 1 to 5 L of water is applied in the planting hole before inserting a seedling. "The use of and/or method of application of the water varies widely between regions and commercial forestry companies. Some companies have a 'no water' policy, others always plant with water, while some only apply water when planting conditions are considered poor (hot weather, dry conditions)" [10]. Applying 2 L of water to young, 7-cm tall, container-grown patula pine seedlings increased survival by 12% at one location but decreased survival of 14-cm tall seedlings by 6% at another site. Adding water had no significant effect at locations where the 100-day survival of non-watered pines was >95%. Photo permission given by Ullrich Hechter—Mondi Forests.

6.18. Adding Fertilizer

There are two ways improper fertilization kills seedlings. When weed growth is stimulated, the extra competition reduces soil moisture, and sometimes weeds overtop seedlings. For example, on a site in Tennessee, applying 114 g of N-P-K fertilizer on the soil surface around each seedling increased weed growth, and grasses overtopped seedlings. As a result, the survival of loblolly pine was reduced by 38% [123]. In Texas, a broadcast treatment (140 kg ha^{-1} of diammonium phosphate) reduced survival by 12% at one location and 27% at another [124]. In Alabama and Virginia, a broadcast treatment (280 kg ha^{-1} of diammonium phosphate) did not reduce the survival of loblolly pine [19].

When fertilizer is placed on roots in the planting hole, the salt effect can kill roots. In Louisiana, an application of 0.6 g of fertilizer per seedling killed 66% to 99% of the treated pines [110]. In California, placing 30 g of fertilizer in the hole reduced pine survival by 24% [125]. Likewise, all container-grown loblolly pine seedlings died when a landowner applied an unknown amount of fertilizer directly into the planting hole just before planting. Even when fertilizer (100 g containing 12 g N) was placed in an adjacent hole, 15 cm distant from the seedling, survival was reduced by 12% [126]. To reduce the chance of increasing weed growth, some researchers place a low rate of fertilizer below ground at the time of planting or several months later. For example, placing a slow-release fertilizer pellet (4.2 g N; $0.15 per pellet) 10–13 cm distant from the roots did not reduce pine survival [127,128].

6.19. Adding Fungicide

At some locations, treating the roots of longleaf pine with benomyl and clay just before planting increased survival by 31% [129]. Likewise, applying benomyl in the planting hole increased the survival of patula pine seedlings by 11% in South Africa [130].

6.20. Adding Insecticide

In Sweden, about 1% of seedlings died when treated with an insecticide, while mortality from *Hylobius abietis* exceeded 50% on control plots [131]. In South Africa, *Hylastes angustatus* and white grubs were responsible for insect-related mortality of patula pine (Figure 18). Applying an insecticide to the planting hole does not increase survival when insect populations are low, but it might improve survival when nematode populations are high. Typically, most researchers assume nematodes do not reduce the survival of pine seedlings.

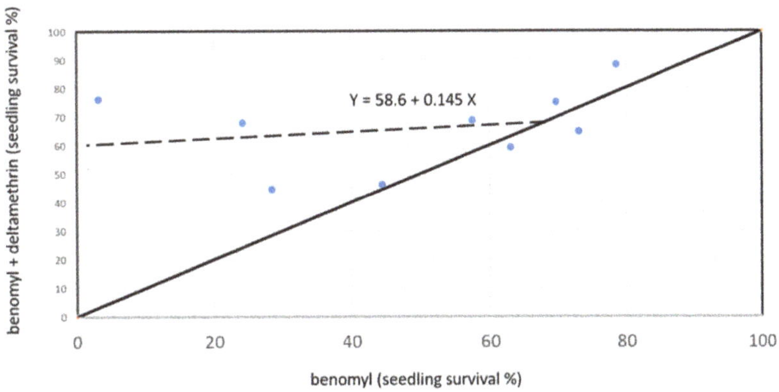

Figure 18. Survival of container-grown patula pine seedlings increased when treated with the insecticide (deltamethrin) at Mpumalanga, South Africa. Data from [130]. The first-year survival was 36%, 49% and 66% for untreated, benomyl only and benomyl + deltamethrin treated seedlings, respectively. White grubs and *Hylastes angustatus* caused most of the insect-related mortality, and *Fusarium circinatum* caused most of the disease-related mortality. Points above the solid line represent sites where the insecticide increased survival. $p > |t| = 0.001$ for intercept and 0.49 for slope; $R^2 = 0.07$.

6.21. Adding Peat Wedge

When rainfall is not optimal, placing a peat wedge at the bottom of a planting hole can increase the survival of bareroot stock planted on shallow soils. Adding one saturated peat wedge increased the survival of 2-0 pine seedlings by 40% or more [122]. Nearly all container seedlings in the SUS are planted with roots already encased in a peat-based plug (cone or rectangle).

6.22. Antitranspirant

Applying an antitranspirant to seedlings before planting will have little effect on survival when soil moisture or rainfall results in more than 85% survival. In Georgia, a 20% increase in survival resulted when a water-emulsifiable organic material (di-1-p-menthene) was applied to pine seedlings in March (Figure 19). This material did not increase survival when seedlings were planted in February (>85% survival) but increased survival by 26% when seedlings were under stress from being planted late in mid-March [132]. Various antitranspirants, are available and some are more effective than others [133].

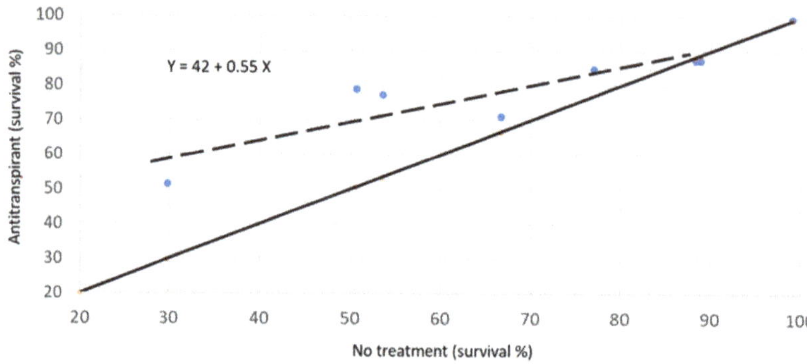

Figure 19. Survival of bareroot pine seedlings after treatment with an antitranspirant (di-1-p-menthene). Data from [132]. Seedlings with >87% survival were planted in February, and the others were planted in March. Points above the solid line represent cases where the antitranspirant increased survival. $p > |t| = 0.0015$ for intercept and 0.0019 for slope; $R^2 = 0.82$.

7. After Planting

7.1. Vegetation Control

In general, competing vegetation reduces the survival of shade-intolerant pines. Therefore, reducing the biomass of weeds can increase survival in areas where moisture is limiting. In some cases, improper use of application equipment and incorrect herbicide calibration has increased seedling mortality. When using a 50% banded herbicide application, some forget and mistakenly apply twice the recommended rate.

The importance of intraspecific competition control of wild natural pine is often overlooked since it can reduce the long-term survival of genetically improved trees. The stocking can be so intense (Appendix A Table A1) as to require a precommercial thinning at age two to three. Without an early thinning, the superior genetics is often diluted as it is nearly impossible to distinguish between wild pines and improved genotypes when trees are five-years-old.

7.1.1. Mowing

When performed correctly, mowing nine times during the first three years after planting can reduce weed biomass without killing planted pine seedlings [134]. However, in some locations, mowing can reduce seedling survival [135].

7.1.2. Herbicides

After planting, about 85% of the area is treated with herbicides in the SUS [8]. Appropriate use of herbicides helps to explain the increase in survival since 1955 (Figure 1). Although errors in application can result in lower survival, the frequency of these mistakes declines when using experienced applicators.

Poor communication can result in herbicide-related seedling mortality. In one case, a contractor was told to treat 10 ha with a 50% band of hexazinone at a rate of 1.1 kg ha^{-1}. The contractor was accustomed to applying a 2-m band of fertilizers (tree rows spaced 4 m apart), and this "fertilizer method" was also used when applying herbicides. Therefore, instead of applying 5.5 kg of hexazinone to 10 gross ha, the contractor purchased 11 kg and applied it in bands to 10 gross ha. This mistake applied twice the amount of herbicide as intended. When herbicides are applied in bands, it is important to clarify how much herbicide to apply to the treated area (instead of the total area). Accurate records are important when herbicide errors kill seedlings.

Seedling mortality may occur due to backpack application errors when herbicide products are mixed in the individual sprayer but not properly agitated to create a uniform tank mixture. Poor agitation may cause some herbicides to settle in the bottom of the sprayer, which applies a potentially toxic rate initially followed by an inadequate rate near the end of the application.

Root-inhibiting herbicides might increase pine mortality when rainfall after planting is below normal. Some organizations apply a tank mix of herbaceous and woody herbicides in October and then plant pine seedlings in November. When the herbaceous herbicide does not inhibit root growth, the reduction in competition can increase survival. However, when rainfall is limited after planting, mortality of newly planted seedlings increases when the root inhibitor reduces the uptake of soil moisture.

7.1.3. Applying a Mulch

Mulch can help reduce loss of soil moisture and, in some years, a mulch will increase survival. In one dry year, adding mulch around pine seedlings increased survival by as much as 26% [136,137]. When rain is near normal, survival in mulched areas increases by 5% [134].

7.1.4. Prescribed Fire

Prescribed burns can reduce fuel loads and can lower the risk of mortality from wildfires. However, if conducted too soon after planting, prescribed fires can reduce the survival of planted pine seedlings [138,139].

7.2. Browse Protection

Browsing can kill pine seedlings when they are pulled out of the planting hole or when the entire shoot is decapitated. Once seedlings become established, however, browsing 50% of the new shoots may not reduce survival. Browsing pines by rabbits can increase mortality [140], but sometime browsing soon after planting will reduce the seedling height and increase survival [141]. When browsing removes the shoot (height after browsing 6 cm), death could occur when weeds overtop pines with suppressed height and root growth. However, at moisture-deficient sites, browsing (height after browsing 16 cm) reduces the rate of transpiration, and this can increase survival [142].

In some areas, deer will browse more on container-grown stock than on bareroot stock [143]. Gopher herbivory of planted seedlings is common in the ponderosa pine range [144].

Cows can kill 10% of pine seedlings by trampling, dislodging and eating pine seedlings [145,146]. At one site that was bedded, cows walked in the furrows and ate all pines that were planted in the furrow but did not injure or dislodge seedlings that were planted on the top of the beds (personal communication John Mexal). Afforestation of pastures following long-term grazing of livestock can result in significant mortality during droughts due to soil compaction, which limits planting dept and root growth.

Carbohydrates in the roots of longleaf pine often attract hogs, which pull up seedlings after they are planted. "Hogs probably have ruined more longleaf plantations than drought, pocket gophers, leaf-cutting ants, and brown spot combined" [110]. In Louisiana, hogs caused the complete destruction of 364 ha of slash pine and recently damaged loblolly pine at two sites in Texas [67].

7.2.1. Tree Tubes

Tree tubes are used to reduce damage from browsing. For eastern white pine, installing plastic tree tubes might increase survival by 11% [147] or decreases it by 27% [148]. The environment in the tube likely explains why mortality can increase when using unvented tubes. The daytime temperature inside the tube is higher than the ambient, and this might explain why survival is low at some locations. Use of tubes can cost more than $3.00 per plant [149] and, therefore, use is limited to governments, universities, wealthy landowners or small areas.

7.2.2. Fences

Deer populations have increased in several areas since 1950. Hunting clubs use fences to keep deer in, and foresters sometimes use fences to keep deer out. In the northeastern region of the USA, fences are commonly used to protect natural regeneration from deer browsing of high-value hardwoods.

7.2.3. Repellents

There are various types of chemical repellents on the market, but most wear out over time. Visual repellents include bud caps, which are fabricated using paper, mesh or cloth [149,150].

7.3. Insect Control

Planting insecticide-treated seedlings adheres to the precautionary principle, and some landowners purchase insecticide-treated seedlings. It might cost $5 to treat 1000 pine seedlings before planting. In contrast, sometimes insecticides are applied after the landowner notices mortality. Insects that kill pine seedlings soon after planting in the SUS include regeneration weevils, grubs and leaf-cutting ants.

7.3.1. Debarking Weevils

Weevils cause pine seedling mortality in Sweden, South Africa, the UK and the SUS [130,131,151,152]. When logging occurs after July in the SUS, planting is usually delayed for 9 months to reduce the risk of mortality from Pales weevil (*Hylobius pales*). When logging occurs in September or later, and seedlings are planted soon after, injury to seedlings can exceed 50% [151,153]. However, even with this practice, seedling mortality due to reproduction weevils can be significant (Figure 20), especially when non-merchantable (generally < 13 cm DBH) pine residuals are present after logging. When non-merchantable pines remain after harvesting, chemical site preparation and burning after July can attract weevils, which can increase the mortality of newly planted seedlings.

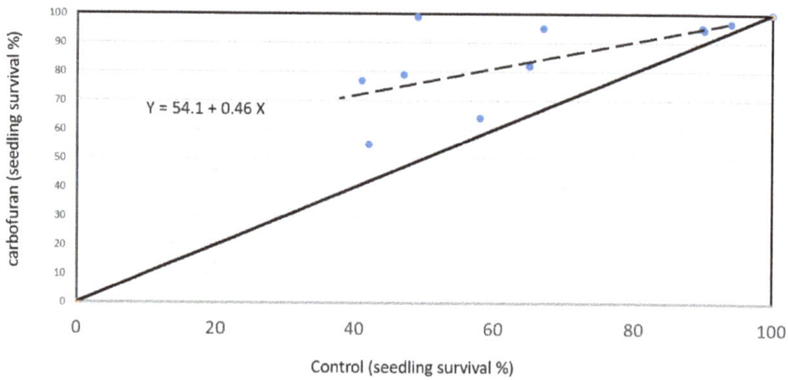

Figure 20. Treating seedlings with an effective soil-applied insecticide at the time of planting increased the survival of bareroot loblolly pine and slash pine. Data from [27,151,154]. Although carbofuran is no longer tested or sold in the USA, a rate of 1.8 g per seedling increased seedling survival. Carbofuran is banned in Canada and the EU, but it currently is used to control weevils in row crops in several countries. Points above the solid line represent sites where carbofuran increased survival. $p > |t| = 0.004$ for intercept and 0.052 for slope; $R^2 = 0.39$.

7.3.2. White Grubs

White grubs is a term that includes several insect species, and when populations are high enough, they can kill newly planted pine seedlings [130,155–157]. Scalping soil in old fields before planting is one method that can improve survival by moving grubs away from planting holes. The chance of mortality may be greater when seedlings are planted on abandoned farmland [27,158], and the reason is partly due to grubs feeding on roots.

7.3.3. Leaf-Cutting Ants

Leaf-cutting ants in the *Atta* and *Acromyrmex* genera have killed pine seedlings in North and South America. In 2016, Texas leafcutter ants (*Atta texana*) were the leading cause of mortality for pines planted on a site in Cherokee County, TX. Mortality ranged from 80% to 99%. "Leafcutter ant damage was observed as early as 1 month after planting and continued through the third year" [67].

7.3.4. Tip-Moths

Insecticides are routinely applied to pine seedlings in order to increase height growth on sites known to contain the Nantucket pine tip moth (*Rhyacionia frustrana* Scudder). On some sites, treatment with an insecticide [159] can increase seedling survival (Figure 21).

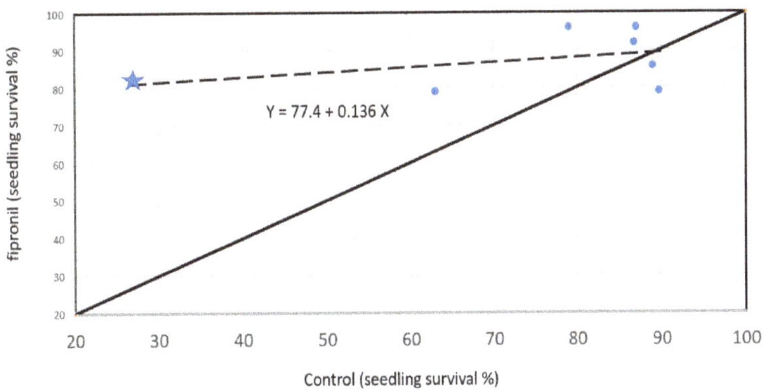

Figure 21. A dilution of fipronil was applied to the root zone within one week of planting bareroot loblolly pine seedlings (March 2008) at seven sites in Virginia [153]. The insecticide increased survival on three sites ($\alpha = 0.10$) and reduced foliar damage to seedlings at six sites. Harvest of pines prior to planting was completed before July 2007 except for one site harvested in October 2007 (blue star). Points above the solid line represent sites where fipronil increased survival. $p > |t| = 0.0007$ for intercept and 0.42 for slope; $R^2 = 0.13$.

7.3.5. Other Insects

Various other insects also feed on pines. Sometimes the redheaded pine sawfly (*Neodiprion lecontei* [Fitch]) can cause 35% defoliation [160], and too much defoliation can kill small seedlings [157].

7.4. Nematodes

Some nematodes are benign, while others can kill pine seedlings in nurseries and in pine stands. Inoculations with the pine wood nematode killed two-year-old pine seedlings in a bareroot nursery [161]. It is possible that when insecticide treatments increased seedling survival on sites without pales weevil, the increase is due to additional root growth due to a reduction in nematodes. At one site in Florida, the endoparasitic pine cystoid nematode (*Meloidodera floridensis* Chitwood, Hannon and Esser) occurred at high levels and treatment with nematicides increased survival by 8 to 10% [162].

7.5. Fungi

At some sites, a lack of ectomycorrhizal fungi can reduce the chance of seedling survival, but healthy pine seedlings have sufficient mycorrhiza when they pass by the nursery gate. At some locations, pathogens such as *Fusarium circinatum* and *Scirrhia acicula* can kill newly planted seedlings [130], and some storage fungi can quickly lower a seedling's root-growth potential.

8. Weather after Planting

Adverse weather can kill pines within a year of transplanting. Dry soil, wet, anaerobic soil, frost heaving, de-acclimation cycles and hail can kill recently planted seedlings. When soil is saturated from rain, strong winds will sometimes topple seedlings during the first five years after planting.

8.1. Rain

Regions with low rainfall typically have few pine nurseries. For example, Alabama and Arizona average 1400 and 320 mm/year, respectively. As a result, Alabama grew 116 million conifer seedlings in 2020 compared to less than a thousand seedlings in Arizona. Average rainfall determines economic returns from planting pines. In dry regions in the West, the US government typically supplies most of the funds to grow and plant pines.

8.1.1. Insufficient Rain

Low seedling survival is expected when limited rain occurs during hot summer months. For example, seedling survival averaged 87% in Mississippi and 25% in Louisiana when rainfall (May to September 1980) averaged 122 and 25 mm/month, respectively (Figure 22).

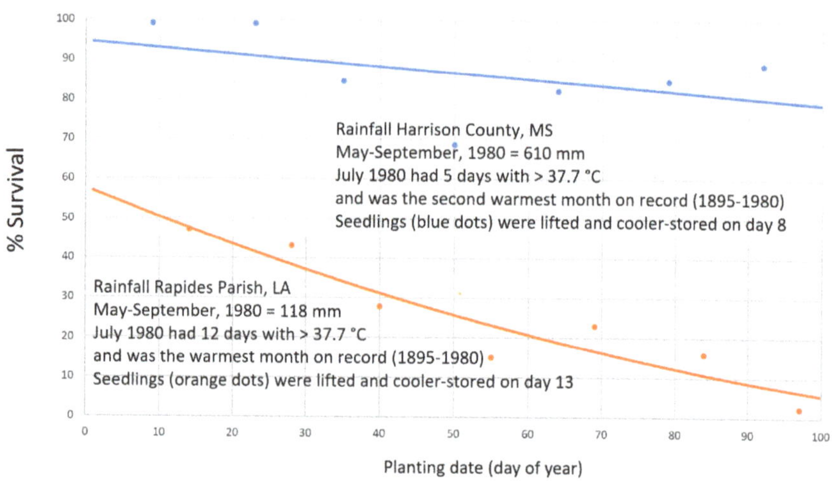

Figure 22. Bareroot longleaf pine seedlings were treated with clay slurry just before hand-planting on sites at the Palustris Experimental Station (Louisiana) and the Harrison Experimental Station (Mississippi). Data from [163] Seedling survival was recorded in October of 1981. For Harrison County, $p > |t| = 0.17$ for intercept and 0.31 for slope; $R^2 = 0.20$. For Rapides Parish, $p > |t| = 0.0001$ for intercept and 0.002 for slope; $R^2 = 0.87$.

In some regions (Figure 23), seedling survival for pine is positively related to the Palmer-Drought-Severity-Index (PDSI). The PDSI uses readily available temperature and precipitation data to estimate relative soil dryness. The PDSI typically has a range of −4 (dry) to +4 (wet), but more extreme values are possible. Current and predicted PDSI values for planting chances are available for use in determining if tree planting should begin or cease (https://www.drought.gov/states/texas) (accessed on 1 December 2022). PDSI at the time of planting is likely a better predictor of pine survival [164] than air temperature recorded at the time seedlings are placed in a planting hole. In East Texas, a three-unit increase in PDSI might increase survival by 10% [164].

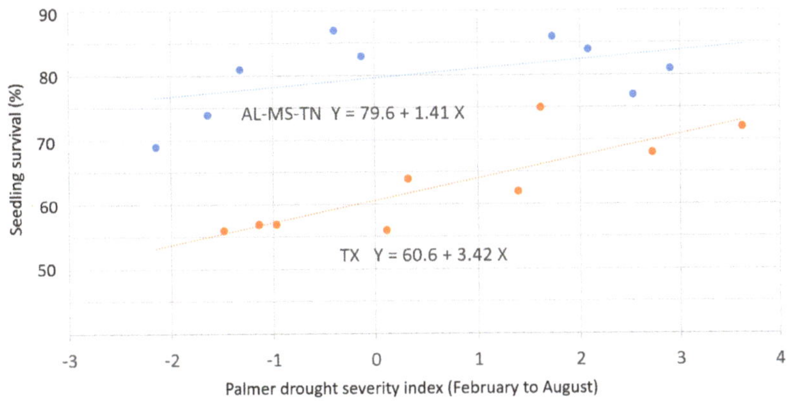

Figure 23. Survival of loblolly pine seedlings is related to the Palmer-Drought-Severity-Index (PDSI) calculated for the period of February to August. Data from [164]. Data from Alabama, Mississippi and Tennessee included company survival data from 690 locations and data from East Texas were obtained from a survey of 3570 sites. All seedlings were planted from 1983 to 1992 (one average per year). In East Texas, the year with the greatest survival (1989) had the highest rainfall (February to August—NOAA climate division 4; 130 mm/month; PDSI = 1.61). The year with the lowest rainfall in East Texas (1988) averaged 68 mm/month (PDSI = −1.49). For East Texas, $p > |t| = 0.0001$ for intercept and 0.0047 for slope; $R^2 = 0.70$. For AL-TN-MS, $p > |t| = 0.0001$ for intercept and 0.21 for slope; $R^2 = 0.21$.

When rainfall is inadequate, shade can slow evapotranspiration and increase survival. Therefore, an interaction exists between soil moisture and benefits from shade. In dry years, shade preserves soil moisture, reduces evapotranspiration and may increase first-year survival. In years with above-average rainfall, shade will reduce photosynthesis and can the reduce survival of newly planted seedlings.

When rainfall is adequate, over 90% of shade-tolerant Douglas-fir seedlings survive transplanting without any shade. In contrast, when rainfall is below normal, shade can slow the loss of soil moisture, and this can increase seedling survival (Figure 24). Since artificial shade is expensive (perhaps $0.40 per screen) and accurately predicting rainfall amounts is difficult, tree planters in western states may select naturally shady planting spots [21,98].

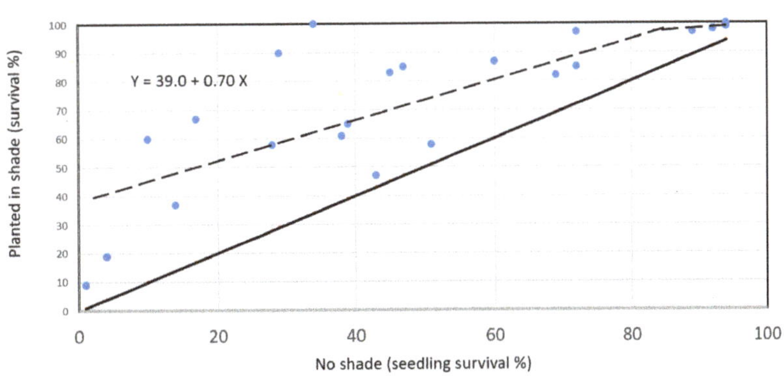

Figure 24. At reforestation sites in Oregon, shade can increase the survival of shade-tolerant Douglas-fir seedlings. Data from [165–168]. The dashed line represents the regression equation ($n = 22$) and the solid line represents no benefit from shade. Points above the solid line represent sites where shade increased survival. $p > |t| = 0.0001$ for intercept and 0.0001 for slope; $R^2 = 0.61$.

In most rainy years, shade-intolerant pines are usually resistant to solar radiation and so artificial shading devices are rarely necessary after planting [169]. East of the Mississippi river, survival of unshaded, bareroot loblolly pines is typically greater than 80%. In years with adequate rainfall, shade can decrease the survival and growth of container-grown longleaf pine [170]. However, in hot and droughty years, partial shade can reduce transpiration and increase seedling survival one year after planting [110,126,143,170–172]. In a Texas study with limited rainfall, shade increased the survival of loblolly pine by 25% in areas with weeds and 30% in areas treated with two herbicide applications [126].

8.1.2. Too Much Rain

Observations over the past four decades indicate that waterlogged conditions during the year of planting lower the survival of pine seedlings, especially soon after planting. Waterlogged soils can develop when frequent rains occur over an extended period. On well-drained soils, above-average rain might not reduce soil oxygen or seedling survival. However, on fine-textured soils, water may accumulate due to low infiltration (Figure 25). For the southeast region, record amounts of rainfall (>430 mm) fell during the last three months of 2009, 2015 and 2019.

Mortality results on some soils when rainfall exceeded 50 mm/week for a period of three weeks or more during the autumn or spring. Anaerobic conditions can occur quickly when warm soils remain saturated for just a few days. Root growth is slowed, and the transpiration of seedlings is reduced.

When anaerobic soil conditions develop, lenticles are produced on the stem of pines near the root-collar or slightly above the soil surface. Although lenticles are not harmful, they can be used to diagnose the cause of low seedling survival (Figure 26). When a regeneration survey is conducted, recording the frequency of lenticles on 100 sampled seedlings may provide a clue as to the cause of unexpected mortality.

In some places, inundation causes complete mortality of loblolly pine [173,174]. In contrast, a mortality of 90% occurred when potted seedlings were flooded for 5 days with salt water [175].

8.2. Temperature

Most SUS foresters plant pines in clear-cut areas or on abandoned agricultural fields where soil temperatures fluctuate. Both low and high temperatures can kill newly planted pine seedlings.

8.2.1. Temperatures above 0 °C

Once pine seedlings leave the nursery, they should be stored in cool environments. Some landowners in Virginia might store seedling bundles in unheated buildings for several weeks in January before planting. Alternatively, seedlings are packed into KP bags or in wax-impregnated cardboard boxes. When stored in an unheated warehouse in Mississippi, the temperature in bags in April can exceed 29 °C (Figure 27). When small pine seedlings (11.6 cm tall) are stored in a cardboard box at 15 °C in Finland, mortality increases after 2 weeks of storage [96].

Figure 25. Scalping a fine-textured soil resulted in standing water which killed longleaf pine seedlings. Data from [68]. Seedlings were machine planted on 29 November 2007, and in April 2008, 243 mm of rain occurred over a 15-day period. A year later, March rain occurred on the 4th (41.1 mm), 7th (81.5 mm) and 12th (4.3 mm) and the photo was taken 13 March 2008. Scalping increased the mortality of bareroot stock (73% and 91% survival on scalped and non-scalped areas, respectively).

Figure 26. Example of two lenticles on the root of a container-grown longleaf pine seedling.

Planting conditions may be marginal for pine seedlings when the air temperature during outplanting is 24 to 29 °C, and relative humidity is 30% to 50% [38,121]. In one trial, mortality increased by more than 18% when the air temperature at planting exceeded 24 °C (Figure 28). Temperatures in bags can be 10 °C higher than air temperatures when bags are stored for 4 to 6 h outside in the direct sun. To reduce the chance of heating seedlings in bags or drying roots before planting, a few foresters suggest tree planting should be stopped when afternoon temperatures reach 30 °C. Many seedlings likely die due to heat buildup when bags are not kept cool.

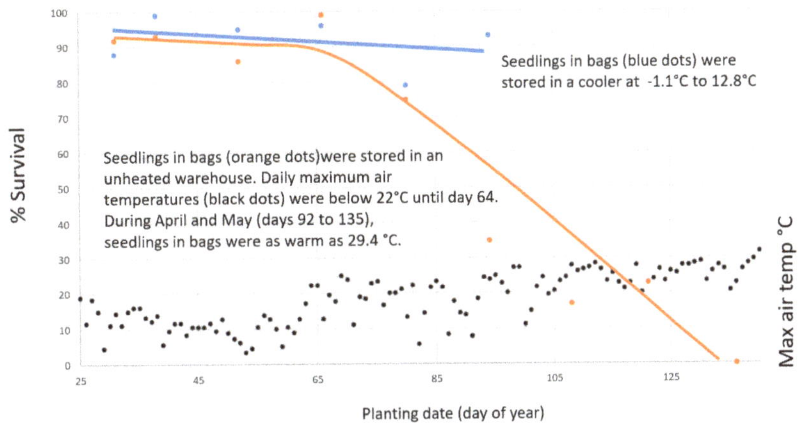

Figure 27. Survival of loblolly pine seedlings is related to the storage method. Data from [176]. Bareroot seedlings lifted on 9 January and cooler-stored for 12 weeks in kraft-polyethylene bags (containing 454 g of sphagnum moss) exhibited good survival when planted at Oxford, Mississippi, on 3 April 1964. The temperature of seedlings stored in an unheated building exceeded 25 °C, and they did not survive well when stored for more than 9 weeks [176]. For the survival of seedlings stored in bags, $p > |t| = 0.0001$ for intercept and 0.0003 for slope; $R^2 = 0.87$.

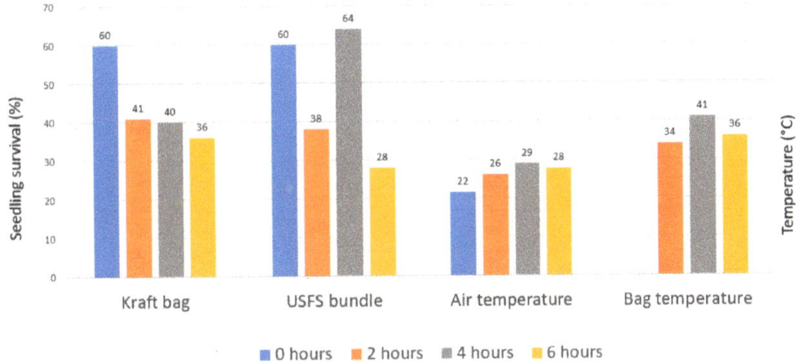

Figure 28. Survival of loblolly pine seedlings lifted on 13 March, stored in a cooler for 47 days at 3.3 °C in either kraft-polyethylene bags or bundles, and planted on 29 April 1970. Data from [177]. Each package contained 1000 bareroot seedlings, and seedlings were extracted and planted from packages at 9:30 a.m. (zero exposure), 11:30 a.m. (2 h), 1:30 p.m. (4 h) and 3:30 p.m. (6 h of exposure while in the bag or bundle). The interaction between the packing method and the exposure time was not significant ($\alpha = 0.05$). The least significant difference ($\alpha = 0.05$) for a one-tail test is −17% and ±32% for a two-tailed test.

The recommendation to stop planting at 30 °C relates to drying exposed roots prior to planting and not to lethal air temperatures after planting stock into moist soil. Five months after planting in February [136], loblolly pine seedlings tolerated a brief air temperature of 67 °C (2.5 cm above a mulch). For some northern pines in a laboratory, mortality occurred when air temperatures exceeded 48 °C for 5 h [178].

Most tree planters in the SUS prefer planting seedlings before May, but sometimes pines are planted during the summer [104,144,179–181]. Planting seedlings in hot and dry soil in July in drought conditions D2 to D4 can kill container-grown pine seedlings [182,183]. When seedlings are handled with care, pines can be planted in moist soil during the summer. For example, at an airport in Pensacola, FL, the month of July 1966 had 80 mm of rain, and each day temperatures in the afternoon exceeded 29 °C (Figure 29). Although the temperature reached 36 °C on July 8th, the survival of bareroot slash pines planted in that month averaged 80%. During 1966, 1.3 million bareroot slash pine seedlings (seed sown in a nursery in mid-November 1965) were planted in the summer in Florida and South Georgia [181].

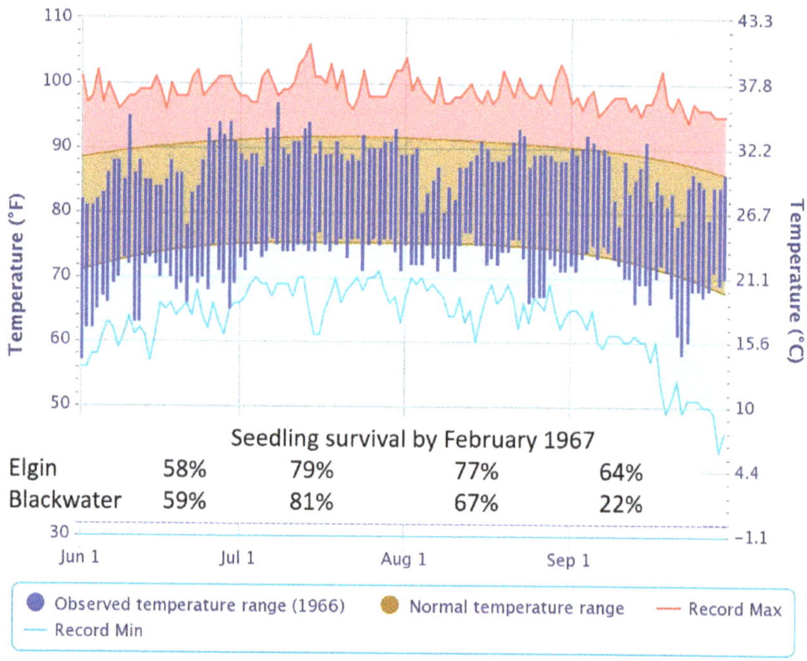

Figure 29. Seedling survivals for bareroot slash pine in Florida vary by planting location and month. Data from [181]. Survival of June planted stock was 58% when planted at the Elgin Air Force Base and 59% when planted at the Blackwater State Forest in Florida. Air temperatures for the Pensacola regional airport for the year 1966 were provided courtesy of the Applied Climate Information System. The highest temperature of the year was 36.1 °C on 8 July 1966. Data for maximum and minimum temperatures began in 1948.

Many pine species will tolerate transitory air temperatures of 45 °C during summer months, and some can tolerate 56 °C for a brief period [184,185]. During a prescribed burn, longleaf pine seedlings can survive brief air temperatures of 400 °C [50]. When moisture is adequate, pine roots can tolerate 45 °C [178].

8.2.2. Temperatures below 0 °C

When soil moisture is adequate, recently planted loblolly or longleaf pines are more likely killed by a sudden December freeze than by air temperatures during planting of 30 to 35 °C. Temperatures below −8 °C have killed seedlings in the SUS from October to March. One week of freezer storage at −6.7 °C can reduce seedling survival by 44% [186], and two weeks of freezer storage at −3 °C killed all the clones of loblolly pine [187]. Likewise, in a growth-room trial, more Douglas-fir seedlings were killed after 2 h of −8 °C temperature than by 24 h of growth-chamber storage at 32 °C [188].

We classify freeze events into three types; preacclimation, acclimation and de-acclimation [189]. In the United States, a preacclimation freeze might occur in October or November when seedlings have not received a sufficient level of freeze tolerance. Acclimated freezes occur after the winter solstice and before warm weather in February (Appendix A Table A1). A de-acclimation freeze occurs when enough warm nighttime temperatures decrease the level of freeze tolerance. The frequency of de-acclimation freezes has likely increased since 1950. In a region stretching from Mississippi to North Carolina, the last hard freeze tends to occur later now when compared to the first part of the 20th century [190].

Unimproved loblolly pine seedlings with sufficient chilling usually tolerate −8 °C freeze events if they have not been de-acclimated due to warm nighttime temperatures (Figure 30).

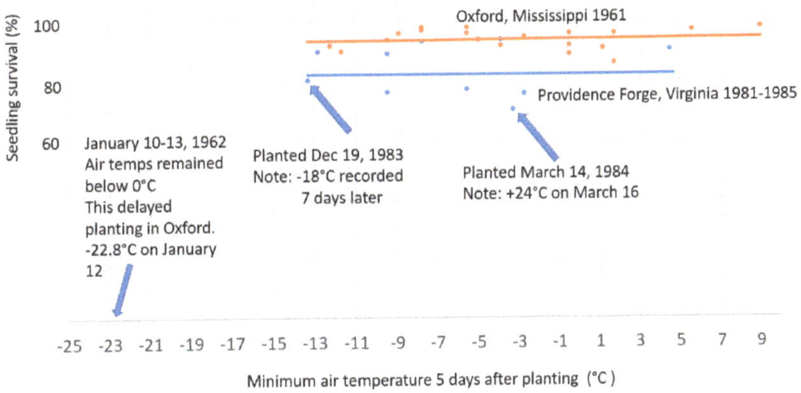

Figure 30. Acclimation freezes above −14 °C had minimal effect on the survival of bareroot loblolly pine seedlings planted from 16 December 1981 to 3 March 1985 (blue dots). Virginia data are from [191], and Mississippi data (orange dots) are from [192]. The cambial injury was not examined, and shoot growth may have been reduced. The X-axis represents the minimum air temperatures during the first 120 h after planting. Air temperatures above −14 °C and below freezing apparently did not reduce survival since the seedlings had sufficient chilling. In contrast, when the soil remains frozen for several days, seedling mortality can increase due to loss of plant water [110]. Blue and orange dots represent first- and third-year survival, respectively. Temperature data from Williamsburg, Virginia and University, Mississippi. Seedlings in Mississippi had little storage, while those planted in March in Virginia were stored in a cooler for 2 to 3 months. The air temperature reached −22.8 °C on the morning of 12 January 1962.

The preferred time to plant pine seedlings in Nebraska is in the spring when the risk of freezing weather is low. In contrast, in Alabama, the preferred time to plant container-grown longleaf pine is mid-September to late November, when −8 °C freezes are rare [100]. In contrast, the survival of longleaf seedling planted in December may average less than 70% due to freezing roots prior to shipping [193] or due to hard freezes just after outplanting (Table 1).

When called to investigate the reason why genetically improved pine seedlings died, we examine the cambium on roots and stems and dead needles and buds. When a de-acclimation freeze injures the cambium, the xylem continues to transport water at a slow rate, and pine needles remain green until warm temperatures arrive. New root growth is reduced, and when evaporation exceeds transpiration, seedlings die. With a de-acclimation freeze event, the landowner might not be convinced a −5 °C freeze (reported on the radio) was the problem. Some landowners were looking to blame planting quality, seedling handling or a nursery cultural practice. In some cases, landowners believed mortality in a frost pocket was not from a −7 °C freeze since they historically achieved good survival with acclimated seedlings that tolerated temperatures below −8 °C (Figure 12).

In Finland, sometimes half of the 155 million container-grown seedlings remain outdoors during winter [194], and a similar amount remains outdoors in winter in the SUS. However, container-grown pine seedlings stored outside can be injured due to freezes [193,195,196]. In Virginia, temperatures also can reach −22 °C in winter and, therefore, seedlings are extracted from containers, placed in boxes in October, and stored in coolers until sold. Some customers purchase seedlings and then plant just prior to freeze events. Others purchase seedlings and then plant in April after seedlings have been stored for 10 weeks in the cooler. In 2019, the survival of seedlings planted just before a freeze ranged from 43% to 67%, while those planted in April had 94% survival [112].

Sometimes landowners store seedlings in unheated buildings before planting. When temperatures remain below −1 °C for a three-day period, seedlings in bags or bundles can freeze solid. Seedlings may survive if allowed to completely thaw before moving (Figure 31).

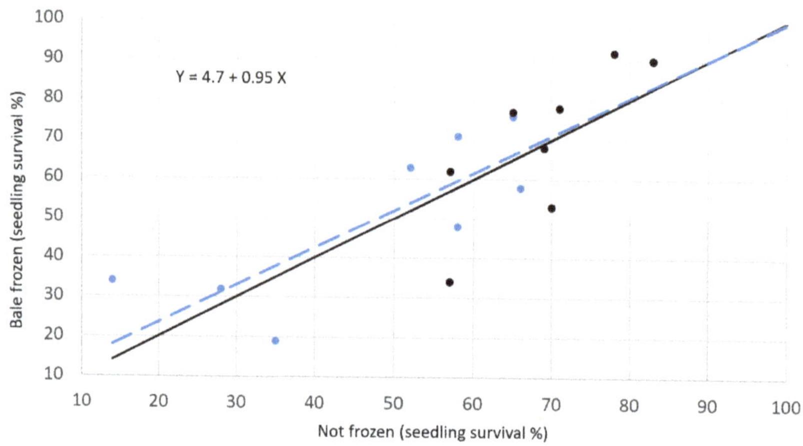

Figure 31. Seedling survivals for genetically unimproved bareroot loblolly pine packed into bundles (2000 seedlings per bail) in January of 1970 at the New Kent nursery in Virginia. Data from [197]. Bales were naturally frozen (−11 to −12 °C) outdoors on January 19 and 20 (blue dots), or February 1 to 3 (black dots). Frozen seedlings were thawed in an unheated building (beginning 23 January or 8 February) and outplanted 34 to 48 days after lifting. When planted in February, survival averaged 69%, and freezing made no difference to survival. When planted in January, non-frozen and frozen seedlings averaged 47% and 50%, respectively (LSD_{05} = 10.8%; two-tailed test). Points >8% below the solid line represent cases where freezing reduced survival. $p > |t|$ = 0.68 for intercept and 0.002 for slope; R^2 = 0.65.

Even after a −13 °C freeze, 80% of sensitive pine genotypes died after roots in the containers froze (Figure 32). For this reason, planting loblolly pine genotypes too far north of their seed origin (more than −5.5 °C in average winter temperature minimum) is not recommended [198].

Some people believe predictions about weather events in 2050 are equivalent to facts [199–201]. In fact, some advocate planting genotypes more than 100 km north of their natural range [202,203]. In fact, more than 29,000 ha of loblolly pine seedlings have been planted north of their natural range. In one trial, loblolly pine planted 700 km north of the origin had 9% survival 37 years after planting [46]. At that age, survival near 50% would be expected for a non-thinned stand planted in Mississippi.

Figure 32. After the wind destroyed a protective plastic cover, rooted cuttings of loblolly pine were killed by a 30 January 2014 freeze (−13 °C). Although average survival was 30%, one northern genotype had 80% survival [204]. Reproduced with permission from Steve McKeand [204].

8.2.3. Winter Desiccation

Winter desiccation can occur when the soil remains frozen for several days and pine seedlings are not protected by a cover of snow. When transpiration is reduced due to frozen soil, a combination of wind and sun can desiccate seedlings [205]. This type of injury is rare in the SUS, but four days of below-freezing weather (16–19 January 1977) desiccated exposed seedlings at a nursery in Virginia [206].

8.2.4. Frost Heaving

Frost heaving in plantations can kill bareroot and container-grown pine seedlings [207,208]. One way to reduce mortality caused by frost heaving is to make a deeper hole and place the root collar 10 to 15 cm below ground and cover the roots with more soil [122,209]. Planting the root-collar of pine only 5 cm below the ground will likely have a minimal effect on frost heave [210]. Even so, in one trial, planting small container seedlings (11 cm taproot) 1.5 cm deeper increased survival by 20% [211] and frost heaving (5 to 12%) did not occur when the root-collar of pine seedlings was planted 11 cm below the soil surface [212]. Frost heaving may be less of a problem with bareroot seedlings with 18-cm taproots [110] because they require deeper holes than seedlings with 11-cm taproots.

8.3. Wind

Naturally regenerated pines typically withstand high winds due to a strong taproot. In contrast, high winds sometimes topple saplings when fast-growing pines are growing in rain-soaked soils [213]. Typically, mortality is near zero when pine lean is only 20° from vertical. However, when pines are prostrate and lying on the ground, mortality increases. For example, hurricane Michael (October 2018) laid 90% of container-grown pines horizontally on the ground [214]. At one site in Brunswick County, North Carolina, the mortality rate exceeded 20%.

8.4. Hail

Hail can damage young pines, and sometimes the damage is minor, and mortality is not increased. When hail is the size of eggs, young seedlings can be killed, and those not killed can attract insects [215]. Sometimes hail might account for 2% or more of early mortality [216,217]. In Australia, the area affected by hailstorms can amount to 300 to 400 ha year^{-1} or about 0.06% of the total plantation area [218].

9. Replanting

When initial mortality is greater than 10% or 20%, some pulpwood companies in Africa will plant more seedlings in the blank spots (a practice known as blanking, beating up or interplanting [219]). Often the second planting is conducted within three months of the first planting [10]. This is because blanking one year later seldom adds to the volume of production at harvest, and the added cost will likely not be recovered [220]. Some landowners would rather focus on maximizing initial seedling survival instead of investing additional time and expense on blanking. For loblolly pine, blanking typically has a benefit/cost ratio less than 1. For example, when 50% mortality occurs two weeks after planting, this might result in 741 live seedlings ha^{-1}. In this case, blanking might increase stocking to 1482 ha^{-1} (at a blanking cost of $150 ha^{-1} at week 2) without increasing wood volume at age 24 years [221]. In other words, the cost of blanking would lower the land expectation value by $150 with no gain in sawtimber production. Typically, most private landowners in the SUS do not blank since sawlog production is near optimal when seedlings are planted at 741 ha^{-1} [222,223].

10. Cost of Improving Survival

Many landowners are willing to spend $100 ha^{-1} to increase seedling growth with either fertilizers or herbicides, but some are reluctant to spend additional money to increase the probability of seedling survival. This is understandable when treatments like mulch or organic amendment costs $2000 ha^{-1} (Appendix A Table A2), but would they be willing to pay $75 ha^{-1} more to plant seedlings deeper with a machine? Machine planting might increase survival by 11% (Figure 10), which might increase initial stocking by 110 trees ha^{-1} (at 1000 ha^{-1}). When it costs $ 1 per plant (75 cents per cutting plus 25 cents for planting), then the benefit/cost ratio for increasing stocking by 110 cuttings ha^{-1} would be 1.46 (i.e., $110/$75). However, when it costs $0.20 to plant a seedling (8 cents plus 12 cents to plant), the benefit/cost ratio is less than 0.3 ($22/$75).

Perhaps 6% of pine-plantation landowners in the SUS are willing to pay $345 ha^{-1} to machine plant ($195) container-grown loblolly pine seedlings ($150). In contrast, about 85% are willing to spend $200 ha^{-1} to hand-plant ($125) bareroot stock ($75).

If a low-cost treatment (Appendix A Table A2) could consistently increase survival by a small amount, would landowners plant fewer trees to offset the added cost? A hypothetical comparison is presented in Table 2, where a relatively cheap treatment increases survival by 4% at site A but has no effect on survival at site B. In this example, the extra $10 ha^{-1} cost is offset by planting 50 fewer seedlings ha^{-1}. The differences in survival and long-term economics are so small that sampling would likely not be able to detect any significant difference, even when using a one-tailed *t*-test.

Table 2. Depending on site, investing an extra $10 ha^{-1} to improve seedling survival of loblolly pine might affect stocking three years after planting. Alternative A increased survival by 4%, while Alternative B did not increase survival. Land expectation value (LEV @6%) estimated from [223].

Regeneration Method	Seedlings ha^{-1}	Treatment Cost ha^{-1}	Seedling Cost ha^{-1}	Hand Planting Cost ha^{-1}	Total Cost ha^{-1}	Third-Year Survival	Third-Year Stocking ha^{-1}	LEV ha^{-1}
No treatment	1300	0	$104	$156	$260	88%	1140	$360
Treatment on site A	1250	$10	$100	$150	$260	92%	1150	$359
Treatment on site B	1250	$10	$100	$150	$260	88%	1100	$364

Table 2 suggests that increasing survival by 4% can reduce the land expectation value by 1%. This small difference would not be noticed when stands A and B are harvested at age 25 years. When too many seedlings are planted, random mortality can increase stand value. This is because, for unthinned plantations, more surviving trees mean more pulpwood, while fewer trees mean more sawlogs. On sites where mortality kills every other planted seedling, stand value can increase by 33% (Figure 33). Many incorrectly assume that increasing survival of planted pines will increase stand value at harvest.

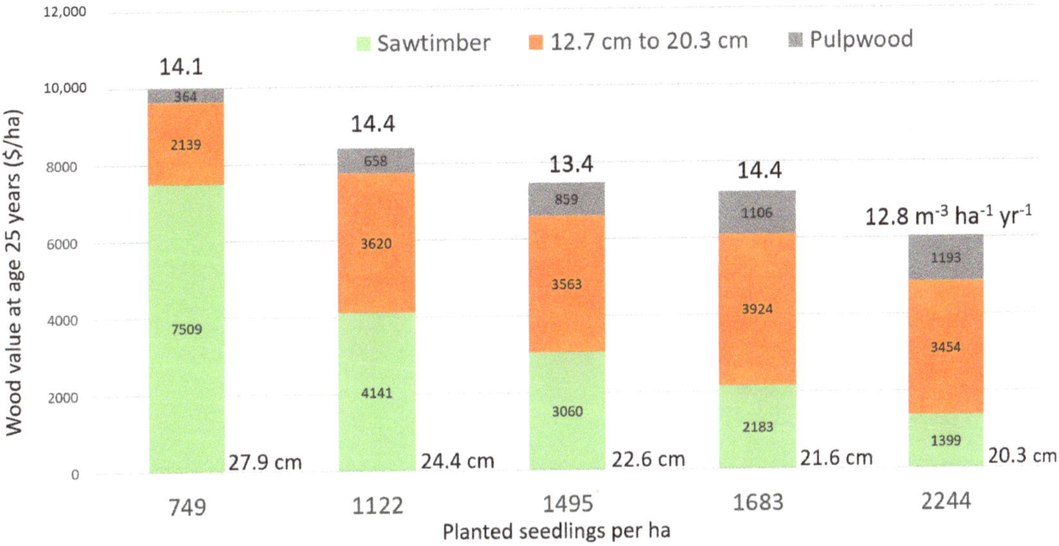

Figure 33. A 25-year old, non-thinned loblolly pine stand (749 planted seedlings ha^{-1}) has 145% more sawlogs than a stand planted with twice as many seedlings. Data from [222]. Too many surviving seedlings can increase wood production while reducing economic returns. In this example, returns per planted seedling varied from $13.30 to $2.70 ($11 m^{-3} for pulpwood and $35 m^{-3} for sawtimber). The mean annual increment is listed at the top of each bar, and the average DBH is listed at the right of each bar. Half of the seedlings planted at 2244 ha^{-1} died before harvest, while mortality was 22% when seedlings were planted at 749 ha^{-1}.

11. Recommendations by Researchers

A main goal of regeneration research is to discover treatments that will provide a real increase in seedling survival. Too often, however, researchers use the precautionary principle and do not recommend effective treatments that, on average, increase survival by 5%. Researchers are trained to avoid making a Type I statistical error (i.e., claiming a treatment works when, in reality, it has no effect on survival). Unfortunately, most of us were not trained in how to install a seedling survival test with high statistical power (i.e., LSD $\leq$ 7% survival). For example, a one-tailed test (α = 0.1) is more powerful than a two-tailed test (α = 0.05), but typically survival means are tested using a low-power, two-tailed test. As a result, many survival trials produce Type II statistical errors (Table 3). When tree planting trials have low statistical power, researchers may consider using more replications [224]. Other methods to increase power include: (1) the use of a one-tailed test [212], (2) the use of a more powerful contrast test (instead of an overall treatment F-test) and (3) conducting survival trials under a roof to reduce soil moisture and increase the onset of mortality [101].

Table 3. When trials have low statistical power [224], a real 10% increase in survival is often declared not significant due to the experimental design and/or method of statistical analysis. Increases listed below were not statistically significant (α = 0.05; two-tailed test). LSD = least significant difference.

Treatment	Survival Increase	LSD α = 0.05	Reference
	%	%	
Subsoiling	42	?	[2]
Ripping	32	?	[23]
No high burn	30	?	[94]
Nursery stock	22	24	[225]
Fertilizer	21	22	[226]
Chilling	20	?	[195]
Root gel	19	21	[87]
Planting date	18	19	[227]
Compost	18	?	[228]
Herbicide	15	17	[229]
3-1 plow	14	?	[230]
Fungicide	14	?	[25]
Root prune	13	?	[53]
Planting date	12	16	[231]
Root gel	11	18	[61]
Herbicide	11	22	[124]
Clipping	11	13	[95]

Examples of Type II errors in reforestation studies can be found in the literature. In a Mississippi study, an alpha level of 0.01 was used to test for significant soil and weather correlations [192]. After analyzing data, the author wrote, "Data in both years suggested higher mortality for seedlings that encountered freezing weather soon after planting, but again trends did not prove: significant." Due to a small alpha value, the power of the statistical test was low, and the following quote from the paper is an example of a Type II statistical error: "Freezing weather immediately after planting is not lethal."

Seedlings planted on 5 February 1962 experienced a −7 °C freeze on February 7 and survival was 68%. In comparison, survival averaged 78% for six other planting dates in February [192]. The claim that no seedlings were killed by the −7 °C freeze was based on (1) not using a one-tailed *t*-test, (2) using a 0.01 alpha value and (3) not knowing the true reason why each seedling died.

12. Discussion

Soon after planting, some landowners want to know why pine mortality was unexpectedly high. Sometimes opinions are provided at no cost without an onsite inspection. When a landowner invites inspectors to quickly examine healthy, dead and dying seedlings, important clues can be documented before they disappear. At several sites, we examined cambium tissue and identified freeze injury as the primary cause of mortality. Weather records can often pinpoint the day of freeze injury. At other sites, lenticels indicated seedlings were exposed to flooded conditions. At several sites, we found piles of roots that indicate new root growth was reduced by trimming roots [69]. At many sites, there were several dead seedlings where the cause of mortality was hard to identify. In Finland, about 24% of mortality is due to unknown factors [232].

Although speculative, we suspect nematodes as a major factor at some old-field sites. Typically, dead and dying roots are not sampled for nematodes, so mortality may be categorized as "unknown." With well-designed studies, future researchers might determine how much of the unknown mortality is due to the feeding on roots by nematodes.

Good seedling survival depends on using practices that are known to not increase mortality. Establishment failure is usually caused by a lack of planning, a lack of adequate supervision, and by not applying simple, inexpensive techniques that are known to increase survival [233].

Supplementary Materials: Information on Standardized Precipitation Index (SPI). https://climatedataguide.ucar.edu/climate-data/standardized-precipitation-index-spi (accessed on 17 December 2022). An example of ArborGen's daily seedling planting log is available online at https://tinyurl.com/2cewmc2c (accessed on 17 December 2022).

Author Contributions: Conceptualization, D.B.S.; data curation, D.B.S. and A.L.; writing—original draft preparation, D.B.S.; writing—review and editing, D.B.S., T.E.S. and A.L.; graphics, D.B.S.; statistical analyses, D.B.S. All authors have read and agreed to the published version of the manuscript.

Funding: This review received no external funding.

Institutional Review Board Statement: Not applicable.

Informed Consent Statement: Not applicable.

Data Availability Statement: Not applicable.

Acknowledgments: This manuscript was developed with input from Steve McKeand. Rainfall data were supplied by the National Oceanic and Atmospheric Administration.

Conflicts of Interest: The authors declare no conflict of interest.

Appendix A

Table A1. Freeze events, type of freeze and type of injury reported on conifers in the eastern United States.

Year	Month	Day	°C	Type of Freeze	Type of Injury	Location
1965	October	30	−8	Preacclimation	Not available	South Carolina
1991	November	5	−7	Preacclimation	Needle burn	Alabama
1950	November	25	−22	Preacclimation	Root cambium	Illinois
1970	November	25	−9	Preacclimation	Needle burn	Georgia
2006	December	9	−10	Preacclimation	Root cambium	Alabama
1962	December	12	−27	Acclimation	Needle burn	Tennessee
1955	December	17	−9	Preacclimation	Needle burn	South Carolina
2022	December	21		———— winter solstice ————		
1989	December	23	−18	Acclimation	Needle burn	Alabama
1983	December	25	−12	Deacclimation	Root cambium	Florida
2004	January	7	−8	Deacclimation	Root cambium	Georgia
2018	January	7	−25	Deacclimation	Root cambium	Virginia
2018	January	7	−18	Deacclimation	Root cambium	North Carolina
1977	January	11	−26	Acclimation	Needle burn	Kentucky
1962	January	12	−23	Acclimation	Root cambium	Mississippi
1994	January	19	−14	Acclimation	Root cambium	Alabama—Mississippi
1957	January	19	−7	Deacclimation	Needle burn	Florida

Table A1. *Cont.*

Year	Month	Day	°C	Type of Freeze	Type of Injury	Location
1996	January	19	−9	Acclimation	Root cambium	Alabama
1985	January	21	−18	Acclimation	Needle burn	Kentucky—Tennessee
1985	January	21	−37	Acclimation	Needle burn	North Carolina
2018	January	21	−12	Acclimation	Root cambium	Virginia
2014	January	30	−13	Acclimation	Root cambium	North Carolina
1996	February	9	−15	Deacclimation	Root cambium	Alabama
1932	March	10	−7	Deacclimation	Shoot cambium	Mississippi
2022	March	20		equinox		
1938	April	7	−4	Deacclimation	Branch cambium	Texas
2007	April	8	−4	Deacclimation	No injury	Tennessee
2002	May	21	−4	Deacclimation	Needle burn	North Carolina
1908	June	3	−4	Deacclimation	No injury	New York

Table A2. Estimated cost of treatments that increase survival at 1000 seedlings ha^{-1}.

Treatment	Cost	Without Treatment	With Treatment	Increase in Survival	Reference
	\$ ha^{-1}	%	%	%	
Correct root gel	0.03	51	81	+30	[234]
Correct root gel—0.33% W/W	0.15	75	88	+13	[87]
Correct clay root dip	0.25	75	89	+14	[57]
No root pruning	3	80	76	+4	[53]
Soaking roots before planting	10	81	88	+7	[83]
Fungicide root dip	10	73	85	+12	[114]
Site preparation burn	75	86	94	+8	[11]
Machine planting—hand plant	75	75	86	+11	[69]
Planted half stem deep—sandhills	75	80	90	+10	[235]
Refrigerator storage—21 days	126	53	77	+24	[37]
Herbicide—herbaceous	130	73	92	+19	[81]
Imidacloprid in planting hole	130	74	86	+12	[153]
Slow careful planting	160	73	83	+10	[3]
Site preparation herbicide	220	89	97	+8	[236]
Carbofuran in planting hole	230	47	85	+38	[50]
Carbofuran treated roots	230	90	95	+5	[154]
Single bedding	253	74	90	+16	[18]
Double bedding	323	75	80	+5	[237]
Subsoiling	370	81	88	+7	[21]
Subsoiling	370	73	90	+17	[2]
3-in-1 combination plow	460	82	96	+14	[230]
Somatic embryogenic stock	500	77	82	+5	[238]
Mulch	655	37	56	+19	[137]
Sewage sludge	2000	84	94	+10	[239]

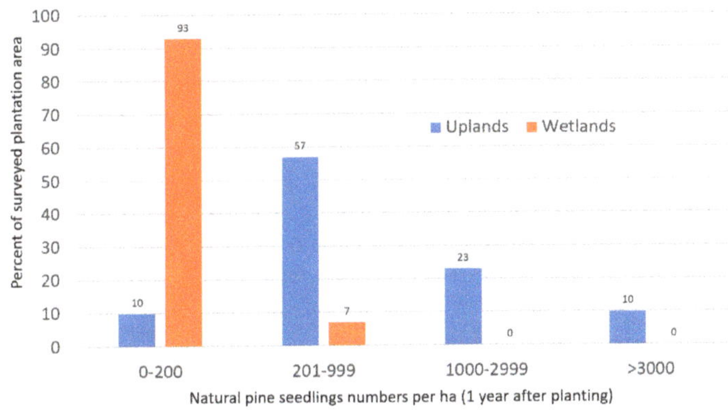

Figure A1. A regeneration survey (>4700 ha) indicates 90% of planted upland areas had more than 200 naturally seeded pines ha^{-1}. As a result, precommercial thinnings are used so managers can harvest a high percentage of genetically improved stock at the end of the rotation. In 2019, over 24,000 ha of stands were precommercially thinned [8].

References

1. South, D.B.; Rakestraw, J.L.; Lowerts, G.A. Early gains from planting large-diameter seedlings and intensive management are additive for loblolly pine. *New For.* **2001**, *22*, 97–110. [CrossRef]
2. Carlson, C.A.; Fox, T.R.; Colbert, S.R.; Kelting, D.L.; Allen, H.L.; Albaugh, T.J. Growth and survival of *Pinus taeda* in response to surface and subsurface tillage in the southeastern United States. *For. Ecol. Manag.* **2006**, *234*, 209–217. [CrossRef]
3. Shriver, B.D.; Borders, B.E.; Page, H.H.; Raper, S.M. Effect of some seedling morphology and planting quality variables on seedling survival in the Georgia Piedmont. *South. J. Appl. For.* **1990**, *14*, 109–114. [CrossRef]
4. Jones, L.; Thacker, S. Survival. In *A Guide to Loblolly and Slash Pine Plantation Management in Southeastern USA*; Wahlenberg, W.G., Ed.; Report 14; Georgia Forest Research Council: Macon, GA, USA, 1968; pp. 68–73.
5. Williston, H.L. The question of adequate stocking _ pine planting programs must allow for seedling survival and plantation losses. *Tree Plant. Notes* **1972**, *23*, 9–10.
6. Schultz, R.P. Intensive Culture of Southern Pines: Maximum Yields on Short Rotations. *Iowa State J. Res.* **1975**, *49*, 325–337. Available online: https://core.ac.uk/download/pdf/224978254.pdf (accessed on 1 December 2022).
7. Weaver, G.H.; Izlar, B.; Zydias, K.; Broerman, F.S. *Southwide Pine Plantation Survival and Nursery Practices*; Publication 82-A-10; American Pulpwood Association: Jackson, MS, USA, 1982; pp. 1–43.
8. Baker, S. *2020 Silviculture Survey: Forest Management Practices of Private Timberland Investors and Managers in the U.S. South*; Forisk Consulting: Atlanta, GA, USA, 2020; pp. 1–15.
9. Schultz, R.P. Loblolly Pine: The Ecology and Culture of Loblolly Pine (*Pinus taeda* L.). In *Agriculture Handbook 713*; U.S. Department of Agriculture, Forest Service: Washington, DC, USA, 1997; p. 493. Available online: https://www.fs.usda.gov/research/treesearch/734 (accessed on 1 December 2022).
10. Rolando, C.A.; Little, K.M. Results from four *Pinus patula* water planting trials in the summer rainfall region of South Africa. *South. Hemisph. For. J.* **2007**, *69*, 9–17. [CrossRef]
11. Carter, M.C.; Dean, T.J.; Wang, Z.; Newbold, R.A. Impacts of harvesting and postharvest treatments on soil bulk density, soil strength, and early growth of *Pinus taeda* in the Gulf Coastal Plain: A long-term soil productivity affiliated study. *Can. J. For. Res.* **2006**, *36*, 601–614. [CrossRef]
12. Schultz, R.P. *Environmental Change after Site Preparation and Slash Pine Planting on a Flatwoods Site*; Research Paper SE-156; Forest Service, United States Department of Agriculture, Southeastern Forest Experiment Station: Asheville, NC, USA, 1976. Available online: https://catalog.hathitrust.org/Record/007422259 (accessed on 1 December 2022).
13. Ryan, M.G.; Covington, W.W. *Effect of a Prescribed Burn in Ponderosa Pine on Inorganic Nitrogen Concentrations of Mineral Soil*; Research Note 464; United States Department of Agriculture Forest Service, Rocky Mountain Forest and Range Experiment Station: Fort Collins, CO, USA, 1986; 5p. Available online: https://books.google.com/books/download/Effect_of_a_Prescribed_Burn_in_Ponderosa.pdf?id=lP5w3DzsWysC&output=pdf (accessed on 1 December 2022).
14. Melvin, M.A. *National Prescribed Fire Use Survey Report*; National Association of State Foresters and the Coalition of Prescribed Fire Councils: Washington, DC, USA, 2020; Available online: https://www.stateforesters.org/wp-content/uploads/2020/12/2020-Prescribed-Fire-Use-Report.pdf (accessed on 1 December 2022).
15. Lowery, R.F.; Gjerstad, D.H. Chemical and mechanical site preparation. In *Forest Regeneration Manual*; Duryea, M.L., Dougherty, P.M., Eds.; Kluwer Academic: Boston, MA, USA, 1991; Chapter 13; pp. 251–261.

16. Derr, H.J.; Mann, W.F., Jr. *Bedding Poorly Drained Sites for Planting Loblolly and Slash Pine in Southwestern Louisiana*; Note SO-134; United States Department of Agriculture Forest Service, Southern Forest Experiment Station Research: New Orleans, LA, USA, 1977; 5p. Available online: https://catalog.hathitrust.org/Record/007407281 (accessed on 1 December 2022).
17. Worst, R.H. A study of effects of site preparation and spacing on planted slash pine in the Coastal Plain of southeast Georgia. *J. For.* **1964**, *62*, 556–560.
18. Cain, M.D. *Planted Loblolly and Slash Pine Response to Bedding and Flat Disking on a Poorly Drained Site: An Update*; Note SO-237; United States Department of Agriculture Forest Service, Southern Forest Experiment Station Research: New Orleans, LA, USA, 1978; 6p. Available online: https://www.srs.fs.usda.gov/pubs/rn/rn_so237.pdf (accessed on 1 December 2022).
19. South, D.B.; Zwolinski, J.B.; Allen, H.L. Economic returns from enhancing loblolly pine establishment on two upland sites: Effects of seedling grade, fertilization, hexazinone, and intensive soil cultivation. *New For.* **1995**, *10*, 239–256. [CrossRef]
20. South, D.B.; Zwolinski, J.B.; Kotze, H. Early growth responses from weed control and planting larger stock of *Pinus radiata* are greater than that obtained from mechanical soil cultivation. *New For.* **2001**, *22*, 199–211. [CrossRef]
21. Wittwer, R.F.; Dougherty, P.M.; Cosby, D. Effects of ripping and herbicide site preparation treatments on loblolly pine seedling growth and survival. *South. J. Appl. For.* **1986**, *10*, 253–257. [CrossRef]
22. Fallis, F.G.; Duzan, H.W. Effects of ripping (deep subsoiling) on loblolly pine plantation establishment and growth—Nineteen years later. In *8th Southern Silvicultural Research Conference Proceedings*; SRS-1; United States Department of Agriculture Forest Service, Southern Research Station: Asheville, NC, USA, 1995; pp. 525–529.
23. Ross, D.W.; Scott, W.; Heninger, R.L.; Walstad, J.D. Effects of site preparation on ponderosa pine (*Pinus ponderosa*), associated vegetation, and soil properties in south central Oregon. *Can. J. For. Res.* **1986**, *16*, 612–618. [CrossRef]
24. Ndlovu, N.; Little, K.; Titshall, L.; Rolando, C.A. Site preparation and vegetation management impacts on *Pinus patula* growth and rotation end productivity in South Africa. *Aust. For.* **2019**, *82*, 107–115. [CrossRef]
25. Hallgren, S.W.; Ferris, D.M. Benomyl applied to roots improves second-year survival and growth of shortleaf pine. *South. J. Appl. For.* **1995**, *19*, 36–41. [CrossRef]
26. Dierauf, T.A.; Scrivani, J.A. *Ripping to Improve Loblolly Seedling Survival and Growth*; Occasional Report 119; Virginia Department of Forestry: Charlottesville, VA, USA, 1995; 3p. Available online: https://dof.virginia.gov/wp-content/uploads/report-0119.pdf (accessed on 1 December 2022).
27. Barnard, E.L.; Dixon, W.N.; Ash, E.C.; Fraedrich, S.W.; Cordell, C.E. Scalping reduces impact of soilborne pests and improves survival and growth of slash pine seedlings on converted agricultural croplands. *South. J. Appl. For.* **1995**, *19*, 49–59. [CrossRef]
28. Creighton, J.; Bowman, W.; Bitoki, O. *Longleaf Pine Establishment—Final Report*; Virginia Department of Forestry: Charlottesville, VA, USA, 2010; p. 4. Available online: https://dof.virginia.gov/wp-content/uploads/FRR-2010-03_pub.pdf (accessed on 1 December 2022).
29. Creighton, J. *Establishment Methods for Shortleaf Pine*; Research Report 146; Virginia Department of Forestry: Charlottesville, VA, USA, 2022; p. 7. Available online: https://dof.virginia.gov/wp-content/uploads/RR-146.pdf (accessed on 1 December 2022).
30. Hainds, M.J. Planting Longleaf Pine in Old Fields and Pastures. 2003. Available online: https://americaslongleaf.org/media/rmqf5ut2/planting-longleaf-in-pastures-_longleaf-alliance_pdf (accessed on 1 December 2022).
31. Moschler, W.W.; Jones, G.D.; Adams, R.E. Effects of loblolly pine fertilization on a Piedmont soil: Growth, foliar composition, and soil nutrients 10 years after establishment. *Soil Sci. Soc. Am. J.* **1970**, *34*, 683–685. [CrossRef]
32. South, D.B.; Carey, W.A.; Johnson, D.A. Copper deficiency in pine plantations in Georgia Coastal Plain. In *12th Biennial Southern Silvicultural Research Conference Proceedings*; GTR SRS-71; United States Department of Agriculture Forest Service, Southern Research Station: Asheville, NC, USA, 2004; pp. 387–390.
33. Walsh, J.R.; Redente, E.F. Comparison of reclamation techniques to re-establish western white pine on smelter-impacted hillsides in Idaho. *Restor. Ecol.* **2011**, *19*, 141–150. [CrossRef]
34. Edgren, J.W. Nursery storage to planting hole: A seedling's hazardous journey. In *Forestry Nursery Manual: Production of Bareroot Seedlings*; Springer: Dordrecht, The Netherlands, 1984; pp. 235–242.
35. Creighton, J.; Bowman, W.F. *Effects of Storage, Handling, and Transportation on Eastern White Pine 1st-Year Survival and Height*; Occasional Report 125; Virginia Department of Forestry: Charlottesville, VA, USA, 2005; 13p. Available online: https://dof.virginia.gov/wp-content/uploads/report-0125.pdf (accessed on 1 December 2022).
36. Carlson, W.C. Lifting, storing and transporting southern pine seedlings. In *Forest Regeneration Manual*; Duryea, M.L., Dougherty, P.M., Eds.; Kluwer Academic: Boston, MA, USA, 1991; Chapter 16; pp. 291–301.
37. Mullins, J.; Buckner, E.; Evans, R.; Moditz, P. Extending Loblolly and Virginia Pine Planting Seasons on Strip Mine Spoils in East Tennessee. *Tenn. Farm Home Sci.* **1989**, *151*, 24–27. Available online: https://digital.lib.utk.edu/collections/islandora/object/agrtfhs%3A2850/datastream/PDF/view (accessed on 1 December 2022).
38. Lantz, C. *A Guide to the Care and Planting of Southern Pine Seedlings*; United States Department of Agriculture Forest Service, Southern Region: Atlanta, GA, USA, 1982. Available online: https://rngr.net/publications/care_and_planting/6.%20Planting/at_download/file (accessed on 1 December 2022).
39. Black, K.; Harper, R.C.; Ryan, S.; O'Reilly, C. Rapid assessments of cold hardiness and quality deterioration during storage of bare root conifer transplants. *Irish For.* **2015**, *72*, 21–41.
40. Barden, C.E.; Feret, P.P. Root growth potential of loblolly pine seedlings held in cold storage under differing moisture treatments. In Proceedings of the Southern Forest Nursery Association Proceedings, Pensacola, FL, USA, 22–24 July 1986; pp. 218–223.

41. DeYoe, D.R. *Guidelines for Handling Seeds and Seedlings to Ensure Vigorous Stock*; Technical Publication 13; College of Forestry, Oregon State University: Corvallis, OR, USA, 1986; Available online: https://ir.library.oregonstate.edu/downloads/gt54kn907 (accessed on 1 December 2022).
42. Stjernberg, E.I. Mechanical shock during transportation: Effects on seedling performance. *New For.* **1997**, *13*, 401–420. [CrossRef]
43. Kauppi, P. Stress, strain and injury: Scots pine transplants from lifting to acclimation on the planting site. *Acta. For. Fenn.* **1984**, *185*, 1–49. [CrossRef]
44. Tabbush, P.M. Rough handling, soil temperature, and root development in outplanted Sitka spruce and Douglas-fir. *Can. J. For. Res.* **1986**, *16*, 1385–1388. [CrossRef]
45. Sharpe, A.; Mason, W.L.; Howes, R.E.J. Early forest performance of roughly handled Sitka spruce and Douglas-fir of different plant types. *Scot. For.* **1990**, *4*, 257–265.
46. Rink, G.; Wells, O.O. Productivity comparison of 37-year-old loblolly-shortleaf pine seed sources in southern Illinois. *North. J. Appl. For.* **1988**, *5*, 155–158. [CrossRef]
47. Creighton, J. *Ten-Year Performance of Eight Longleaf Pine Provenances in Virginia*; Research Report 136; Virginia Department of Forestry: Charlottesville, VA, USA, 2017; 6p. Available online: https://dof.virginia.gov/wp-content/uploads/RR-136.pdf (accessed on 1 December 2022).
48. McKeand, S.; Peter, G.; Byram, T. Trends in deployment of advanced loblolly pine germplasm. In *PINEMAP Year 4 Annual Report*; United States Department of Agriculture: Washington, DC, USA, 2015; Chapter 11; pp. 26–27. Available online: http://pinemap.org/reports/annual-reports/PINEMAP_AnnualReport_press4v2.pdf (accessed on 1 December 2022).
49. Dierauf, T.A.; Hannah, H. *Root Pruning White Pine Seedlings Prior to Planting*; Occasional Report 82; Virginia Division of Forestry: Charlottesville, VA, USA, 1989; 4p. Available online: https://dof.virginia.gov/wp-content/uploads/report-0082.pdf (accessed on 1 December 2022).
50. Yu, F.; Guo, X.; Xu, X. Shoot and root pruning and exposure treatments affect root growth potential of Chinese fir and Masson pine bare-root seedlings after lifting. *N. Z. J. For. Sci.* **2002**, *32*, 299–308.
51. South, D.B.; Stumpff, N.J. Root stripping reduces root growth potential of loblolly pine seedlings. *South. J. Appl. For.* **1990**, *14*, 196–199. [CrossRef]
52. Dierauf, T.A.; Garner, J.W. *Root Pruning Loblolly Pine Seedlings Effect on Survival and Growth*; Occasional Report 52; Virginia Division of Forestry: Charlottesville, VA, USA, 1978; 4p. Available online: https://dof.virginia.gov/wp-content/uploads/report-0052.pdf (accessed on 1 December 2022).
53. Harrington, T.B.; Howell, K.D. Planting cost, survival, and growth one to three years after establishing loblolly pine seedlings with straight, deformed, or pruned taproots. *New For.* **1998**, *15*, 193–204. [CrossRef]
54. Marx, D.H.; Hatchell, G.E. Root stripping of ectomycorrhizae decreases field performance of loblolly and longleaf pine seedlings. *South. J. Appl. For.* **1986**, *10*, 173–179.
55. McKay, H.M. A review of the effect of stresses between lifting and planting on nursery stock quality and performance. *New For.* **1997**, *13*, 369–399. [CrossRef]
56. Feret, P.P.; Kreh, R.E.; Mulligan, C. Effects of air drying on survival, height, and root growth potential of loblolly pine seedlings. *South. J. Appl. For.* **1985**, *9*, 125–128. [CrossRef]
57. Tabor, C.A.; Davey, C.B. Clay-suspension root-coating as antidesiccants and rhizospheric nutrient sources. *Soil Sci. Soc. Am. J.* **1966**, *30*, 516–520. [CrossRef]
58. Cummings, W.H. Exposure of roots of shortleaf pine stock. *J. For.* **1942**, *40*, 490–492.
59. Slocum, G.K.; Maki, T.E. Exposure of loblolly pine planting stock. *J. For.* **1956**, *54*, 313–315.
60. Switzer, G.L. Exposure and planting depth effects on loblolly pine planting stock on poorly drained sites. *J. For.* **1960**, *58*, 390–391.
61. Alm, A.; Stanton, J. Field trials of root dipping for red pine, jack pine, and white spruce nursery stock in Minnesota. *Tree Plant. Notes* **1990**, *41*, 18–20.
62. Dierauf, T.A.; Marler, R.L. *Effect of Exposure, Clay Treatment, and Storage on Survival and Growth of Loblolly Pine Seedlings*; Occasional Report 34; Virginia Division of Forestry: Charlottesville, VA, USA, 1971; 10p. Available online: https://dof.virginia.gov/wp-content/uploads/report-0034.pdf (accessed on 1 December 2022).
63. Nowak, J.; McCarthy, W.; Holland, R. *Silviculture Best Management Practices 2019 Implementation Survey Report*; Florida Division of Forestry: Tallahassee, FL, USA, 2020; 51p. Available online: https://www.fdacs.gov/content/download/78966/file/silviculture-bmp-2019-implementation-survey-report.pdf (accessed on 1 December 2022).
64. Kluender, R.A.; Burger, J.A.; Izlar, B.L.; Stuart, W.B.; Walbridge, T.A. *Southwide Pine Regeneration Survey*; American Pulpwood Association: Washington, DC, USA, 1985; 37p.
65. Senior, M.T.; Hassan, A.E. Field Evaluation of Tree Transplanting Methods; Paper No. 83-1606. In Proceedings of the American Society of Agricultural Engineers, Chicago, IL, USA, 13–16 December 1983.
66. Nieuwenhuis, M.; Egan, D. An Evaluation and Comparison of Mechanised and Manual Tree Planting on Afforestation and Reforestation Sites in Ireland. *Int. J. For. Eng.* **2002**, *13*, 11–23. Available online: https://www.tandfonline.com/doi/abs/10.1080/14942119.2002.10702459 (accessed on 1 December 2022). [CrossRef]
67. Hooker, J.M.; Oswald, B.P.; Stovall, J.P.; Weng, Y.; Williams, H.M.; Grogan, J. Third year survival, growth, and water relations of West Gulf Coastal Plain pines in East Texas. *For. Sci.* **2021**, *67*, 347–355.
68. South, D.B. Scalping improves early growth of longleaf pine seedlings. *Native Plants J.* **2011**, *12*, 18–26. [CrossRef]

69. South, D.B.; Mitchell, B. Survival of the Fittest: Pine Seedling Survival Increased by Machine Planting Large Seedlings. *Highlights Agric. Res.* **1999**, *46*, 16–18. Available online: https://rngr.net/publications/fnn/2008-winter/new-nursery-literature/survival-of-the-fittest-pine-seedlings-survival-increased-by-machine-planting-large-seedlings/at_download/file (accessed on 1 December 2022).
70. Cameron, D.A. *A Guide to Planting Machine Operation in the Boreal Forest of Ontario*; Great Lakes Forest Research Center: Sault Ste Marie, ON, Canada, 1975; Available online: https://cfs.nrcan.gc.ca/pubwarehouse/pdfs/8911.pdf (accessed on 1 December 2022).
71. Thompson, M.A. *Evaluation of a Mechanized Tree-Planting Operation*; Research Paper NC-258; United States Department of Agriculture Forest Service, North Central Forest Experiment Station: St. Paul, MN, USA, 1984; 13p. Available online: https://www.nrs.fs.usda.gov/pubs/rp/rp_nc258.pdf (accessed on 1 December 2022).
72. Ghaffariyan, M.R. A Short Review on Studies on Work Productivity of Mechanical Tree Planting. *Silva Balcanica* **2021**, *22*, 25–32. Available online: https://doi.org/10.3897/silvabalcanica.22.e64233 (accessed on 1 December 2022). [CrossRef]
73. Laine, T.; Rantala, J. Mechanized tree planting with an excavator mounted M-Planter planting device. *Int. J. For. Eng.* **2013**, *24*, 183–193. [CrossRef]
74. Blake, J.I.; South, D.B. *Planting Morphologically Improved Seedlings with Shovels*; School of Forestry Series 13; Auburn University: Auburn, AL, USA, 1991; 7p, Available online: http://aurora.auburn.edu/bitstream/handle/11200/955/0044FORE.pdf (accessed on 1 December 2022).
75. Granzow, R.F.; Schall, M.C.; Smidt, M.F.; Chen, H.; Fethke, N.B.; Huangfu, R. Characterizing exposure to physical risk factors among reforestation hand planters in the Southeastern United States. *Appl. Ergon.* **2018**, *66*, 1–8. [CrossRef] [PubMed]
76. Beineke, W.F. Genetic Variation in the Ability to Withstand Transplanting Shock in Loblolly Pine (*Pinus taeda* L.). Ph.D. Dissertation, North Carolina State University, Raleigh, NC, USA, 1966.
77. Schmidtling, R.C. *Southern Pine Seed Sources*; GTR SRS-44; United States Department of Agriculture Forest Service, Southern Research Station: Asheville, NC, USA, 2001; 25p. Available online: https://babel.hathitrust.org/cgi/pt?id=umn.31951d029884323&view=1up&seq=1 (accessed on 1 December 2022).
78. Garriou, D.; Girard, S.; Guehl, J.M.; Généré, B. Effect of desiccation during cold storage on planting stock quality and field performance in forest species. *Ann. For. Sci.* **2000**, *57*, 101–111. [CrossRef]
79. Mullen, R.E. Some Effects of Root Dipping, Root Exposure and Extended Planting Dates with White Spruce. *For. Chron.* **1970**, *74*, 90–93. Available online: https://pubs.cif-ifc.org/doi/pdf/10.5558/tfc47090-2 (accessed on 1 December 2022). [CrossRef]
80. Maki, T.E.; Marshall, H. Effects of soaking with indolebutyric acid on root development and survival of tree seedlings. *Bot. Gaz.* **1945**, *107*, 268–276. [CrossRef]
81. Ursic, S.J. Modifications of planting technique not recommended for loblolly on eroded soils. *Tree Plant. Notes* **1963**, *57*, 13–17.
82. Dierauf, T.A.; Edwards, L.C. *A Three-Year Study of Immersing Stored Loblolly Pine Seedlings in Water before Planting*; Occasional Report 66; Virginia Division of Forestry, Department of Conservation and Economic Development: Charlottesville, VA, USA, 1985; 7p. Available online: https://dof.virginia.gov/wp-content/uploads/report-0066.pdf (accessed on 1 December 2022).
83. Davey, C.B. Current studies on clay slurry root dips. In *Region 8 Forest Nurseryman's Conference Proceedings*; United States Department of Agriculture Forest Service, State and Private Forestry: Atlanta, GA, USA, 1964; pp. 53–59. Available online: https://rngr.net/publications/proceedings/1964/current-studies-on-clay-slurry-root-dips/at_download/file (accessed on 1 December 2022).
84. Carey, W.A.; South, D.B.; Williford, M.; Britt, J. Washing roots reduces vigor of loblolly pine seedlings. *South. J. Appl. For.* **2001**, *25*, 25–30. [CrossRef]
85. Sloan, J.P. The use of rootdips on North American conifer seedlings: A review of the literature. *Tree Plant. Notes* **1994**, *45*, 26–31.
86. Starkey, T.E.; Enebak, S.A.; South, D.B.; Cross, R.E. Particle Size and Composition of Polymer Root Gels Affect Loblolly Pine Seedling Survival. *Native Plants J.* **2012**, *20*, 19–26. Available online: https://muse.jhu.edu/article/470878 (accessed on 1 December 2022). [CrossRef]
87. Landis, T.D.; Haase, D.L. Applications of hydrogels in the nursery and during outplanting. In *Forest and Conservation Nursery Associations-2011 Proceedings*; RMRS-P-68; United States Department of Agriculture Forest Service, Rocky Mountain Forest and Range Experiment Station: Fort Collins, CO, USA, 2012; pp. 53–58.
88. Alm, A.; Stanton, J. Polymer root dip increases survival of stressed bareroot seedlings. *North. J. Appl. For.* **1993**, *10*, 90–92. [CrossRef]
89. Crous, J.W. Use of hydrogels in the planting of industrial wood plantations. *South. For.* **2017**, *79*, 197–213. [CrossRef]
90. Boyer, J.N.; South, D.B. Excessive seedling height, high shoot-to-root ratio, and benomyl root dip reduce survival of stored loblolly pine seedlings. *Tree Plant. Notes* **1987**, *38*, 19–22.
91. Grossnickle, S.C. Why seedlings survive: Influence of plant attributes. *New For.* **2012**, *43*, 711–738. [CrossRef]
92. Allen, R.M. Foliage treatments improve survival of longleaf pine plantings. *J. For.* **1955**, *10*, 724–727.
93. Knapp, B.O.; Pile, L.S.; Walker, J.L.; Wang, G.G. Fire Effects on a Fire-Adapted Species: Response of Grass Stage Longleaf Pine Seedlings to Experimental Burning. *Fire Ecol.* **2018**, *14*, 1–6. Available online: https://link.springer.com/content/pdf/10.1186/s42408-018-0003-y.pdf (accessed on 1 December 2022). [CrossRef]
94. South, D.B. Needle-clipping longleaf pine and top-pruning loblolly pine in bareroot nurseries. *South. J. Appl. For.* **1998**, *22*, 235–240. [CrossRef]

95. Luoranen, J.; Pikkarainen, L.; Poteri, M.; Peltola, H.; Riikonen, J. Duration limits on field storage in closed cardboard boxes before planting of Norway spruce and Scots pine container seedlings in different planting seasons. *Forests* **2019**, *10*, 1126. [CrossRef]
96. Pikkarainen, L.; Luoranen, J.; Peltola, H. Early field performance of small-sized silver birch and Scots pine container seedlings at different planting depths. *Forests* **2021**, *12*, 519. [CrossRef]
97. Greaves, R.D. Planting. In *Regenerating Oregon's Forests: A Guide for the Regeneration Forester*; Cleary, B.D., Greaves, R.D., Hermann, R.K., Eds.; Oregon State University: Corvallis, OR, USA, 1978; Chapter 8; pp. 134–147.
98. Abbott, J.E. Operational planting of container grown slash pine seedlings on problem sites. In *Southern Containerized Forest Tree Seedling Conference Proceedings*; United States Department of Agriculture Forest Service, Southern Forest Experiment Station: New Orleans, LA, USA, 1982. Available online: https://www.srs.fs.usda.gov/pubs/8383 (accessed on 1 December 2022).
99. Larson, D.R. Field planting containerized longleaf pine seedlings. In *Workshops on Growing Longleaf Pine in Containers—1999 and 2001 Proceedings*; GTR SRS-56; United States Department of Agriculture Forest Service, Southern Forest Experiment Station: Asheville, NC, USA, 2002; pp. 62–63. Available online: https://www.fs.usda.gov/research/treesearch/5046 (accessed on 1 December 2022).
100. Betts, J.W. A discussion of tubed seedlings. In *Proceedings of the North American Containerized Forest Tree Seedling Symposium, Denver, CO, USA, 26–29 August 1974*; Tinus, R.W., Stein, W.I., Balmer, W.E., Eds.; Publication No. 68; Great Plains Agricultural Council: Lincoln, NE, USA, 1974; pp. 139–140. Available online: https://rngr.net/publications/proceedings/1974/a-discussion-of-tubed-seedlings/at_download/file (accessed on 1 December 2022).
101. South, D.B.; Jackson, D.P.; Starkey, T.E.; Enebak, S.A. Planting Deep Increases Early Survival and Growth of *Pinus echinata* Seedlings. *Open For. Sci. J.* **2012**, *5*, 33–41. Available online: https://benthamopen.com/contents/pdf/TOFSCIJ/TOFSCIJ-5-33.pdf (accessed on 1 December 2022). [CrossRef]
102. South, D.B. A Review of the "Pull-Up" and "Leave-Down" Methods of Planting Loblolly Pine. *Tree Plant. Notes* **2005**, *31*, 53–67. Available online: https://rngr.net/publications/tpn/51-1/a-review-of-the-pull-up-and-leave-down-methods-of-planting-loblolly-pine-1/at_download/file (accessed on 1 December 2022).
103. Wakeley, P.C. *Artificial Reforestation in the Southern Pine Region*; Technical Bulletin 492; United States Department of Agriculture: Washington, DC, USA, 1935. Available online: https://ageconsearch.umn.edu/record/164666/files/tb492.pdf (accessed on 1 December 2022).
104. Hainds, M.J. Determining the correct planting depth for container-grown longleaf pine seedlings. In *12th Biennial Southern Silvicultural Research Conference Proceedings*; GTR SRS-71; United States Department of Agriculture Forest Service, Southern Research Station: Asheville, NC, USA, 2004; pp. 317–318.
105. Pickens, B.; Blackburn, D. *Effect of Planting Depth on the Survival and Height Growth of Container-Grown Longleaf Seedlings*; Technical Resource Bulletin TRB-006; North Carolina Forest Service: Raleigh, NC, USA, 2006; 4p.
106. Sutton, R.F. Influence of planting depth on early growth of conifers. *Commonw. For. Rev.* **1967**, *46*, 282–295.
107. VanderSchaaf, C.L.; South, D.B. Effect of planting depth on growth of open-rooted *Pinus elliottii* and *Pinus taeda* seedlings in the United States. *South. Afr. For. J.* **2003**, *198*, 63–73.
108. Starkey, T.E.; Enebak, S.A.; South, D.B. Forest seedling nursery practices in the southern United States: Container nurseries. *Tree Plant. Notes* **2015**, *58*, 4–17.
109. South, D.B. Chilling hours: Myths and facts. In *Forest and Conservation Nursery Associations-2012 Proceedings*; RMRS-P-69; United States Department of Agriculture Forest Service, Southern Forest Experiment Station: Fort Collins, CO, USA, 2013; pp. 3–10.
110. Wakeley, P.C. *Planting the Southern Pines*; Agricultural Monograph 18; United States Department of Agriculture: Washington, DC, USA, 1954. Available online: https://www.srs.fs.usda.gov/pubs/57741 (accessed on 1 December 2022).
111. Riffle, J.W. Field performance of ponderosa, Scots, and Austrian pines with *Pisolithus tinctorius* ectomycorrhizae in prairie soils. *For. Sci.* **1989**, *35*, 935–945.
112. Creighton, J. *Comparison of Early Performance of Containerized and Bare-Root Loblolly Pine Seedlings Planted from October through April*; Research Report 140; Virginia Department of Forestry: Charlottesville, VA, USA, 2021; 4p. Available online: https://dof.virginia.gov/wp-content/uploads/RR-140.pdf (accessed on 1 December 2022).
113. Omi, S.K.; Yoder, B.; Rose, R. Fall lifting and long-term freezer storage of ponderosa pine seedlings: Effects on post-storage leaf water potential, stomatal conductance, and root growth potential. *Tree Physiol.* **1991**, *8*, 315–325. [CrossRef] [PubMed]
114. Brissette, J.C.; Barnett, J.P.; Jones, J.P. Fungicides improve field performance of stored loblolly and longleaf pine seedlings. *South. J. Appl. For.* **1996**, *20*, 5–9. [CrossRef]
115. Tolbert, J.L.; Burger, J.L.; Daniels, W.L. Effect of overburden type and organic amendments on the growth of pines on a reclaimed surface mine. In *3rd Biennial Southern Silvicultural Research Conference Proceedings*; GTR SO-54; United States Department of Agriculture Forest Service, Southern Forest Experiment Station: New Orleans, LA, USA, 1985; pp. 369–374.
116. Clabo, D.; Clatterbuck, W. Restoration of shortleaf pine (*Pinus echinata*)-hardwood mixtures in low quality mixed upland hardwood stands using cluster planting and natural regeneration. *Forests* **2020**, *11*, 457. [CrossRef]
117. South, D.B. The December dip of loblolly pine. In *Forest and Conservation Nursery Associations Proceedings*; GTR SRS-25; United States Department of Agriculture Forest Service, Southern Research Station: Asheville, NC, USA, 1999; pp. 14–17. Available online: https://rngr.net/publications/proceedings/1998/South/?searchterm=december%20dip (accessed on 1 December 2022).

118. Pickens, B. *Effects of Extended Cold Storage on the Survival and Performance of Container Grown Longleaf Seedlings*; Technical Resource Bulletin TRB-007; North Carolina Forest Service: Raleigh, NC, USA, 2012; 3p. Available online: https://www.ncforestservice.gov/publications/techBulletins/TRB007.pdf (accessed on 1 December 2022).
119. Ramantswana, M.; Guerra, S.P.S.; Ersson, B.T. Advances in the Mechanization of Regenerating Plantation Forests: A review. *Curr. For. Rep.* **2020**, *6*, 143–158. Available online: https://link.springer.com/content/pdf/10.1007/s40725-020-00114-7.pdf (accessed on 1 December 2022). [CrossRef]
120. de Luis, M.; Novak, K.; Raventós, J.; Gričar, J.; Prislan, P.; Čufar, K. Cambial activity, wood formation and sapling survival of *Pinus halepensis* exposed to different irrigation regimes. *For. Ecol. Manag.* **2011**, *262*, 1630–1638. [CrossRef]
121. Long, A.J. Proper planting improves performance. In *Forest Regeneration Manual*; Duryea, M.L., Dougherty, P.M., Eds.; Kluwer Academic: Boston, MA, USA, 1991; Chapter 17; pp. 303–331.
122. Stroempl, G. Deeper planting of seedlings and transplants increases plantation survival. *Tree Plant. Notes* **1990**, *41*, 17–21.
123. Walker, R.F.; West, D.C.; McLaughlin, S.B. *Pisolithus tinctorius* ectomycorrhizae enhance the survival and growth of *Pinus taeda* on a southern Appalachian coal spoil. In *1st Biennial Southern Silvicultural Research Conference Proceedings*; GTR SO-1; United States Department of Agriculture Forest Service, Southern Forest Experiment Station: New Orleans, LA, USA, 1981; pp. 29–33.
124. Yeiser, J.L. Weed control and seedling performance using Oust, Velpar and Velpar+Oust impregnated diammonium phosphate. In *11th Biennial Southern Silvicultural Research Conference Proceedings*; GTR SRS-48; United States Department of Agriculture Forest Service, Southern Research Station: Asheville, NC, USA, 2002; pp. 260–265.
125. Walker, R.F. Reforestation of an Eastern Sierra Nevada surface mine with containerized Jeffrey pine. *J. Sustain. For.* **1999**, *9*, 127–147. [CrossRef]
126. Messina, M.G. Herbicide, fertilizer and shade influence loblolly pine growth and survival on harsh Texas sites. In *6th Biennial Southern Silvicultural Research Conference Proceedings*; GTR SE-70; United States Department of Agriculture Forest Service, Southern Forest Experiment Station: Asheville, NC, USA, 1991; pp. 155–162.
127. Schoenholtz, S.H. Survival, and Growth of Pine Seedlings on Strip-Mined Sites. Master's Thesis, Virginia Polytechnic Institute and State University, Blacksburg, VA, USA, 1983. Available online: https://vtechworks.lib.vt.edu/bitstream/handle/10919/31469/LD5655.V855_1983.S363.pdf (accessed on 1 December 2022).
128. Gleason, J.F.; Duryea, M.; Rose, R.; Atkinson, M. Nursery and field fertilization of 2+0 ponderosa pine seedlings: The effect on morphology, physiology, and field performance. *Can. J. For. Res.* **1990**, *20*, 1766–1772. [CrossRef]
129. Kais, A.G.; Cordell, C.E.; Affeltranger, C.E. Benomyl root treatment controls brown-spot disease on longleaf pine in the southern United States. *For. Sci.* **1986**, *32*, 506–511.
130. Crous, J.W. Post establishment survival of *Pinus patula* in Mpumalanga, one year after planting. *South. Afr. For. J.* **2005**, *205*, 3–11.
131. Örlander, G.; Nilsson, U. Effect of reforestation methods on pine weevil (*Hylobius abietis*) damage and seedling survival. *Scand. J. For. Res.* **1999**, *14*, 341–354. [CrossRef]
132. Rowan, S.J. Vapor-gard affects survival and growth of outplanted pine seedlings. In Proceedings of the Southern Forest Nursery Association Proceedings, Charleston, SC, USA, 25–28 July 1988; pp. 27–32. Available online: https://rngr.net/publications/proceedings/1988/vapor-gard-affects-survival-and-growth-of-outplanted-pine-seedlings/at_download/file (accessed on 1 December 2022).
133. Williams, P.A.; Gordon, A.M.; Moeller, A.W. Effects of five antitranspirants on white spruce and white pine seedlings subjected to greenhouse drought. *Tree Plant. Notes* **1990**, *41*, 34–38.
134. Haywood, J.D. *Controlling Herbaceous Competition in Pasture Planted with Loblolly Pine Seedlings*; Research Note SO-381; United States Department of Agriculture Forest Service, Southern Forest Experiment Station: New Orleans, LA, USA, 1995; 4p. Available online: https://www.srs.fs.usda.gov/pubs/rn/rn_so381.pdf (accessed on 1 December 2022).
135. Pruett, E.W.; Gatherum, G.E. Control of herbaceous vegetation in forest plantings. *Iowa Acad. Sci.* **1961**, *68*, 153–161.
136. Ursic, S.J. Bale storage effective for loblolly pine seedlings. *J. For.* **1956**, *54*, 815–816.
137. Ringe, J.M. Economic aspects of sawmill residue use for tree seedling establishment on surface mines. *Int. J. Surf. Min. Reclam. Environ.* **1988**, *2*, 129–133. [CrossRef]
138. Knapp, B.O.; Wang, G.G.; Hu, H.; Walker, J.L.; Tennant, C. Restoring longleaf pine (*Pinus palustris* Mill.) in loblolly pine (*Pinus taeda* L.) stands: Effects of restoration treatments on natural loblolly pine regeneration. *For. Ecol. Manag.* **2011**, *262*, 1157–1167. [CrossRef]
139. Chirico, D.A. Effects of Shortleaf Pine Seedling Stock (bareroot vs. Containerized) on Growth and Survival in the Absence and Presence of Fire. Master's Thesis, North Carolina State University, Raleigh, NC, USA, 2019. Available online: https://repository.lib.ncsu.edu/bitstream/handle/1840.20/36802/Chirico,%20Dominic%20final.pdf?sequence=1 (accessed on 1 December 2022).
140. Hunt, E.V. How serious is rabbit damage on loblolly pine seedlings? *J. For.* **1968**, *66*, 853.
141. Wakeley, P.C. Long-time effects of damage by rabbits to newly planted southern pines. *Tree Plant. Notes* **1970**, *21*, 6–9.
142. South, D.B.; Starkey, T.E.; Jackson, D.P. Needle-clipping of longleaf pine (*Pinus palustris* Mill.) can increase seedling survival while reducing transpiration and root growth potential. *J. Hort. For.* **2011**, *3*, 159–162.
143. Schnake, D.K. Underplanted Shortleaf Pine Seedling Survival and Growth in the North Carolina Piedmont. Master's Thesis, Mississippi State University, Mississippi State, MS, USA, 2016. Available online: https://scholarsjunction.msstate.edu/cgi/viewcontent.cgi?article=5815&context=td (accessed on 1 December 2022).

144. Vickers, L.A.; Houser, J.; Rooni, J.; Guldin, J.M. Some lessons learned on early survival and growth of containerized, locally-sourced ponderosa pine seedlings in the Davis Mountains of western Texas, US. *Forests* **2019**, *10*, 267. [CrossRef]
145. Lewis, C.E. Simulated Cattle Injury to Planted Slash Pine: Combinations of Defoliation, Browsing, and Trampling. *J. Range Manag.* **1980**, *33*, 340–345. Available online: https://journals.uair.arizona.edu/index.php/jrm/article/download/7083/6695 (accessed on 1 December 2022). [CrossRef]
146. Pearson, H.A.; Baldwin, V.C.; Barnett, J.P. Cattle grazing and pine survival and growth in subterranean clover pasture. *Agrofor. Syst.* **1990**, *10*, 161–168. [CrossRef]
147. Ward, J.S.; Gent, M.P.; Stephens, G.R. Effects of planting stock quality and browse protection-type on height growth of northern red oak and eastern white pine. *For. Ecol. Manag.* **2000**, *127*, 205–216.
148. Keeton, W.S. Evaluation of tree seedling mortality and protective strategies in riparian forest restoration. *North. J. Appl. For.* **2008**, *25*, 117–123. [CrossRef]
149. Schaap, W.; DeYoe, D. *Seedling Protectors for Preventing Deer Browse*; Research Bulletin 54; Forest Research Laboratory, Oregon State University: Corvallis, OR, USA, 1986; Available online: https://ir.library.oregonstate.edu/downloads/p8418p58q (accessed on 1 December 2022).
150. Ward, J.S.; Mervosh, T.L. Strategies to reduce browse damage on eastern white pine (*Pinus strobus*) in southern New England, USA. *For. Ecol. Manag.* **2008**, *255*, 1559–1567. [CrossRef]
151. Walstad, J.D.; Hart, J.B.; Cade, S.C. Carbofuran-clay root dip protects loblolly pine seedlings from debarking weevils. *J. Econ. Entomol.* **1973**, *66*, 1219–1220.
152. Hardy, C.; Sayyed, I.; Leslie, A.D.; Dittrich, A.D. Effectiveness of insecticides, physical barriers and size of planting stock against damage by the pine weevil (*Hylobius abietis*). *Crop Prot.* **2020**, *137*, 105307. [CrossRef]
153. Asaro, C.; Creighton, J. Use of systemic fipronil and imidacloprid to control regeneration pests of loblolly pine. *J. Econ. Entomol.* **2011**, *104*, 1272–1279. [CrossRef] [PubMed]
154. Mitchell, R.J.; Runion, G.B.; Kelley, W.D.; Gjerstad, D.H.; Brewer, C.H. Factors associated with loblolly pine mortality on former agricultural sites in the Conservation Reserve Program. *J. Soil Water Conserv.* **1991**, *46*, 306–311.
155. Speers, C.V.; Schmiege, D.C. *White Grubs in Forest Tree Nurseries and Plantations*; Forest Pest Leaflet 63; United States Department of Agriculture, Forest Service: Asheville, NC, USA, 1971; 5p. Available online: https://www.fs.usda.gov/Internet/FSE_DOCUMENTS/fsbdev2_043588.pdf (accessed on 1 December 2022).
156. Sutton, R.F.; Stone, E.L. *White Grubs: A Description for Foresters, and an Evaluation of Their Silvicultural Significance*; Information Report. O-X-212; Canadian Forestry Service: Sault Ste Marie, ON, Canada, 1974; 21p.
157. Mangini, A.C. Regeneration insect pests: Protecting southern pine seedlings after outplanting. *Tree Plant. Notes* **2017**, *62*, 70–80.
158. Zwolinski, J.B.; South, D.B.; Droomer, E.A.P. Pine mortality after planting on post-agricultural lands in South Africa. *Silva fennica* **1998**, *32*, 271–280. [CrossRef]
159. King, J.S.; Kelley, A.M.; Rees, R. Systemic control of Nantucket pine tip moth (*Rhyacionia frustrana* Scudder in Comstock, 1880) enhances seedling vigor, plantation establishment, and early stand-level productivity in *Pinus taeda* L. *For. Sci.* **2014**, *60*, 97–108. [CrossRef]
160. Schnake, D.K.; Roberts, S.D.; Willis, J.L.; Kushla, J.D.; Munn, I.A. Overstory retention and stock type impact survival and growth of underplanted shortleaf pine beneath a hardwood canopy. *For. Sci.* **2021**, *67*, 219–230. [CrossRef]
161. Yamanobe, T. Relationships between morphological traits and resistance to pine wood nematode in two Japanese pines. *Eur. J. Plant Path.* **2009**, *124*, 543–552. [CrossRef]
162. Bengtson, G.W.; Smart, G.C. Slash pine growth and response to fertilizer after application of pesticides to the planting site. *For. Sci.* **1981**, *27*, 487–502.
163. Kais, A.G.; Barnett, J.P. Longleaf pine growth following storage and benomyl root-dip. *Tree Plant. Notes* **1984**, *35*, 30–33.
164. Zwolinski, J.; South, D.B.; Barber, B.L. Drought and survival of loblolly pine seedlings after planting. In *8th Biennial Southern Silvicultural Research Conference Proceedings*; GTR SRS-1; United States Department of Agriculture Forest Service, Southern Research Station: Asheville, NC, USA, 1995; pp. 419–423.
165. Petersen, G.J. The effects of artificial shade on seedling survival on Western Cascade harsh sites [USA]. *Tree Plant. Notes* **1982**, *33*, 20–23.
166. Childs, S.W.; Flint, L.E. Effect of shadecards, shelterwoods, and clearcuts on temperature and moisture environments. *For. Ecol. Manag.* **1987**, *18*, 205–217. [CrossRef]
167. Helgerson, O.T.; Bunker, J.D. Alternate types of artificial shade increase survival of Douglas-fir (*Pseudotsuga menziesii* (Mirb.) Franco) seedlings in clearcuts. *Tree Plant. Notes* **1985**, *36*, 7–12.
168. Oester, P.T. Ten-year response of western larch and Douglas-fir seedlings to mulch mats, sulfometuron, and shade in northeast Oregon. *Tree Plant. Notes* **2008**, *53*, 29–36.
169. Rynearson, B. Planting. In *Reforestation Practices for California Conifers*; Stewart, W., Ed.; University of California Agriculture and Natural Resources: Oakland, CA, USA, 2021; Chapter 9; pp. 421–470. Available online: https://ucanr.edu/sites/forestry/files/337245.pdf (accessed on 1 December 2022).
170. Hu, H.; Wang, G.G.; Walker, J.L.; Knapp, B.O. Silvicultural treatments for converting loblolly pine to longleaf pine dominance: Effects on planted longleaf pine seedlings. *For. Ecol. Manag.* **2012**, *276*, 209–216. [CrossRef]

171. Messina, M.G.; Jenkins, J. Loblolly pine stand early development under reserve-tree silvicultural systems in East Texas. *South. J. Appl. For.* **2000**, *24*, 11–16. [CrossRef]
172. Rodríguez-Trejo, D.A.; Duryea, M.L.; White, T.L.; English, J.R.; McGuire, J. Artificially regenerating longleaf pine in canopy gaps: Initial survival and growth during a year of drought. *For. Ecol. Manag.* **2003**, *180*, 25–36. [CrossRef]
173. Walker, L.C. Flooding and drainage effects on slash pine and loblolly pine seedlings. *For. Sci.* **1961**, *7*, 2–15.
174. Peairs, S.E. First year height performance of planted loblolly pine following site preparation and herbaceous release treatments on a lower Coastal Plain site. In *21st Biennial Southern Silvicultural Research Conference Proceedings*; GTR SRS-268; United States Department of Agriculture Forest Service, Southern Research Station: Asheville, NC, USA, 2022; pp. 78–82.
175. Conner, W.H.; Askew, G.R. Response of baldcypress and loblolly pine seedlings to short-term saltwater flooding. *Wetlands* **1992**, *12*, 230–233. [CrossRef]
176. Williston, H.L. Moss not needed in kraft-polyethylene bags during loblolly pine seedling transport and cold storage. *Tree Plant. Notes* **1965**, *72*, 10–11.
177. Dierauf, T.A.; Garner, J.W. *Survival of Loblolly Pine Seedlings Packed in Kraft Paper Bags Compared to Conventional Packaging*; Occasional Report 45; Virginia Division of Forestry: Charlottesville, VA, USA, 1974; 6p. Available online: https://dof.virginia.gov/wp-content/uploads/report-0045.pdf (accessed on 1 December 2022).
178. Shirley, H.L. Lethal high temperatures for conifers, and the cooling effect of transpiration. *J. Agric. Res.* **1936**, *53*, 239–258.
179. Alm, A.A.; Schantz-Hansen, R. Five-year results from tubeling plantings in Minnesota. *J. For.* **1972**, *70*, 616–619.
180. Lambert, B.B.; Harper, S.J.; Robinson, S.D. Effect of container size at time of planting on tree growth rates for baldcypress [*Taxodium distichum* (L.) Rich], red maple (*Acer rubrum* L.), and longleaf pine (*Pinus palustris* Mill.). *J. Arboric.* **2010**, *36*, 93–99. [CrossRef]
181. Schultz, R.P.; Wilhite, L.P. *Operational Summer Planting of Slash Pine*; SE-80; United States Department of Agriculture Forest Service, Southeastern Forest Experiment Station: Asheville, NC, USA, 1967; 3p.
182. Kolb, T.; Dixit, A.; Burney, O. Challenges and opportunities for maintaining ponderosa pine forests in the southwestern United States. *Tree Plant. Notes* **2019**, *62*, 104–112.
183. Dixit, A.; Kolb, T.; Burney, O.; Mock, K.; Grady, K. Provenance variation in early survival, growth, and carbon isotope discrimination of southwestern ponderosa pine growing in three common gardens across an elevational gradient. *Forests* **2021**, *12*, 1561. [CrossRef]
184. Roeser, J. Transpiration capacity of coniferous seedlings and the problem of heat injury. *J. For.* **1932**, *30*, 381–395.
185. Kolb, P.F.; Robberecht, R. High temperature and drought stress effects on survival of *Pinus ponderosa* seedlings. *Tree Physiol.* **1996**, *16*, 665–672. [CrossRef]
186. Byrd, R.R. Freezing temperatures affect survival of planted loblolly and slash pine seedlings. *Tree Plant. Notes* **1963**, *58*, 18–19.
187. Grossnickle, S.C.; South, D.B. Fall Acclimation and the Lift/Store Pathway: Effect on Reforestation. *Open For. Sci. J.* **2014**, *7*, 1–20. Available online: https://benthamopen.com/contents/pdf/TOFSCIJ/TOFSCIJ-7-1.pdf (accessed on 1 December 2022). [CrossRef]
188. McCreary, D.D.; Duryea, M.L. OSU Vigor Test: Principles, procedures, and predictive ability, 85–92. In *Evaluating Seedling Quality: Principles, Procedures, and Predictive Abilities of Major Tests*; Duryea, M.L., Ed.; Oregon State University: Corvallis, OR, USA, 1985.
189. South, D.B. Freeze injury to roots of southern pine seedlings in the USA. *South. Hemisph. For. J.* **2007**, *69*, 151–156. [CrossRef]
190. Marino, G.P.; Kaiser, D.P.; Gu, L.; Ricciuto, D.M. Reconstruction of false spring occurrences over the southeastern United States, 1901–2007: An increasing risk of spring freeze damage? *Environ. Res. Lett.* **2011**, *6*, 024015. [CrossRef]
191. Dierauf, T.A. *A Five-Year Study of Different Sawdust and Nitrogen Rates in a Loblolly Pine Nursery*; Occasional Report 94; Virginia Division of Forestry: Charlottesville, VA, USA, 1991; 19p. Available online: https://dof.virginia.gov/wp-content/uploads/report-0094.pdf (accessed on 1 December 2022).
192. McClurkin, D.C. Survival of planted loblolly pine seedlings: Moisture, temperature, and soil as influences. *J. For.* **1966**, *64*, 731–734.
193. Tinus, R.W.; Sword, M.A.; Barnett, J.P. Prevention of cold damage to container-grown longleaf pine roots. In *Workshops on Growing Longleaf Pine in Containers—1999 and 2001 Proceedings*; GTR SRS-56; United States Department of Agriculture Forest Service, Southern Research Station: Asheville, NC, USA, 2002; pp. 55–57. Available online: https://www.fs.usda.gov/research/treesearch/5051 (accessed on 1 December 2022).
194. Di, B.; Luoranen, J.; Lehto, T.; Himanen, K.; Silvennoinen, M.; Silvennoinen, R.; Repo, T. Biophysical changes in the roots of Scots pine seedlings during cold acclimation and after frost damage. *For. Ecol. Manag.* **2019**, *431*, 63–72. [CrossRef]
195. Mexal, J.G.; Timmis, R.; Morris, W.G. Cold-hardiness of containerized loblolly pine seedlings: Its effect on field survival and growth. *South. J. Appl. For.* **1979**, *3*, 15–19. [CrossRef]
196. Bigras, F.J.; Dumas, D. Root-freezing damage in the containerized nursery: Impact on plantation sites—A review. *New For.* **2005**, *30*, 167–184. [CrossRef]
197. Garner, J.W.; Dierauf, T.A. *Effect of Freezing on Survival of Loblolly Pine Seedlings*; Occasional Report 44; Virginia Division of Forestry: Charlottesville, VA, USA, 1974; 4p. Available online: https://dof.virginia.gov/wp-content/uploads/report-0044.pdf (accessed on 1 December 2022).
198. Lambeth, C.; McKeand, S.; Rousseau, R.; Schmidtling, R. Planting nonlocal seed sources of loblolly pine–managing benefits and risks. *South. J. Appl. For.* **2005**, *29*, 96–104.

199. Thomas, C.D.; Cameron, A.; Green, R.E.; Bakkenes, M.; Beaumont, L.J.; Collingham, Y.C.; Erasmus, B.F.; De Siqueira, M.F.; Grainger, A.; Hannah, L. Extinction risk from climate change. *Nature* **2004**, *427*, 145–148. [CrossRef]
200. South, D.B. A few thoughts about assisted migration. *J. For.* **2013**, *111*, 382, 440.
201. Reich, P.B.; Oleksyn, J. Climate warming will reduce growth and survival of Scots pine except in the far north. *Ecol. Lett.* **2008**, *11*, 588–597.
202. Gray, L.K.; Rweyongeza, D.; Hamann, A.; John, S.; Thomas, B.R. Developing management strategies for tree improvement programs under climate change: Insights gained from long-term field trials with lodgepole pine. *For. Ecol. Manag.* **2016**, *377*, 128–138. [CrossRef]
203. Sáenz-Romero, C.; O'Neill, G.; Aitken, S.N.; Lindig-Cisneros, R. Assisted migration field tests in Canada and Mexico: Lessons, limitations, and challenges. *Forests* **2020**, *12*, 9. [CrossRef]
204. McKeand, S. Freeze damage and cold tolerance of clones in the northern and piedmont elite breeding population. In *NCSU Tree Improvement Program 59th Annual Report 2015*; North Carolina State University: Raleigh, NC, USA, 2015; p. 22. Available online: https://www.treeimprovement.org/s/59thAnnualReport2015.pdf (accessed on 1 December 2022).
205. Baig, M.N.; Tranquillini, W. The effects of wind and temperature on cuticular transpiration of *Picea abies* and *Pinus cembra* and their significance in dessication damage at the alpine treeline. *Oecologia* **1980**, *47*, 252–256. [CrossRef]
206. Dierauf, T.A.; Olinger, H.L. *Cold Damage to Loblolly Seedlings at New Kent Nursery*; Occasional Report 51; Virginia Division of Forestry: Charlottesville, VA, USA, 1977; 4p. Available online: https://dof.virginia.gov/wp-content/uploads/report-0051.pdf (accessed on 1 December 2022).
207. Goulet, F. Frost heaving of forest tree seedlings: A review. *New For.* **1995**, *9*, 67–94. [CrossRef]
208. Heidmann, L.J.; Haase, S.M. Causes of mortality in outplanted ponderosa pine container seedlings in the southwest. *Tree Plant. Notes* **1989**, *40*, 16–19.
209. de Chantal, M.; Rita, H.; Bergsten, U.; Ottosson Löfvenius, M.; Grip, H. Frost heaving of *Picea abies* seedlings as influenced by soil preparation, planting technique, and location along gap-shelterwood gradients. *Silva Fennica* **2009**, *43*, 39–50. [CrossRef]
210. Sahlén, K.; Goulet, F. Reduction of frost heaving of Norway spruce and Scots pine seedlings by planting in mounds or in humus. *New For.* **2002**, *24*, 175–182. [CrossRef]
211. Heidmann, L.J.; Thorud, D.B. *Controlling Frost Heaving of Ponderosa Pine Seedlings in Arizona*; RM-172; United States Department of Agriculture Forest Service, Rocky Mountain Forest and Range Experiment Station: Fort Collins, CO, USA, 1976; 12p.
212. Paquette, A.; Girard, J.P.; Walsh, D. Deep planting has no short-or long-term effect on the survival and growth of white spruce, black spruce, and jack pine. *North. J. Appl. For.* **2011**, *28*, 146–151. [CrossRef]
213. South, D.B. Questions about toppling of pine seedlings. *Reforesta* **2022**, *14*, 63–106.
214. Caisley, L.L. The Effects of Nitrogen Fertilization on Soil pH and Calcium Fertilization on Growth and Form of Loblolly Pine. Ph.D. Dissertation, North Carolina State University, Raleigh, NC, USA, 2021.
215. Holmes, J.S. Damage to forests by hail in North Carolina. *Mon. Weather. Rev.* **1921**, *49*, 333. [CrossRef]
216. Dobbs, R.C.; McMinn, R.G. Hail damage to a new white spruce and lodgepole pine plantation in central British Columbia. *For. Chron.* **1973**, *49*, 174–175. [CrossRef]
217. Jokela, E.J.; Wilson, D.S.; Allen, J.E. Early growth responses of slash and loblolly pine following fertilization and herbaceous weed control treatments at establishment. *South. J. Appl. For.* **2000**, *24*, 23–30. [CrossRef]
218. Cremer, K.W. Nature and impact of damage by wind, hail and snow in Australia's pine plantations. *Aust. For.* **1984**, *47*, 28–38. [CrossRef]
219. Helms, J.A. *Dictionary of Forestry*; Society of American Foresters: Bethesda, MD, USA, 1998; p. 99.
220. Morris, A.R. Five-year results from a blanking trial with *Pinus patula*. *S. Afr. For. J.* **1997**, *178*, 1–7.
221. Zhao, D.; Bullock, B.P.; Montes, C.R.; Wang, M. Production, tree size inequality and growth dominance in loblolly pine plantations under different silvicultural management regimes. *For. Ecol. Manag.* **2020**, *526*, 120594. [CrossRef]
222. Amateis, R.L.; Burkhart, H.E. Rotation-age results from a loblolly pine spacing trial. *South. J. Appl. For.* **2012**, *36*, 11–18. [CrossRef]
223. Caulfield, J.P.; South, D.B.; Somers, G.L. The price-size curve and planting density decisions. *South. J. Appl. For.* **1992**, *16*, 24–29. [CrossRef]
224. South, D.B.; VanderSchaaf, C.L. Should Forest Regeneration Studies Have More Replications? *Reforesta* **2017**, *3*, 19–30. Available online: https://journal.reforestationchallenges.org/index.php/REFOR/article/view/51/pdf (accessed on 1 December 2022). [CrossRef]
225. Larsen, H.S.; South, D.B.; Williams, H.M.; Boyer, J.N. Survival and early growth of loblolly pine seedlings from six nurseries. In *5th Biennial Southern Silvicultural Research Conference Proceedings*; GTR SO-74; United States Department of Agriculture Forest Service, Southern Forest Experiment Station: New Orleans, LA, USA, 1989; pp. 151–154.
226. Oliet, J.A.; Planelles, R.; Artero, F.; Valverde, R.; Jacobs, D.F.; Segura, M.L. Field performance of *Pinus halepensis* planted in Mediterranean arid conditions: Relative influence of seedling morphology and mineral nutrition. *New For.* **2009**, *37*, 313–331. [CrossRef]
227. Pickens, B.; Rohr, H. *The Effects of Fall Planting on the Performance of Containerized Longleaf Pine Seedlings*; Technical Resource Bulletin TRB-008; North Carolina Forest Service: Raleigh, NC, USA, 2012; 4p. Available online: https://www.ncforestservice.gov/publications/techBulletins/TRB008.pdf (accessed on 1 December 2022).

228. Slesak, R.A.; Kelso, S.G.; Windmuller-Campione, M.A. Effect of biochar and manual vegetation control on early growth and survival of planted jack pine (*Pinus banksiana* Lamb.) seedlings in northern Minnesota. *For. Sci.* **2022**, *68*, 104–112. [CrossRef]
229. Gardiner, E.S.; Yeiser, J.L. First-year survival and growth of loblolly pine seedlings released from perennial weeds in old fields of northeastern Arkansas. *Tree Plant. Notes* **1993**, *44*, 25–32.
230. Carlson, C.A.F.; Thomas, R.; Allen, H.L.; Albaugh, T.J.; Stape, J.L.; Rubilar, R.A. Effects of combination plowing on the survival and growth of loblolly pine. *Tree Plant. Notes* **2014**, *57*, 18–22.
231. Venator, C.R.; Mexal, J.G. *The Effect of Wrenching and Planting Date on the Survival of Loblolly Seedlings [Pinus taeda]*; GTR SO-34; United States Department of Agriculture Forest Service, Southern Forest Experiment Station: New Orleans, LA, USA, 1981; pp. 20–24.
232. Pikkarainen, L.; Luoranen, J.; Kilpeläinen, A.; Oijala, T.; Peltola, H. Comparison of planting success in one-year-old spring, summer and autumn plantings of Norway spruce and Scots pine under boreal conditions. *Silva Feinnica* **2020**, *54*, 10243. [CrossRef]
233. Jackson, J.K. Why do forest plantation fail. In *Strategies and Designs for Afforestation, Reforestation and Tree Planting*; PUDOC: Wageningen, The Netherlands, 1984; pp. 277–285. Available online: https://library.wur.nl/WebQuery/wurpubs/fulltext/313656#page=277 (accessed on 1 December 2022).
234. Kulhavy, D.; Watterson, K.G.; Kroll, J.; Meeker, J.R. Management and pest management considerations on droughty soils: Four years results. In *4th Biennial Southern Silvicultural Research Conference Proceedings*; GTR SE-42; United States Department of Agriculture Forest Service, Southeastern Forest Experiment Station: Asheville, NC, USA, 1987; pp. 559–562.
235. McGee, C.E.; Hatcher, J.B. Deep-planting small slash pine on old field sites in the Carolina sandhills. *J. For.* **1963**, *61*, 382–383.
236. Reynolds, J.P.; Greene, T.A.; Britt, J.R. Effects of lifting method, seedling size, and herbaceous weed control on first-year growth of loblolly pine seedlings. In *11th Biennial Southern Silvicultural Research Conference Proceedings*; GTR SRS-48; United States Department of Agriculture Forest Service, Southern Research Station: Asheville, NC, USA, 2002; pp. 51–54.
237. South, D.B.; Mitchell, R.J. Determining the "optimum" slash pine seedling size for use with four levels of vegetation management on a flatwoods site in Georgia, USA. *Can. J. For. Res.* **1999**, *29*, 1039–1046. [CrossRef]
238. Bryan, N.T. Impacts on growth and quality of interplanting loblolly pine (*Pinus taeda* L.) seedlings with clonal material in the lower coastal plain of North Carolina. Master's Thesis, North Carolina State University, Raleigh, NC, USA, 2014.
239. Hatchell, G.E.; Marx, D.H. Response of longleaf, sand, and loblolly pines to *Pisolithus* ectomycorrhizae and fertilizer on a sandhills site in South Carolina. *For. Sci.* **1987**, *33*, 301–315.

forests

MDPI

Article

Potential Use of Two Forest Species (*Salix alba* and *Casuarina glauca*) in the Rhizofiltration of Heavy-Metal-Contaminated Industrial Wastewater

Malek Bousbih [1,2], Mohammed S. Lamhamedi [3], Mejda Abassi [2], Damase P. Khasa [4] and Zoubeir Béjaoui [1,2,*]

1 Faculty of Sciences of Bizerte, University of Carthage, Jarzouna 7021, Tunisia
2 Laboratory of Forest Ecology (LR11INRGREF03), National Institute of Research in Rural Engineering, Water and Forests (INRGREF), University of Carthage, Hédi Elkarray Street, Elmenzah IV, BP 10, Ariana 2080, Tunisia
3 Direction de la Recherche Forestière, Ministère des Ressources Naturelles et des Forêts, 2700, Rue Einstein, Quebec, QC G1P 3W8, Canada
4 Centre for Forest Research and Institute for Systems and Integrative Biology, Université Laval, 1030 Avenue de la Médecine, Quebec, QC G1V 0A6, Canada
* Correspondence: zoubeir.bejaoui@fsb.ucar.tn

Abstract: The discharge of raw industrial wastewater (IWW) into ecosystems is a major environmental problem that adversely affects water quality, soil physicochemical properties, the food chain and, therefore, human health. Injection of treated IWW into irrigation and "fertigation" systems is an ecological, sustainable and economical approach for its appropriate disposal. Seedlings of two forest species (*Salix alba, Casuarina glauca)* were grown hydroponically and subjected to 25% diluted IWW and control (tap water) treatments for 35 days. Morphological and physiological traits were evaluated, including leaf symptoms, stem and root dry masses, leaf water potential, relative water content, chlorophyll content, photosystem II efficiency, hydrogen peroxide, thiobarbituric acid reactive substances, bioaccumulation and translocation factor estimates and removal efficiency for various heavy metals. Application of 25% IWW stress affected many aspects of plant morphology: chlorosis and necrosis in leaves, epinasty, leaf curling, early leaf senescence and root browning. In both species, the 25% IWW treatment reduced leaf, stem and root dry masses relative to controls. *S. alba* exhibited greater removal capacity for heavy metal ions and could be effective as a remediator of toxic-metal-polluted industrial effluent water.

Keywords: industrial wastewater; toxicity; heavy metals; *Salix alba*; *Casuarina glauca*; rhizofiltration

Citation: Bousbih, M.; Lamhamedi, M.S.; Abassi, M.; Khasa, D.P.; Béjaoui, Z. Potential Use of Two Forest Species (*Salix alba* and *Casuarina glauca*) in the Rhizofiltration of Heavy-Metal-Contaminated Industrial Wastewater. *Forests* **2023**, *14*, 654. https://doi.org/10.3390/f14030654

Academic Editor: Tiina Maileena Nieminen

Received: 9 February 2023
Revised: 14 March 2023
Accepted: 21 March 2023
Published: 22 March 2023

1. Introduction

Intensification of global industrialization has led to the release of large amounts of industrial wastewater (IWW). In North African countries, enormous volumes of IWW are dumped into the environment without sufficient treatment [1]. The main industrial users of water are electrical power stations, oil refineries, manufacturing, cooling tower usage and supplying boilers. Wastewater is discharged by industries following the extraction and smelting of metals, the production of beverages, textiles, paints, paper, plastic and tanneries, and the manufacture of pesticides and phosphate fertilizers [2]. It has been widely reported that industrial wastewater, especially discharges from the petrochemical industry, often contains high levels of the heavy metals Cd, Co, Fe, Mn, Ni and Zn, with concentrations of 0.015, 0.13, 2.98, 0.95, 1.57 and 8.07 mg L^{-1}, respectively, observed [3]. Along with heavy metals, IWW carries small quantities of various organic chemical contaminants, such as detergents and fossil fuels from the petroleum industry. Indeed, they are loaded with many constituents: oils, fats, suspended solids and organic matter of various origin, and they are often characterized by high levels of salts [4]. Heavy metals in IWW are important pollutants given their persistence in the environment and at high concentrations can induce toxicity through the formation of free radicals that cause oxidative stress [5]. These heavy metals are by nature non-biodegradable and chemically non-oxidizable and tend to reduce soil quality by inhibiting the development of microflora and through their accumulation in superficial horizons, or they are transported in runoff to groundwater and surface water bodies, contaminating them [6]. Heavy metals can also be bioconcentrated with increasing trophic levels [6].

Faced with these negative effects of IWW and the increasing public awareness of environmental issues, it is essential to find solutions that would limit the risks associated with pollutants in wastewater. Different electrochemical, physical and chemical methods have been developed to treat IWW [7]. These methods are robust but face major disadvantages, such as low selectivity, limiting the removal of metals from industrial wastewater [8]. In fact, the overall goal of wastewater treatment is to treat it in the most cost-effective manner as possible in an environmentally friendly manner [9]. Indeed, scientists continually search for the best way to purify IWW at minimum expense and with maximum efficiency. Currently, biological techniques are strongly recommended as they are more in line with the principles of sustainable development. Rhizofiltration is an ecological and promising technology applicable to the remediation of contaminated water by using the plant root system to fix, extract, immobilize and adsorb toxic ions from soil or water solution [10]. This strategy combines phytoextraction and phytostabilization [10]. It is a relatively inexpensive phytotechnology, and serves as a clean and effective alternative solution for reducing water pollution [11]. Thus, phytoremediation methods seem to overcome heavy metal pollution even at low concentrations [12,13].

More than 450 plant species have been identified as being useful in rhizofiltration, ranging from floating aquatic plants such as *Eichhornia crassipes* [14], *Pistia stratiotes* [14], *Nuphar* and *Limnobium spongia* [15] to emergent aquatics such as *Scirpus*, *Phalaris arundinacea* and *Juncus* [15] to submerged aquatics such as *Elodea* [15]. This identification has continued with herbaceous terrestrial species that are grown in hydroponic systems, such as *Brassica juncea* [16], *Helianthus annuus* [17] and *Zea mays* [15]. These herbaceous plants are characterized by low biomass production [18]. In this study, interest is directed towards woody forest species that would facilitate accumulation of high levels of pollutants in their biomass [18]. Therefore, these species offer an opportunity for ecologically mediated disposal of wastewater as a source of irrigation and fertilization. This can be achieved via their ability to transpire large amounts of water, the production of large quantities of biomass, long lifespans, deeply penetrating root systems [18], rapid sap rise, rapid vegetative growth [19], ease of handling and harvesting [19] and their regrowth potential [19].

The use of poplars (*Populus*) and willows (*Salix*) has been shown to recycle water that is high in salinity and rich in boron [20] to treat larger amounts of dissolved pollutants including Cd and Zn, and to absorb excess nitrates and phosphates. For example, the spontaneous hybrid of *Salix viminalis* L. and *S. caprea* (*Salix* × *smithiana Willd.*) is characterized by a high capacity to accumulate Cd and Zn in the shoots [21]. *Paulownia* [22] and *Eucalyptus* [23] are promising genera, given the ability of their species to tolerate high concentrations of metals [24]. Seedling roots of *Casuarina glauca* Sieb ex Spreng can scavenge Cd, Pb, Ni and Zn ions at concentrations of 0.13, 1.31, 0.53 and 6.92 mg L^{-1}, and with removal efficiencies reaching 92%, 77%, 83% and 73%, respectively [13]. These plant species, especially their roots, show a wide range of tolerance to wastewater and heavy metal accumulation [13] via inhibiting the translocation of metals from roots to shoots (rhizoaccumulation), which is an essential process for phytostabilization purposes [25]. Nevertheless, it is widely accepted that rhizofiltration with hyperaccumulator woody species is a very interesting and feasible approach. These species can treat several types of contaminants, even at low concentrations [12].

Therefore, the present study was conducted to evaluate (*i*) the ecotoxicity of industrial wastewater and its effect on the ecophysiology of two tree species, i.e., *Casuarina glauca* and *Salix alba*, and (*ii*) to evaluate their rhizofiltration potential for possible applications in phytoremediation.

2. Materials and Methods

2.1. Plant Material

Casuarina glauca seedlings of the same size (mean ± SD: height, 80.0 ± 3.0 cm; root collar diameter, 5.0 ± 1.2 mm, n = 36) were produced from seeds at the forest nursery of El Agba, Manouba, Tunisia (36°46′34′′ N, 8°41′05′′ E). *Salix alba* seedlings (height, 20.0 ± 1.4 cm; diameter, 10.8 ± 1.4 mm, n = 36) were produced from cuttings in Ain Drahem, Tunisia (36°46′34′′ N, 8°41′05′′ E). Techniques and cultural practices for seedling and cutting production in modern and traditional forest nurseries are described in detail in our previous publications [26–28].

2.2. Experimental Design

The experimental design was installed at the National Institute for Research in Rural Engineering, Water and Forests (INRGREF), Tunisia (36°50′ N, 10°14′ E, 3 m asl). *Casuarina glauca* and *Salix alba* seedlings were placed under hydroponic conditions in plastic pots (20 cm diameter × 17 cm deep; volume, 5.3 L) containing tap water. The experimental design consisted of six complete random blocks (four pots per block and twelve pots per species). In each pot, three seedlings of the same species were fixed in holes in the lids of the pots. Seedlings were kept suspended by strips of Styrofoam, while half of their length was subjected to external environmental conditions (photoperiod, 14 h; daily temperature, 29.6 ± 4 °C; relative humidity, 64 ± 5%) and the other half was submerged. The medium was renewed three times each week. Both species were subjected to an acclimatization period of 3 months or 90 days (March–June 2021) to allow full development of leaves and roots. Five different concentrations of IWW (0, 25, 50, 75, 100% IWW) were used to evaluate the ecotoxicity of industrial wastewater and its effect on the ecophysiology of *Casuarina glauca* and *Salix alba*, but a concentration of IWW higher than 25% was found to be lethal to seedlings. Thus, only two concentrations (treatments) were tested: 0% IWW (control: C) and 25% IWW (stressed treatment: S). Six pots of each species remained filled with tap water, and the plants served as controls (control: C, tap water). For the other six pots of each species, tap water was replaced with 25% IWW, whereby the seedlings were stressed (stress: S, 25% IWW). A total of 72 seedlings were deployed as 3 seedlings × 2 species × 2 treatments × 6 blocks. The treatments and the species were randomly distributed within each block. Once the roots were saturated with heavy metals [29], the morphology and growth were affected; they were then harvested and removed [29]. The treatment lasted 35 days. The experiments were conducted in a greenhouse and the pots were continuously aerated using two types of oxygen pumps (one pump per four pots). In assessing effects of industrial wastewater on growth and the appearance of several symptoms in *Casuarina glauca* and *Salix alba* seedlings, visual checks of shoot and root morphology were made every three days.

2.3. Composition of Industrial Wastewater

The wastewater that was used in this study was collected from the Tunisian Electricity and Gas Company (STEG) in La Goulette, Tunisia (36°49′09′′ N, 10°18′22′′ E). The initial physicochemical composition of industrial wastewater prior to dilution (100% IWW) is summarized in Table 1, which is compared with World Health Organization standards [30]. The physicochemical characterization of diluted industrial wastewater (25% IWW) was carried out on two dates: before purification (25% IWW) and at the end of the experiment after 35 days of purification (treated diluted industrial wastewater: T 25% IWW). T 25% IWW was collected in sterilized plastic bottles and transported to STEG for further laboratory analyses. Heavy metal concentrations were determined via flame atomic absorption spectrometry (ContrAA 300, ANALYTIK JENA, France). The determination of nitrite (NO_2^-), nitrate (NO_3^-) and suspended matter (SM) was conducted using visible spectrophotometry (DR 2800 model, HACH, Loveland, CO, USA). Electrical conductivity (EC, $\mu S\ cm^{-1}$) and the hydrogen ion potential (pH) were measured using a multiparameter probe (HQ30D model, HACH).

Table 1. Composition of initial industrial wastewater before dilution (100% IWW) and World Health Organization (WHO) standard. Three replications ± SD of each variable.

Variable	100% IWW	WHO Standard 2021
pH $_{H_2O_2}$	7.46 ± 0.01	6.5–8.4
CE (mS cm^{-1})	5.31 ± 0.02 *	3
MES (mg L^{-1})	200.0 ± 0.009 *	100
Si (mg L^{-1})	15.6 ± 0.13	–
Ca (mg L^{-1})	660.0 ± 0.02 *	200
Mg (mg L^{-1})	240.0 ± 0.005 *	150
Zn (mg L^{-1})	1.56 ± 0.002	2
Na (mg L^{-1})	2300.0 ± 0.9 *	200
Mn (mg L^{-1})	1.12 ± 0.09 *	0.2
K (mg L^{-1})	60.8 ± 0.04	–
Co (mg L^{-1})	1.28 ± 0.11 *	0.05
Fe (mg L^{-1})	0.894 ± 0.017 *	0.2
Cu (mg L^{-1})	0.58 ± 0.007	2
Ni (mg L^{-1})	0.008 ± 0.0005	0.07
Cr (mg L^{-1})	0.012 ± 0.008	0.05
Cd (mg L^{-1})	0.001 ± 0.0001	0.003
Mo (mg L^{-1})	0.018 ± 0.0007	0.07
NO_2^- (mg L^{-1})	156.0 ± 0.005 *	3
NO_3^- (mg L^{-1})	148.0 ± 0.002 *	30
$C_2H_6O_2$ (mg L^{-1})	128.0 ± 0.001 *	26
RD11 (mg L^{-1})	156.0 ± 0.005 *	20

pH $_{H_2O_2}$ hydrogen ion potential of water, CE electrical conductivity, MES suspended matter, Si silicon, Ca calcium, Mg magnesium, Zn zinc, Na sodium, Mn manganese, K potassium, Co cobalt, Fe iron, Cu copper, Ni nickel, Cr chromium, Cd cadmium, Mo molybdenum, NO_2^- nitrite, NO_3^- nitrate, $C_2H_6O_2$ ethylene glycol. RD11 detergents used as antifreeze and anticorrosion to protect installations. * Higher than the limits of the World Health Organization standard of wastewater.

The efficiency of the 25% IWW treatment with *C. glauca* and *S. alba* seedlings was evaluated as described by Wang et al. [31] using the following formula:

$$E\ (\%) = (C_i - C_f/C_i) \times 100 \quad (1)$$

where E (%) is the efficiency of treatment, C_i is the initial elemental concentration in the 25% IWW and C_f is the final elemental concentration after 35 days in the T 25% IWW.

2.4. Determination of Leaf, Stem and Root Dry Masses

Dry mass was obtained after rinsing the roots with tap water and then reweighing the seedlings before and after 48 h at 70 °C. The seedlings were separated into stems, leaves and roots. Dry mass (g DM/seedling) was recorded using a precision balance with three decimal places (0.001g).

2.5. Determination of Chlorophyll Concentration (SPAD Values) and Fluorescence

Chlorophyll concentrations were measured on the leaves (or needles) that had been most exposed to natural environmental conditions (photoperiod, 14 h; daily temperature, 29.6 ± 4 °C; relative humidity, 64 ± 5%) with a chlorophyll meter (SPAD-502 model, Konica Minolta, Tokyo, Japan) with a measurement accuracy within ±1.0 SPAD unit [32]. SPAD-502 m provides an alternative means of measuring relative leaf chlorophyll levels that overcomes the drawbacks of the organic solvent pigment extraction method [33]. Indeed, SPAD values (generally between 0.0 and 50.0) correlate strongly with direct photometric measurements of chlorophyll (nmol cm^{-2}) that is extracted from the leaf [33]. Five randomly selected leaves (or needles of each twig) were used from each treatment combination (control: C, 0% IWW vs. stress: S, 25% IWW) per block, with five replicates per measurement.

Initial fluorescence (F_0), maximum fluorescence (F_m), variable fluorescence ($F_v = F_m - F_0$) and maximum quantum yield of PSII (F_v/F_m) were measured using a portable infrared fluorometer (LCpro-SD, ADC Bioscientific Ltd., Hoddesdon, Herts, UK). Five randomly selected leaves (or needles of each twig) from each seedling/species/treatment (C and S)/block were used, with five replicates per measurement. All measurements were taken on leaves (or needles) after dark adaptation overnight. The potential quantum yield of photosystem II (PSII), which was expressed as F_v/F_m, was calculated as follows [34]:

$$F_v/F_m = (F_m - F_0)/F_m \quad (2)$$

Chlorophyll concentrations (SPAD values) and fluorescence were measured on the middle part of leaves (or needles) and on the uppermost portion of seedlings that were fully developed [35], after 12 days from the start of the experiment.

2.6. Determination of Leaf Water Potential and Relative Water Content

Leaf water potential was measured via the Scholander pressure chamber technique (600 model, PMS Instrument Company, Albany, OR, USA) on young, fully developed leaves. The freshly harvested leaf (or needle) was inserted into a gas-tight stopper with the cut end protruding a few millimeters from the stopper. Once sealed in the chamber, pressure was applied and monitored until sap was expressed from the cut stem's surface.

Relative water content (RWC) was determined according to the protocol described by Scotti-Campos et al. [36]. Six randomly selected leaves (or needles) from each seedling/species/treatment (C and S)/block were used. The leaves (or needles) were first weighed to determine the fresh weight (FW). They were then cut at the petiole and floated in test tubes containing 15 mL of distilled water for 24 h at ambient temperature in the dark. The samples were dried on filter paper and reweighed to determine the turgor weight (TW). They were then oven-dried (80 °C for 48 h) and reweighed to obtain the dry weight (DW). RWC was calculated according to the following formula [37]:

$$RWC (\%) = [(FW - DW)/(TW - DW)] \times 100 \quad (3)$$

2.7. Determination of the Integrity of Membrane Structures, Hydrogen Peroxide (H_2O_2) and Thiobarbituric Acid Reactive Substances (TBARSs)

Membrane permeability was measured by determining electrolyte leakage induced by metallic stress. Electrolyte leakage was measured according to the protocol described by Thiaw [38]. One gram of fresh leaf material/species/treatment was cut into pieces 1 cm in length and rinsed three times with distilled water (DW) in Petri dishes, and then floated in glass tubes containing 15 mL of DW in order to eliminate the electrolytes released following excision. The samples were then placed in a water bath at 40 °C in the dark for 1 h and cooled to room temperature to measure the free conductivity (FC) of the supernatant using a conductivity meter (Cellox 325 model, Multiline P3 PH/LF-SET, WTW GmbH, Weilheim, Germany) and expressed in $\mu S\ cm^{-1}$. The value that was obtained corresponds to the residual tonoplast permeability to ions. After this measurement, these samples that had been soaked in water were placed in a water bath boiling at 100 °C for 20 min to destroy the leaf tissue. After cooling to ambient temperature, total conductivity (TC) was measured. The test was repeated five times; we determined the percentage of damage (PD%) for each treatment (C, S) according to the following formula:

$$PD (\%) = 100 - PRI \quad (4)$$

$$PRI (\%) = (PAI\ stress/PAI\ control) \times 100 \quad (5)$$

$$PAI (\%) = (1 - FC/TC) \times 100 \quad (6)$$

where PRI is the percentage of relative integrity and PAI is the percentage of absolute integrity.

Hydrogen peroxide (H_2O_2) determination was carried out according to the method described by Sergiev et al. [39] using 20 randomly selected seedlings (5 leaves or needles/bloc/treatment (C and S)/species). To do this, 0.5 g of fresh material was homogenized with 5 mL of TCA (0.1%) with a cold (4 °C) mortar and pestle (i.e., 10 mL/g of fresh material). The homogenate was centrifuged at 12,000× *g* for 15 min at 4 °C. A 0.5 mL aliquot of the supernatant was mixed with 0.5 mL of phosphate buffer (10 mM, pH = 7) and 1mL of potassium iodide (KI) (1 M). Under the same conditions, a range of suitable concentrations of hydrogen peroxide were used for calibration. Absorbance was determined at 390 nm using a UV spectrophotometer (UV-3100 Model, Xian Yima Optoelec Co., Ltd., Xi'An, Shaanxi, China). The concentration of H_2O_2 was calculated using its molar extinction coefficient (ε = 39.4 mM^{-1} cm^{-1}).

Lipid peroxidation was performed by determining the concentration of thiobarbituric acid reactive substances (TBARSs) using the method of Heath et al. [40]. Twenty seedlings (5 leaves or needles/bloc/treatment (C and S)/species) were selected, and 1 g was ground in 10 mL of extraction buffer consisting of thiobarbituric acid (TBA: 0.5%) (w/v), trichloroacetic acid (TCA: 10%) (w/v) and 0.2 mM EDTA. The ground material was heated in a water bath at 95 °C for 30 min. The reaction between TBA and endoperoxides, primarily MDA (malondialdehyde), leads to the formation of a TBA-MDA complex [41]. The enzymatic reaction was stopped by immediate cooling in an ice bath at 4 °C. After centrifugation at 1000 rpm for 10 min, absorbance of the supernatants was measured at 532 nm and 600 nm against a blank using a UV spectrophotometer (UV-1200 Model, Tomos Life Science Group Pte, Ltd., China). TBARS concentrations were calculated using the extinction coefficient (ε = 155 mM^{-1} cm^{-1}).

2.8. Potential for Rhizofiltration

At the end of the experiment, the shoot and root parts of the seedlings of *C. glauca* and *S. alba* were harvested, rinsed with distilled water, dried separately at 105 °C for 72 h and ground. For each species, five seedlings per treatment (C and S) were randomly selected, after which 0.2 g samples were transferred into numbered Teflon tubes to be digested in HNO_3-H_2SO_4-$HClO_4$ (4:4:2) for 2 h at 300 °C. After digestion and mineralization, the mixture was filtered, and then the volume was adjusted to 100 mL with distilled water. Quantification of metallic trace elements was conducted using a flame mode atomic absorption spectrometer (ContrAA 300, ANALYTIK JENA, France). Nitrite and nitrate assays were carried out with a visible spectrophotometer (DR 2800, HACH). Results are expressed in mg g^{-1} DM.

The phytoremediation potential of *Salix alba* and *Casuarina glauca* was assessed at the end of the experiment, using bioaccumulation (BAF) and translocation (TF) factors [42]:

$$\text{BAF} = \text{ion concentration in the seedling/ion concentration in 25\% IWW} \quad (7)$$

$$\text{TF} = \text{ion concentration in shoots/ion concentration in roots} \quad (8)$$

2.9. Statistical Analyses

Statistical analyses were performed with SPSS 22.0 software (IBM, Armonk, NY, USA). One-way ANOVA [43] was conducted on each response variable. Means comparisons of the measurements (dry mass seedling growth, water status, chlorophyll concentration and fluorescence, TBARSs, H_2O_2 and integrity of membrane structures) relating to the two treatments (control C vs. stress S) in the two species (*S. alba* and *C. glauca*) was carried out with Student–Newnan–Keuls (SNK) tests at a significance level of 5%. Each value is presented as the mean ± standard deviation (SD).

3. Results

3.1. Physicochemical Composition of Initial Industrial Wastewater before Dilution (100% IWW)

The compositions of 100% IWW and World Health Organization (WHO) standard values [30] are summarized in Table 1. This 100% IWW exhibited a high electrical conductivity and a high load of suspended matter relative to the standard reference values demanded by the WHO (Table 1). Concentrations of Ca, Mg, Na, Mn, Fe, Co, NO_2^- and NO_3^- were elevated in 100% IWW (Table 1). The 100% IWW included two types of detergents, ethylene glycol and RD11, whose concentrations were almost 5-fold and 8-fold higher, respectively, than the standard WHO values (Table 1).

3.2. Morphological Aspects and Growth of Seedlings

For *Salix alba* seedlings that were treated with 25% IWW (stressed treatment: S), the appearance of chlorosis and progressive necrotic patches on the edges of the leaf blades were recorded after 6 days (Figure 1B). In addition, preferential downward inclination of the leaves (=epinasty) (Figure 1D) that was accompanied by leaf rolling was observed after six days (Figure 1C). For *Casuarina glauca* seedlings that were treated with 25% IWW (stressed treatment: S), progressive drying of the needles was recorded beyond day 15 (Figure 1F).

The stressed treatment (S) caused browning of the initial root system in the two species from day 3 onward. After 12 days, we noted the emergence of new white roots in seedlings of *S. alba* treated with 25% IWW (Figure 2C), followed by budding of the buds at about day 15 (Figure 2A). Additionally, beyond 21 days, the emergence of new leaves of a smaller size was noted in *Salix alba* that was treated with 25% IWW (S) (Figure 2B). At the end of the treatments, the stressed treatment that was imposed on the seedlings of *S. alba* (S) and *C. glauca* (S) caused a significant loss of leaf dry mass (Figure 3A,D; $p < 0.001$, $p < 0.017$), stem dry mass (Figure 3B,E; $p < 0.043$, $p < 0.051$) and root dry mass (Figure 3C,F; $p < 0.01$, $p < 0.297$) compared to controls. Subsequent reductions in the dry mass of leaves, stems and roots compared to the control were 58, 37.7 and 74.6% in *S. alba* (S), yet only 34.6, 28.3 and 24.4%, respectively, in *Casuarina glauca* (S) (Figure 3A–F).

(A) (B)

Figure 1. *Cont.*

Figure 1. Appearance of chlorosis and necrotic areas on the edges of the limbus (B), leaf rolling (C) and epinasty (D) after 6 days in seedlings of S. alba treated with 25% industrial wastewater (25% IWW) compared to control (A), and desiccation of the needles of C. glauca treated with 25% IWW after 15 days (F) compared to control (E).

Figure 2. The appearance of a recovery phase in S. alba treated with 25% of industrial wastewater (25% IWW), which manifested as bud burst after 15 days (A), emergence of new leaves after 21 days (B) and formation of white roots after 12 days (C).

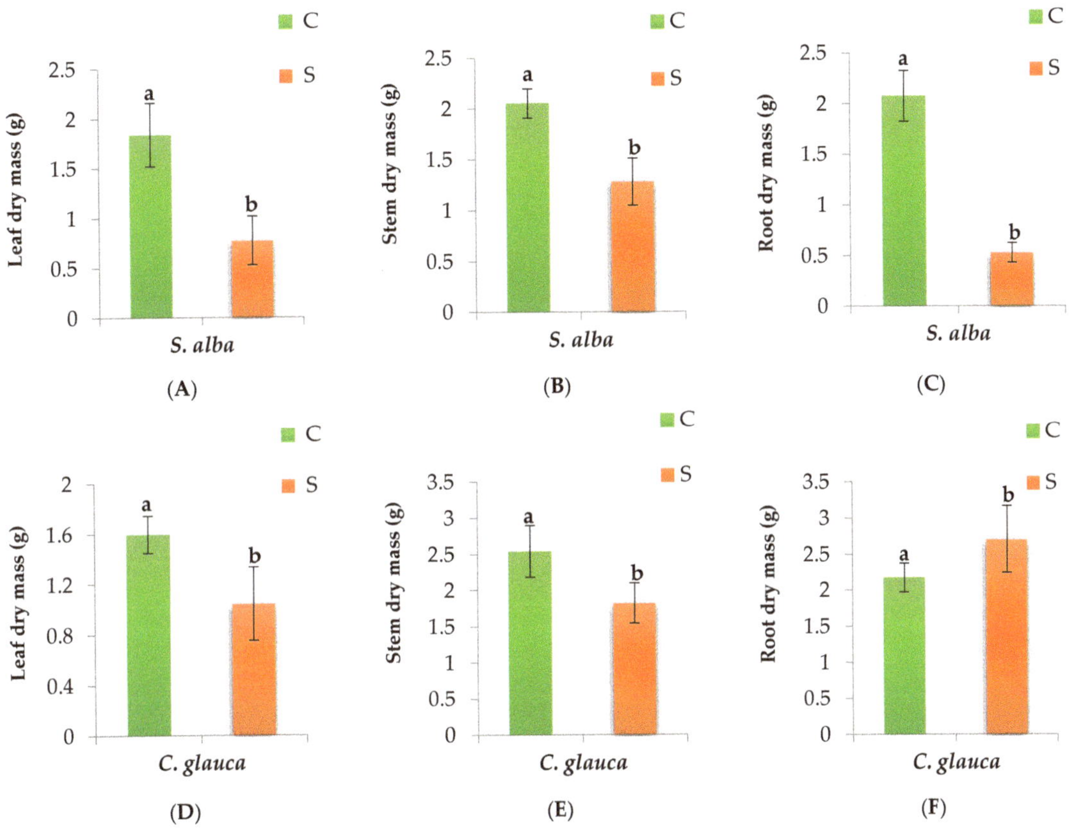

Figure 3. Comparison of leaf (**A**,**D**), stem (**B**,**E**) and root dry masses (**C**,**F**) in seedlings of *S. alba* and *C. glauca* treated with tap water (control, C) or with 25% industrial wastewater (25% IWW; stressed, S) after 35 days of treatment, (n = 5, mean ± SD). Different letters above the means indicate a significant difference at $p < 0.05$ according to Student–Newnan–Keuls (SNK) tests.

3.3. Leaf Water Potential and Relative Water Content

Tissue relative water content (RWC) was significantly reduced during the treatment of the two species, *S. alba* (S) ($p < 0.045$) and *C. glauca* (S) ($p < 0.027$) (Figure 4C). After 30 days, these reductions were 12% and 16% in *S. alba* (S) and *C. glauca* (S), respectively (Figure 4C). Treatment with 25% industrial wastewater (25% IWW) (S) that was imposed on the seedlings of *S. alba* (S) and *C. glauca* (S) significantly lowered leaf water potential (Figure 4A). After 30 days, these values reached 0.6 and 1.21 MPa in *S. alba* (S) ($p < 0.005$) and *C. glauca* (S) ($p < 0.0001$), respectively (Figure 4A).

3.4. Chlorophyll Fluorescence and the Concentration of Chlorophyll (SPAD Value)

The (Fm-Fo)/Fm value reflects the photosynthetic efficiency of PS II in using light for photochemical conversion. Application of 25% industrial wastewater (25% IWW) significantly reduced chlorophyll fluorescence in *C. glauca* (S) ($p < 0.0001$), reaching 21% that of the control (Figure 4D). Moreover, the application of this treatment for 12 days caused a significant degradation of chlorophyll, which is explained by the decrease in SPAD values in *S. alba* (S) ($p < 0.0001$) and in *C. glauca* (S) ($p < 0.0001$) as compared to the control (Figure 4B). This reduction was 56% and 42%, respectively (Figure 4B).

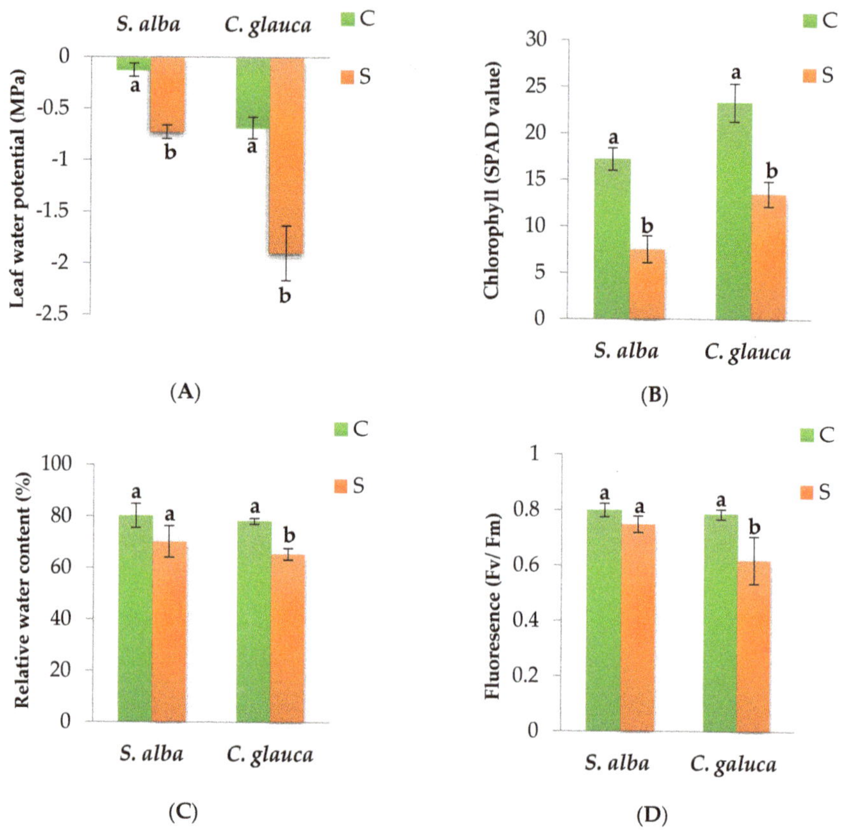

Figure 4. Variation in leaf water potential (**A**), SPAD value (**B**), relative water content (**C**) and chlorophyll fluorescence (**D**) measured after 12 days in *S. alba* and *C. glauca* treated with 25% industrial wastewater (25% IWW; stress, S) or with tap water (control, C), (n = 5, mean ± SD). Different letters above the means indicate a significant difference at $p < 0.05$ according to SNK tests.

3.5. Concentration of Thiobarbituric Acid Reactive Substances (TBARSs), Hydrogen Peroxide H_2O_2 and Percentage of Damage to Membrane Structures

The stress treatment (25% IWW) significantly increased *TBARS* concentrations in *S. alba* (S) ($p < 0.042$) and in *C. glauca* (S) ($p < 0.006$) (Figure 5B). These rates of increase were similar and averaged 50%. These results show that stress induced a significant increase in H_2O_2 levels in *S. alba* (S) ($p < 0.0001$) and in *C. glauca* (S) ($p < 0.0001$) (Figure 5A). The rates of increase were 1000% and 130%, respectively. The stress treatment significantly affected the stability of membrane structures in both species (Figure 5C). The percentage of damage to the membrane structures compared to the control was 15% and 65% in *S. alba* (S) and *C. glauca* (S), respectively (Figure 5C).

3.6. Accumulation and Compartimentation

Accumulation of ions (Na, Fe, Co and Mn) varied according to the species and the component being considered (shoot or root), as indicated in Table 2. In stressed seedlings of *Salix alba* (S), compared to the control, Fe and Mn levels decreased in leaves and increased in roots, while the opposite trend was observed for Na. With respect to Co concentrations, an increase was recorded in the shoots and roots of stressed seedlings (Table 2). In *C. glauca* (S), the accumulation of Co ions was greater in shoots than that observed for the roots, while the opposite was true for Na, Fe and Mn (Table 2). It should be noted that in *S. alba* (S), the ions that were measured (Na, Mn, Co and Fe) in old senescent leaves showed clear accumulation, reaching concentrations of 0.4, 0.015, 0.03 and 0.08 mg g^{-1}, respectively (Table 2).

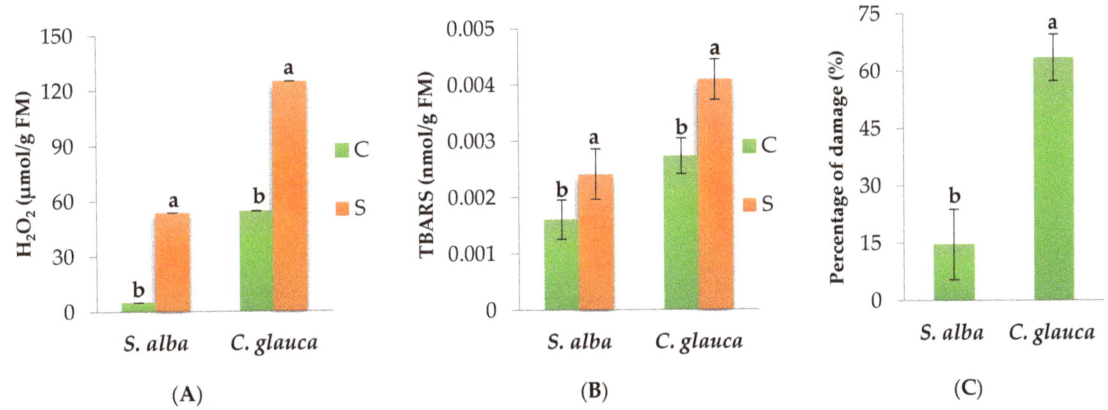

Figure 5. Variation in the concentrations of hydrogen peroxide H_2O_2 (**A**), *thiobarbituric acid reactive substances (TBARSs)* (**B**) and percentage damage to membrane structures (**C**) in *S. alba* and *C. glauca* treated with IWW (S, stress) or with tap water (C, control), (n = 5; mean ± SD). Different letters above the means indicate a significant difference at $p < 0.05$ according to SNK tests.

For the bioaccumulation (BAF) and translocation (TF) factors (Table 3) of the toxic ions, the BAF followed the order Na < Mn < Co < Fe and Na < Co < Mn < Fe for *S. alba* (S) and *C. glauca* (S), respectively (Table 3).The BAF was greater than 1 for Fe in *C. glauca* (S) and the young leaves of *S. alba* (S) and for Co only in the young leaves of *S. alba* (S) (Table 3). The TF is an indicator that aids in understanding the mobility of ions in the seedling. The TF followed the order Fe < Na < Mn < Co in newly formed *S. alba* (S) leaves, with the TF being greater than 1 for Co (Table 3). At the level of senescent leaves in *S. alba* (S), the TF followed the order Fe < Na < Co < Mn, where all values were greater than 1 (Table 3). In *C. glauca* (S), the TF followed the order Fe < Mn < Na < Co (Table 3).

Table 2. Variation in Na, Mn, Co and Fe ion concentrations in the shoot and roots of *S. alba* and *C. glauca* seedlings subjected to stress for 35 days imposed by 25% of industrial wastewater (S, stressed treatment) or in optimal conditions with tap water (C, control). Three replications ± SD of each parameter were carried out.

			Ions (mg g^{-1} DM)			
			Na	Mn	Co	Fe
S. alba	Root	C	0.47 ± 0.002 a	0.006 ± 0.0003 a	0.007 ± 0.0002 a	0.03 ± 0.008 a
		S	0.44 ± 0.002 a	0.008 ± 0.0003 b	0.02 ± 0.001 b	0.12 ± 0.002 b
	Shoot	C	0.15 ±0.001 a	0.01 ± 0.005 a	0.008 ± 0.0003 a	0.04 ± 0.005 a
		S	0.18 ± 0.001 b	0.006 ±0.005 a	0.02 ± 0.003 b	0.03 ± 0.005 a
C. glauca	Root	C	0.62 ± 0.003 a	0.008 ± 0.0007 a	0.002 ± 0.0008 a	0.07 ± 0.003 a
		S	0.67 ± 0.003 b	0.01 ± 0.003 b	0.005 ± 0.0006 b	0.18 ± 0.002 b
	Shoot	C	0.54 ± 0.001 a	0.005 ± 0.0003 a	0.008 ± 0.0004 a	0.08 ± 0.005 a
		S	0.56 ± 0.001 a	0.008 ± 0.0003 b	0.009 ± 0.0003 b	0.05 ± 0.002 a

Na sodium, Mn manganese, Co cobalt, Fe iron. Note: for each element, means followed by distinct letters within the root or aerial parts show significant differences at the 5% level according to SNK tests.

Table 3. Bioaccumulation factor (BAF) and translocation factor (TF) in seedlings of *C. glauca* (S, stress) and *S. alba* (S, stress) treated with 25% industrial wastewater (S, stressed treatment) over 35 days.

Ions (mg g^{-1} DM)	*S. alba* (S)				*C. glauca* (S)	
	Young Leaves		Senescent Leaves			
	TF	BAF	TF	BAF	TF	BAF
Na	0.41	0.001	2.26	0.0007	0.84	0.002
Mn	0.75	0.7	5	0.05	0.8	0.11
Co	1.15	1.12	4	0.13	0.86	0.14
Fe	0.2	1.02	1.1	0.4	0.3	1.05

Na sodium, Mn manganese, Co cobalt, Fe iron. Note: early senescence was only recorded in *S.alba* seedlings (S).

3.7. Rhizofiltration Potential

Physicochemical and heavy metal data of diluted industrial wastewater (25% IWW) and treated diluted industrial wastewater (T 25% IWW) are summarized in Table 4. The pH indicated a strong reduction, reaching 23.9% for *C. glauca* (S), while a low increase of 6.5% was attained for *S. alba (S)* (Table 4). Electrical conductivity (EC) was 44.8% and 42.1% lower after T 25% IWW treatment for *S. alba* (S) and *C. glauca* (S), respectively, as compared to the EC of 25% IWW treated plants (Table 4). Efficiency of treatment (E) significantly increased in *S. alba* (S) seedlings, compared to *C. glauca* (S), for Ca, Mg, Na, K, NO_2^-, NO_3^-, Mn, Co, Fe, Cu, Ni, Cr, Cd and Mo (Table 4). All ion concentrations were significantly reduced in T 25% IWW, indicating the efficiency of these two forest species, which are used to reduce or remove toxic ions from 25% IWW (Table 4).

Table 4. Physicochemical parameters (mean ± SD, n = 3) of diluted untreated industrial wastewater (25% IWW) over time: d = 0 day, of the diluted treated industrial wastewater (T 25% IWW) with *S. alba* or *C. glauca* after 35 days of treatment and the efficiency of this treatment (E) for these 2 species. Three replications ± SD of each parameter were carried out.

Elements	25% IWW (d = 0 day)	T 25% IWW + *S. alba*, T 25% IWW + *C. glauca* (d = 35 days)			
		T 25% IWW	E (%)	T 25% IWW	E (%)
$pH_{H_2O_2}$	7.46 ± 0.01	7.95 ± 0.01	–	5.67 ± 0.01	–
CE (mS cm^{-1})	5.31 ± 0.02 *	2.93 ± 0.02	–	3.07 ± 0.02	–
Ca (mg L^{-1})	165.0 ± 0.02	116.0 ± 0.02	29.6	133.5 ± 0.02	19
Mg (mg L^{-1})	60.0 ± 0.005	38.5 ± 0.005	35.8	42.0 ± 0.005	30
Zn (mg L^{-1})	0.389 ± 0.002	<LQ	100	0.16 ± 0.002	58.8
Na (mg L^{-1})	575.0 ± 0.9 *	460.0 ± 0.9 *	20	495.0 ± 0.9 *	13.9
Mn (mg L^{-1})	0.28 ± 0.09 *	0.025 ± 0.003	89.3	0.155 ± 0.030	42.9
K (mg L^{-1})	15.2 ± 0.04	12.8 ± 0.04	15.7	12.7 ± 0.04	16.4
Co (mg L^{-1})	0.32 ± 0.11 *	0.042 ± 0.010	87.5	0.1 ± 0.01 *	68.8
Fe (mg L^{-1})	0.223 ± 0.017 *	0.149 ± 0.017	31.8	0.168 ± 0.017	22.7
Cu (mg L^{-1})	0.144 ± 0.007	<LQ	100	<LQ	100
Ni (mg L^{-1})	0.002 ± 0.0005	<LQ	100	<LQ	100
Cr (mg L^{-1})	0.002 ± 0.0008	<LQ	100	<LQ	100
Cd (mg L^{-1})	0.0004 ± 0.0001	<LQ	100	<LQ	100
Mo (mg L^{-1})	0.005 ± 0.0007	<LQ	100	<LQ	100
NO_2^- (mg L^{-1})	39.0 ± 0.005 *	5.8 ± 0.005 *	85.1	6.1 ± 0.005 *	84.3
NO_3^- (mg L^{-1})	37.0 ± 0.002 *	4.2 ± 0.002	88.6	5.9 ± 0.002	84.3

pH $_{H_2O_2}$ hydrogen potential of water, CE electrical conductivity, Ca calcium, Mg magnesium, Zn zinc, Na sodium, Mn manganese, K potassium, Co cobalt, Fe iron, Cu copper, Ni nickel, Cr chromium, Cd cadmium, Mo molybdenum, NO_2^- nitrite, NO_3^- nitrate. LQ limit of quantification * Higher than the limits of the World Health Organization standard.

4. Discussion

4.1. Morphological Changes and Seedling Growth

The industrial wastewater (IWW) applied at 25% IWW to *S. alba* (S, stress) and *C. glauca* seedlings (S) caused the onset of several symptoms related to the toxicity of metal ions (pronounced chlorosis, progressive necrosis, early senescence and root browning) (Figure 1B–D,F). This adverse effect of heavy metals appears at the whole-seedling level and was attributed to excessive accumulation of Mn, Co, Fe and Na (Table 2), which is associated with physiological disturbances [44] and nutritional imbalance [45]. In addition, leaf rolling in *S. alba* (S) decreases leaf area and can lead to a decrease in photosynthetic activity [46]. Similar signs were observed in *S. viminalis* [47]. Moreover, exposure to 25% IWW affected root morphology and caused progressive root browning, leading to a reduction in water and mineral uptake by the roots. Such morphological disturbances could be attributed to a calcium deficiency [48] or to an accumulation of potentially toxic elements, thereby limiting their transport to the leaves [49]. This highlights the premature senescence of older leaves and the appearance of a recovery phase with the emergence of new leaves and white roots, markedly so in *S. alba* (S) (Figure 2A–C). This response has been reported to be an adaptive strategy following vacuolar sequestration of the most toxic heavy metals [50].

The 25% IWW caused a significant reduction in the growth of *S. alba* (S) and *C. glauca* seedlings (S) (Figure 3A–F). Similar results were observed with *Populus* sp. [51]. In *S. alba* (S) that was treated with 25% IWW, heavy metals were preferentially accumulated in older leaves during the process of leaf senescence (Figure 3A). The decrease in whole-plant dry mass could be a direct consequence of excessive accumulation and effects of heavy metal ions in the cells [52,53], and can also occur with higher levels of Na. Our results (Table 4) are consistent with other studies showing the stress-reducing effects of excess Na on plant growth through negative impacts on net assimilation in *Salix viminalis* plants that were irrigated with wastewater containing high Na concentrations [45].

4.2. Water Status, Chlorophyll Concentration and Chlorophyll Fluorescence

The significant decreases in relative water content (RWC) and water potential in both species (Figure 4A,C) testify to the severity of the stress that was imposed by 25% IWW. In effect, osmotic stress incurred by the excess of ions would result in a reduction in the supply of water to the leaf tissues. Beyer et al. [43] showed that excess metallic elements limit absorption of essential ions and can cause a hydromineral imbalance.

This growth retardation in seedlings could be explained by a dysfunction of the photosynthetic system due to an impoverishment (depletion) of chlorophyll pigments (Figure 4B). This reduction has been attributed to an inhibition of the chlorophyll biosynthetic pathway [54], as seen in *Populus* [55], and the formation of a proteolytic enzyme responsible for chloroplast degradation [54]. Regardless, under the effect of 25% IWW treatment, degradation of chlorophyll pigments (Figure 4B) could be attributed to a destruction of the chloroplast membranes, an increase in the activity of chlorophyllase or a reduction in the absorption of Mg ions [56] or Fe, which is an essential cofactor for the synthesis of chlorophyll [57]. In *S. alba* (S) and *C. glauca* (S), this decline in chlorophyll (Figure 4B) led to a decrease in the functioning of the photosynthetic system II (PSII) (Figure 4D). This loss of PS II effectiveness would generally result in stomatal closure [58], which is often considered an indicator of root stress [59]. It should be noted that the leaf rolling recorded in *S. alba* (S) (Figure 1C) contributes to limiting light interception and would result in a reduction in transpiration [60]. It was shown that irrigation with wastewater caused a significant reduction in photosynthetic characteristics in two cultivars of *Triticum aestivum* L. (wheat), i.e., Chamran and Behrang, most notably in terms of chlorophyll fluorescence (Fv/Fm) and photosynthetic pigments [3].

4.3. Traits Promoting Metal Stress Tolerance

Our results show that 25% IWW toxicity caused significant tissue accumulation of hydrogen peroxide (H_2O_2) (Figure 5A). Heavy metals induce the overproduction of reactive oxygen species (ROS), in particular H_2O_2 [61], which are likely to cause membrane disruption [12] and an alteration in the photosynthetic electron transport chain [62]. Other observations showed that high H_2O_2 concentrations induced severe disturbances at the cellular level and generally translate into electrolyte leakage [63]. Furthermore, H_2O_2 is a cellular response that signals the presence of stress [64]. At low concentrations, H_2O_2 can function as a secondary messenger that activates antioxidant defenses [65], as is the case in *S. alba* seedlings (S) (Figure 5A). This defensive behavior in *S. alba* (S) resulted in greater stability of membrane structures (Figure 5C). Membranes are selective barriers that control the diffusion of ions. They must overcome the harmful effects of heavy metals [66], generally leading to the loss of membrane integrity [67]. Our results show a significant increase in thiobarbituric acid reactive substances (TBARSs) in both species (Figure 5B), i.e., a cytotoxic product of membrane lipid peroxidation [68].

4.4. Metal Bioaccumulation and Rhizofiltration Potential

For dosed ions (Na, Mn, Co, Fe), seedlings of both species showed variability in extraction capacity (Table 3). This behavior is correlated with a strategy for excluding heavy metals opted by seedlings that minimizes their harmful effects [69]. Weak translocation of heavy metals towards aerial parts is considered as a defense mechanism in seedlings to protect the photosynthetic apparatus [70]. About 75 to 90% of the heavy metals that are adsorbed by the seedling are blocked at the level of the roots [71]. This limitation of the mobility of heavy metals results from their complexation, transport and sequestration in root vacuoles [72]. Thus, the roots act as a trap organ that reduces transfer of heavy metals to aerial tissues, thereby limiting their toxicity and improving tolerance in the plant [73]. This has been considered as a resistance mechanism where an heavy metals excluding capacity is manifested in the face of high external heavy metal concentrations [73]. In stressed plants, the decrease in the concentrations of certain nutrients can be attributed to antagonistic interactions between nutrients and heavy metal exclusion [73]. As was seen in the work of Hajihashemi et al. [3], toxic levels of mineral elements in wastewater resulted in a significant drop in the content of K and Zn from the leaves in two cultivars of Triticum aestivum L. (wheat), i.e., Chamran and Behrang. The values of translocation factor TF < 1 in *C. glauca* (S) and *S. alba* (S) for Na, Mn and Fe testify to their aptitude for phytostabilization (Table 3). Comparable results have been reported by many authors [54,74]. In addition, other species limit the absorption of heavy metals in the roots due to avoidance or tolerance strategies [75]. Plants reorganize their root architecture to avoid growth in contaminated soils [76] that would maintain low ion concentrations in aerial parts [77]. Our results also showed that for senescent leaves of *S. alba* (S), TF values were >1 for all toxic ions. All concentrations of toxic ions were significantly reduced in treated diluted industrial wastewater (T 25% IWW) with *S. alba* (S) and *C. glauca* (S) compared to the 25% IWW, confirming the effectiveness of the two species in the removal of toxic ions (Table 4). Slaimi et al. [13] showed similar results with *C. glauca* seedlings (S). The preferential order of toxic ion accumulation in *C. glauca* (S) was noted as Na < Mn < Co < Fe and in *S. alba* as Na < Co < Mn < Fe. The bioaccumulation factor (BAF) was always < 1 for both species, except in *C. glauca* (S) for the Fe ion (Table 3). This high BAF value confirmed that *C. glauca* (S) could be a suitable seedling species for Fe phytoextraction as accumulator seedlings showed BAF values > 1 [13]. The BAF is an index of a plant's ability to accumulate a toxic ion depending on its concentration in the medium [78]. The high values recorded in this experiment are confirmed by Zacchini et al. [79].

The high removal efficiency (E), ranging from about 13% to 99%, which was observed in both species, testifies to their capability to extract toxic ions from 25% IWW (Table 4). Several studies have reported the use of rhizofiltration systems for the purification of water contaminated with heavy metals [80].

5. Conclusions and Research Needs

In the present study, two woody forest species, *S. alba* and *C. glauca*, were explored for the purification via rhizofiltration of diluted industrial wastewater (25% IWW) contaminated with heavy metals. This treated diluted industrial wastewater (T 25% IWW) can be used as a source of irrigation and fertilization. Heavy metals impose negative effects on seedling growth by decreasing leaf water potential, inducing morphological disturbances (chlorosis, necrosis, epinasty, leaf rolling and early senescence) and changes in physiological processes (photosynthetic dysfunction activity, a reduction in chlorophyll fluorescence, osmotic adjustment, lipid peroxidation and loss of selective permeability of membranes). The two species tested in this study have revealed great potential for removing toxic heavy metals and filtering polluted wastewater. Our results show the performance of *S. alba* seedlings for rhizofiltration applications with a preferential accumulation of heavy metals in senescent older leaves. In addition, *S. alba* seedlings showed a recovery period with the emergence of new leaves during the experiment, allowing better physiological functioning. The results make it possible to continue this study on a larger operational scale in situ by associating the two species and applying the treatment over a longer period.

Author Contributions: Conceptualization, M.B. and Z.B.; Methodology, experimental design, data collection, laboratory analyses: M.B., Z.B., M.S.L., M.A. and D.P.K.; Statistical analyses: M.B.; writing—original draft preparation, M.B., Z.B. and M.S.L.; writing—review and editing, M.B., Z.B., M.S.L., M.A. and D.P.K. All authors have read and agreed to the published version of the manuscript.

Funding: This research was funded by the University of Carthage, Ministry of Higher Education and Scientific Research, Tunisia, the National Institute of Research in Rural Engineering, Water and Forests (INRGREF), Tunisia, and the NSERC Discovery Grant to DPK to cover publication costs.

Data Availability Statement: Not applicable.

Acknowledgments: We are thankful to William F.J. Parsons for English editing. We would also like to extend our thanks to the technicians of the forest ecology laboratory of INGRREF for their valuable technical support to this project.

Conflicts of Interest: The authors declare no conflict of interest.

References

1. Alaya, M.A.; Zwiers, F.; Zhang, X. Probable Maximum Precipitation: Its Estimation and Uncertainty Quantification Using Bivariate Extreme Value Analysis. *Hydrometeorology* **2018**, *19*, 679–694. [CrossRef]
2. Chen, D.; Yu, X.; Song, C.; Pang, X.; Huang, J.; Li, Y. Effect of pyrolysis temperature on the chemical oxidation stability of bamboo biochar. *Bioresour. Technol.* **2016**, *218*, 1303–1306. [CrossRef]
3. Hajihashemi, S.; Mbarki, S.; Skalicky, M.; Noedoost, F.; Raeisi, M.; Brestic, M. Effect of Wastewater Irrigation on Photosynthesis, Growth, and Anatomical Features of Two Wheat Cultivars (*Triticum aestivum* L.). *Water* **2020**, *12*, 607. [CrossRef]
4. Ladjel, F. *Exploitation D'une Station D'épuration à Boue Active Niveau 02*; Centre de Formation au Métier de L'assainissement CFMA: Boumerdès, Algeria, 2006; p. 80.
5. Ashraf, S.; Ali, Q.; Ahmad Zahir, Z.; Ashraf, S.; Asghar, H.N. Phytoremediation: Environmentally sustainable way for reclamation of heavy metal polluted soils. *Ecotoxicol. Environ. Saf.* **2019**, *174*, 714–727. [CrossRef]
6. Alghobar, M.; Suresha, S. Evaluation of metal accumulation in soil and tomatoes irrigated with sewage water from Mysore city, Karnataka, India. *Saudi Soc. Agric. Sci.* **2017**, *16*, 49–59. [CrossRef]
7. Delgado-González, C.R.; Madariaga-Nacarrete, A.; Fernandz-Cortes, J.M.; Islas-Pelcastre, M.; Oza, G.; Iqbal, H.M.N.; Sharma, A. Advances and Applications of Water Phytoremediation: A Potential Biotechnological Approach for the Treatment of Heavy Metals from Contaminated Water. *Int. J. Environ. Res. Public Health* **2021**, *18*, 5215. [CrossRef]
8. Yan, A.; Wang, Y.; Tan, S.N.; Yusof, M.L.M.; Ghosh, S.; Chen, Z. Phytoremediation: A Promising Approach for Revegetation of Heavy Metal-Polluted Land. *Front. Plant Sci.* **2020**, *11*, 359. [CrossRef]
9. Javed, M.T.; Tanwir, K.; SohailAkram, M.; Shahid, M.; KhanNiazi, N.; Lindberg, S. Phytoremediation of Cadmium-Polluted Water/Sediment by Aquatic Macrophytes: Role of Plant-Induced pH Changes. In *Cadmium Toxicity and Tolerance in Plants from Physiology to Remediation*; Academic Press: Cambridge, MA, USA, 2019; pp. 495–529. [CrossRef]
10. Glick, B. Using soil bacteria to facilitate phytoremediation. *Biotechnol. Adv.* **2010**, *28*, 367–374. [CrossRef]
11. Khalid, K.M.; Ganjo, D.G.A. Native aquatic plants for phytoremediation of metals in outdoor experiments: Implications of metal accumulation mechanisms, Soran City-Erbil, Iraq. *Int. J. Phytoremediation* **2021**, *23*, 374–386. [CrossRef]

12. Dalcorso, G.; Fasani, E.; Manara, A.; Visioli, G.; Furini, A. Heavy Metal Pollutions: State of the Art and Innovation in Phytoremediation. *Int. J. Mol. Sci.* **2019**, *20*, 3412. [CrossRef]
13. Slaimi, R.; Abassi, M.; Béjaoui, Z. Assessment of *Casuarina glauca* as biofiltration model of secondary treated urban wastewater: Effect on growth performances and heavy metals tolerance. *Environ. Monit. Assess.* **2021**, *193*, 653. [CrossRef]
14. Gagnon, V.; Brisson, J. *Les Marais Filtrants, Fiches Techniques de la Société Québécoise de Phytotechnologie*; Société québécoise de phytotechnologie: Montréal, QC, Canada, 2014; Available online: https://phytotechno.com/fiches-techniques/ (accessed on 8 February 2023).
15. Sahay, S.; Inam, A.; Iqbal, S. Risk analysis by bioaccumulation of Cr, Cu, Ni, Pb and Cd from wastewater-irrigated soil to Brassica species. *Int. J. Environ. Sci. Technol.* **2020**, *17*, 2889–2906. [CrossRef]
16. Dhanwal, P.; Kumar, A.; Dudejau, S. Recent Advances in Phytoremediation Technology. In *Advances in Environmental Biotechnology*; Springer: Singapore, 2017; pp. 227–241. [CrossRef]
17. Hassan, S.; Bhat, S.A.; Kumar, V.; Ganai, B.A.; Ameen, F. Phytoremediation of Heavy Metals: An Indispensable Contrivance in Green Remediation Technology. *Plants* **2022**, *11*, 1255. [CrossRef]
18. Kuzovkina, Y. Compilation of the Checklist for Cultivars of *Salix* L. (Willow). *HortScience* **2015**, *50*, 1608–1609. [CrossRef]
19. Cloutier-Hurteau, B.; Turmel, M.C.; Mercier, C.; Courchesne, F. The sequestration of trace elements by willow (*Salix purpurea*) which soil properties favor uptake and accumulation. *Environ. Sci. Pollut. Res.* **2014**, *21*, 4759–4771. [CrossRef]
20. Bañuelos, G.S.; LeDuc, D.; Johnson, J. Evaluating the Tolerance of Young Hybrid Poplar Trees to Recycled Waters High in Salinity and Boron. *Phytoremediation* **2010**, *12*, 419–439. [CrossRef]
21. Tlustoš, P.; Száková, J.; Vysloužilová, M.; Pavlíková, D.; Weger, J.; Javorská, H. Variation in the uptake of Arsenic, Cadmium, Lead, and Zinc by different species of willows *Salix* spp. grown in contaminated soils. *Open Life Sci.* **2007**, *2*, 254–275. [CrossRef]
22. Azzarello, E.; Pandolfi, C.; Pollastri, S.; Masi, E.; Mugnai, S.; Mancuso, S. The use of trees in phytoremediation. *Plant Sci.* **2012**, *135*. [CrossRef]
23. Rockwood, D.L.; Rudie, A.W.; Ralph, S.A.; Zhu, J.Y.; Winandy, J.E. Energy product options for Eucalyptus species grown as short rotation woody crops. *Int. J. Mol. Sci.* **2008**, *9*, 1361–1378. [CrossRef]
24. Burges, A.; Alkorta, I.; Epelde, L.; Garbisu, C. From phytoremediation of soil contaminants to phytomanagement of ecosystem services in metal contaminated sites. *Phytoremediation* **2018**, *20*, 384–397. [CrossRef]
25. Ghazouani, S.; Béjaoui, Z.; Spiers, G.; Beckett, P. Effects of Rhizobioaugmentation with N-Fixing Actinobacteria *Frankia* on Metal Mobility in *Casuarina glauca*-Soil System Irrigated with Industrial Wastewater: High Level of Metal Exclusion of *C. glauca*. *Water Air Soil Pollut.* **2020**, *231*, 395. [CrossRef]
26. Lamhamedi, M.S.; Fecteau, B.; Godin, L.; Gingras, C. *Guide Pratique de Production en Hors Sol de Plants Forestiers, Pastoraux et Ornementaux en Tunisie*; Pampev Internationale: Manseau, QC, Canada, 2006; pp. 1–80.
27. Lamhamedi, M.S.; Abourouh, M.; Fortin, J.A. Technological transfer: The use of ectomycorrhizal fungi in conventional and modern forest tree nurseries in northern Africa. In *Advances in Mycorrhizal Science and Technology*; Khasa, D.Y., Piché, Y., Coughlan, A.P., Eds.; NRC Research Press: Ottawa, ON, Canada, 2009; pp. 139–152.
28. Sbay, H.; Lamhamedi, M.S. *Guide Pratique de Multiplication Végétative des Espèces Forestières et Agroforestières: Techniques de Valorisation et de Conservation des Espèces à Usages Multiples Face Aux Changements Climatiques en Afrique du Nord*; Royaume du Maroc, Haut Commissariat aux Eaux et Forêts et à la Lutte Contre la Désertification, Centre de Recherche Forestière: Rabat, Morocco, 2015; pp. 1–124.
29. Wuana, R.A.; Okieimen, F.E. Heavy Metals in Contaminated Soils: A Review of Sources, Chemistry, Risks and Best Available Strategies for Remediation. *Int. Sch. Res. Netw.* **2011**, *2011*, 402647. [CrossRef]
30. WHO. *World Health Organization*; WHO: Geneva, Switzerland, 2021.
31. Wang, H.; Wang, T.; Zhang, B.; Li, F.; Toure, B.; Omosa, I.B.; Chiramba, T.; Abdel-Monem, M.; Pradhan, M. Water and wastewater treatment in Africa–current practices and challenges. *Clean-Soil Air Water* **2014**, *42*, 1029–1035. [CrossRef]
32. Mendoza-Tafolla, R.O.; Juarez-Lopez, P.; Ontiveros-Capurata, R.; Sandovalvilla, M.; Alia-Tejacal, I.; Alejo-Santiago, G. Estimating Nitrogen and Chlorophyll Status of Romaine Lettuce Using SPAD and at LEAF Readings. *Not. Bot. Horti Agrobot. Cluj-Napoca* **2019**, *47*, 751–756. [CrossRef]
33. Ling, Q.; Huang, W.; Jarvis, P. Use of a SPAD-502 meter to measure leaf chlorophyll concentration in Arabidopsis thaliana. *Photosynth. Res.* **2011**, *108*, 89. [CrossRef]
34. Krause, G.H.; Weis, E. Chlorophyll fluorescence and photosynthesis: The basics. *Annu. Rev. Plant Physiol. Plant Mol. Biol.* **1991**, *42*, 313–349. [CrossRef]
35. Yuan, Z.; Cao, Q.; Zhang, K.; Ata-Ul-Karim, S.T.; Tian, Y.; Zhu, Y.; Cao, W.; Liu, X. Optimal Leaf Positions for SPAD Meter Measurement in Rice. *Front. Plant Sci.* **2016**, *7*, 719. [CrossRef]
36. Scotti-Campos, P.; Duro, N.; da Costa, M.; Pais, I.P.; Rodrigues, A.P.; Batista-Santos, P.; Semedo, J.N.; Leitao, A.E.; Lidon, F.C.; Pawlowski, K.; et al. Antioxidative ability and membrane integrity in salt-induced responses of *Casuarina glauca* Sieber ex Spreng. in symbiosis with N_2-fixing *Frankia* Thr or supplemented with mineral nitrogen. *Plant Physiol.* **2016**, *196*, 60–69. [CrossRef]
37. Turner, N.C.; Begg, J.E. Plant-water relations and adaptation to stress. *Plant Soil* **1981**, *58*, 97–131. [CrossRef]
38. Thiaw, S. *Association between Slow Leaf-Electrolyte-Leakage under Heat Stress and Heat Tolerance during Reproductive Development in Cowpea*; University of California: Riverside, CA, USA, 2003; Available online: Search.proquest.com (accessed on 8 February 2023).

39. Sergiev, P.V.; Lavrik, I.N.; Wlasoff, V.A.; Dokudovskaya, S.S.; Dontsova, O.A.; Bogdanov, A.A.; Brimacombe, R. The path of mRNA through the bacterial ribosome: A site-directed crosslinking study using new photoreactive derivatives of guanosine and uridine. *RNA* **1997**, *3*, 464–475.
40. Heath, R.L.; Packer, L. Photoperoxidation in isolated chloroplasts: I. Kinetics and stoichiometry of fatty acid peroxidation. *Arch. Biochem. Biophys.* **1968**, *125*, 189–198. [CrossRef]
41. Buege, J.A.; Aust, S.D. Microsomal lipid peroxidation. *Methods Enzymol.* **1978**, *52*, 302–310. [CrossRef]
42. Rani, P.; Kumar, A.; Arya, R.C. Stabilization of tannery sludge amended soil using Ricinus communis, *Brassica juncea* and *Nerium oleander*. *Soils Sediments* **2017**, *17*, 1449–1458. [CrossRef]
43. Steel, G.D.; Torrie, J.H.; Dickey, D.A. *Principles and Procedures of Statistics: A Biometrical Approach*, 3rd ed.; The McGraw-Hill Companies Inc.: New York, NY, USA, 1997.
44. Beyer, W.; Green, C.; Beyer, M.; Chaney, R. Phytotoxicity of zinc and manganese to seedlings grown in soil contaminated by zinc smelting. *Environ. Pollut.* **2013**, *179*, 167–176. [CrossRef]
45. Sivasankar, R.; Kalaikandhan, R.; Vijayarengan, P. Phytoremediating capability of four plant spécifs under zinc stress. *Int. J. Res. Environ. Sci. Technol.* **2012**, *2*, 1–9.
46. Martineau, C. *La Fertilisation des Plantes Ligneuses Cultivées en Contenant*; IQDHO Institut Québécois du Développement de L'horticulture Ornementale: Saint-Hyacinthe, QC, Canada, 2004; p. 57.
47. Dimitriou, L.; Aronsson, P.; Weih, M. Stress tolerance of five willow clones after irrigation with different amounts of landfill leachate. *Bioresour. Technol.* **2006**, *97*, 150–157. [CrossRef]
48. Greger, M.; Lindberg, S. Effects of Cd^{2+} and EDTA on young sugar beets (*Beta vulgaris*). II-Net uptake and distribution of Mg^{2+}, Ca^{2+}, and Fe^{2+}/Fe^{3+}. *Physiol. Plant* **1987**, *69*, 81–86. [CrossRef]
49. Breckle, S.W. Growth under stress: Heavy metals. In *Plant Roots: The Hidden Half*; Waisel, Y., Eshel, A., Kafkafi, U., Eds.; Marcel Dekker: New York, NY, USA, 1991; pp. 351–373.
50. Thakur, S.; Singh, L.; Wahid, Z.; Siddiqui, M.; Atnaw, S.; Din, M. Plant-driven removal of heavy metals from soil: Uptake, translocation, tolerance mechanism, challenges, and future perspectives. *Environ. Monit. Assess.* **2016**, *188*, 206. [CrossRef]
51. DiBaccio, D.; Tognetti, R.; Minnocci, A.; Sebastiani, L. Responses of the *Populus* × *euramericana* clone I-214 to excess zinc: Carbon assimilation, structural modifications, metal distribution and cellular localization. *Environ. Exp. Bot.* **2009**, *67*, 153–163. [CrossRef]
52. Ouelhadj, A.; Kaminski, M.; Mittag, M.; Humbeck, K. Receptor-like protein kinase HvLysMR1 of barley (*Hordeum vulgare* L.) is induced during leaf senescence and heavy metal stress. *Exp. Bot.* **2007**, *58*, 1381–1396. [CrossRef]
53. Abbasia, M.; Kamalan, H.R. Waste Management Planning in Amirkabir Petrochemical Complex. *Environ. Energy Econ. Res.* **2018**, *2*, 63–74. [CrossRef]
54. Touati, M.; Bottega, S.R.; Castiglione, M.; Sorce, C.; Béjaoui, Z.; Spanò, C. Modulation of the defence responses against Cd in willow species through a multifaceted analysis. *Plant Physiol. Biochem.* **2019**, *142*, 125–136. [CrossRef] [PubMed]
55. Clemns, S. Toxic metal accumulation, responses to exposure and mechanisms of tolerance in plants. *Biochimie* **2006**, *88*, 1707–1719. [CrossRef]
56. Sharma, P.; Dubey, R.S. Lead toxicity in plants. *Braz. J. Plant Physiol.* **2005**, *17*, 35–52. [CrossRef]
57. Pätsikkä, E.; Kairavuo, M.; Sěrsěn, F.; Aro, E.M.; Tyystjärvi, E. Excess Copper Predisposes Photosystem II to Photoinhibition in Vivo by Outcompeting Iron and Causing Decrease in Leaf Chlorophyll. *Plant Physiol.* **2002**, *129*, 1359–1367. [CrossRef]
58. Sidhu, G.P.S.; Singh, P.H.; Batish, D.R.; Kohli, R.K. Tolerance and hyperaccumulation of cadmium by a wild, unpalatable herb *Coronopus didymus* (L.) Sm. (Brassicaceae). *Ecotoxicol. Environ. Saf.* **2017**, *135*, 209–215. [CrossRef]
59. Béjaoui, Z.; Albouchi, A.; Lamhamedi, M.S.; Abassi, M.; El Aouni, M.H. Adaptation and morpho-physiology of three Populus deltoids Marsh. *P. nigra* L. clones after preconditioning to prolonged waterlogging. *Agrofor. Syst.* **2012**, *86*, 433–442. [CrossRef]
60. Woodrow, L.; Jiao, J.; Tsujita, M.J.; Grodzinski, B. Whole Plant and Leaf Steady State Gas Exchange during Ethylene Exposure in *Xanthium strumarium* L. *Plant Physiol.* **1989**, *90*, 85–90. [CrossRef]
61. Sánchez-Pardo, B.; Fernández-Pascual, M.; Zornoza, P. Copper microlocalisation, ultrastructural alterations and antioxidant responses in the nodules of white lupin and soybean plants grown under conditions of copper excess. *Environ. Exp. Bot.* **2012**, *84*, 52–60. [CrossRef]
62. Kupper, H.; Setlik, I.; Setlikova, E.; Ferimazova, N.; Spiller, M.; Kupper, F.C. Copper –induced inhibition of photosynthesis: Limiting steps of in vivo copper chlorophyll formation in Scenedesmus quadricauda. *Funct. Plant Biol.* **2003**, *30*, 1187–1196. [CrossRef]
63. Abassi, M.; Lamhamedi, M.S.; Hachani, C.; Béjaoui, Z. Morphophysiological evaluation of three *Salix* clones from semi-arid zones subjected to an excess of zinc under hydroponic conditions. *Can. J. For. Res.* **2018**, *48*, 599–609. [CrossRef]
64. Maksymiec, W. Signaling responses in plants to heavy metal stress. *Acta Physiol. Plant.* **2007**, *29*, 177–187. [CrossRef]
65. Dat, J.F.; Capelli, N.; Folzer, H.; Bourgeade, P.; Badot, P.M. Sensing and signaling during plant flooding. *Plant Physiol. Biochem.* **2004**, *42*, 273–282. [CrossRef]
66. Yruela, I.; Alfonso, M.; Barón, M.; Picorel, R. Copper effect on the protein composition of photosystem II. *Physiol. Plant.* **2008**, *110*, 551–557. [CrossRef]
67. Jozefczak, M.; Remans, T.; Vangronsveld, J.; Cuypers, A. Glutathione is a key player in metal-induced oxidative stress defenses. *Int. J. Mol. Sci.* **2012**, *13*, 3145–3175. [CrossRef]

68. Kaur, S.; Singh, H.P.; Rani Batish, D.; Kumar Kohli, R. *Artemisia scoparia* essential oil inhibited root growth involves reactive oxygen species (ROS)-mediated disruption of oxidative metabolism: In vivo ROS detection and alterations in antioxidant enzymes. *Biochem. Syst. Ecol.* **2012**, *44*, 390–399. [CrossRef]
69. Baker, A.J.M. Accumulators and excluders-strategies in the response of plants to heavy metals. *Plant Nutr.* **1981**, *3*, 643–654. [CrossRef]
70. Arduino, P.; Verra, F.; Cianchi, R.; Rossi, W.; Corrias, B.; Bullini, L. Genetic variation and natural hybridization between *Orchis laxiflora* and *Orchis palustris* (*Orchidaceae*). *Plant Syst. Evol.*.**1996**, *202*, 87–109. [CrossRef]
71. Florijin, P.; Van Beusichem, M.L. Cadmium distribution in maize inbred lines: Effects of pH and level of Cd supply. *Plant Soil* **1993**, *153*, 79–84. [CrossRef]
72. Mendoza-Cózatl, D.G.; Zhai, Z.; Jobe, T.O.; Vatamaniuk, O.K.; Russell, P.; Schroeder, J.I. Tonoplast-localized Abc2 Transporter Mediates Phytochelatin Accumulation in Vacuoles and Confers Cadmium Tolerance. *Biol. Chem.* **2010**, *285*, 40416–40426. [CrossRef] [PubMed]
73. Kahle, H. Response of roots of trees to heavy metals. *Environ. Exp. Bot.* **1993**, *33*, 99–119. [CrossRef]
74. Zouari, M.; Souguir, D.; Bloem, E.; Schnug, E.; Hanchi, B.; Hachicha, M. Saline soil reclamation by agroforestry species under Kalaât Landelous conditions and irrigation with treated wastewater in Tunisia. *Environ. Sci. Pollut. Res.* **2019**, *26*, 28829–28841. [CrossRef]
75. Singh, A.; Sharma, R.K.; Agrawal, M.; Marshall, F.M. Risk assessment of heavy metal toxicity through contaminated vegetables from waste water irrigated area of Varanasi, India. *Trop. Ecol.* **2010**, *51*, 375–387.
76. Khare, V.; Nema, S.; Baredar, P. Solar–wind hybrid renewable energy system. *Renew. Sustain. Energy Rev.* **2016**, *58*, 23–33. [CrossRef]
77. Baker, A.J.M.; Proctor, J. The influence of cadmium, copper, lead, and zinc on the distribution and evolution of metallophytes in the British Isles. *Plant Syst. Evol.* **1990**, *173*, 91–108. [CrossRef]
78. Bakar, A.F.A.; Yusoff, I.; Fatt, N.T.; Othman, F.; Ashraf, M.A. Arsenic, Zinc, and Aluminium Removal from Gold Mine Wastewater Effluents and Accumulation by Submerged Aquatic Plants (*Cabomba piauhyensis*, *Egeria densa*, and *Hydrilla verticillata*). *BioMed Res. Int.* **2013**, *2013*, 890803. [CrossRef] [PubMed]
79. Zacchini, M.; Pietrini, F.; Scarascia Mugnozza, G.; Lori, V.; Pietrosanti, L.; Massacci, A. Metal Tolerance, Accumulation and Translocation in Poplar and Willow Clones Treated with Cadmium in Hydroponics. *Water Air Soil Pollut.* **2009**, *197*, 23–34. [CrossRef]
80. Rana, V.; Maiti, S.K. Municipal wastewater treatment potential and metal accumulation strategies of *Colocasia esculenta* (L.) Schott and *Typha latifolia* L. in a constructed wetland. *Environ. Monit. Assess.* **2018**, *190*, 328. [CrossRef]

forests

MDPI

Article

Nutritional Assessment and Comparison of the Composition of Oil Extracted from Argan Nuts Collected from a Plantation and Two Natural Forest Stands of ARGAN Trees

Chaimaa Sabiri [1], Bouchra Tazi [2], Nadia Maata [3], Souad Rahim [1], Hassan Taki [1], Ahmed Bennamara [4], Lhoussaine Saad [3] and Abdelfettah Derouiche [1,*]

1 Research Unit, Human Nutrition, Laboratory of Biology and Health, Faculty of Sciences Ben Msik, Hassan II University of Casablanca, Avenue Driss El Harti B.P 7955, Sidi Othmane, Casablanca 20660, Morocco
2 Department of Basic Sciences, National School of Agriculture (ENAM), km 10, Route Haj Kaddour, B.P. S/40, Meknes 50000, Morocco
3 Laboratory of Analysis and Chemical Research (LOARC), 25, Rue Rahal Nichakra, Casablanca 20100, Morocco
4 Laboratory of Biomolecules and Organic Synthesis, Department of Chemistry, Faculty of Sciences Ben M'Sik, University Hassan II of Casablanca, Avenue Driss El Harti B.P 7955, Sidi Othmane, Casablanca 20660, Morocco
* Correspondence: derouicheaf@yahoo.com; Tel.: +212-638-083834

Abstract: Argan oil (AO), extracted from the argan tree's fruits, is principally composed of mono-unsaturated fatty acids, polyphenols, tocopherols, and sterols. This unique chemical composition is likely to be responsible for its beneficial effects. The argan tree (*Argania spinosa*) grows endemically in the southwest of Morocco. This study aimed to evaluate the chemical composition of three types of argan oil from three geographical locations: argan oil extracted from argan nuts collected from a plantation (Casablanca, AOC) and two forest stands of argan trees growing naturally in their native environment of the south-west of Morocco ((regions of Essaouira (AOE) and Taroudant (AOT)). The composition of the three oils corresponds to the known composition of argan oil in terms of fatty acids and unsaponifiable fraction. The chemical analyses revealed that the argan oil extracted from the plantations (AOC) is significantly richer in linoleic acid, linolenic acid, and tocopherols compared to the oil from the two natural stands (AOE and AOT) of argan trees. These results suggest that it is possible to facilitate an assisted migration of the argan tree outside its natural area into sites exposed to sea spray without affecting the quality of its argan oil.

Keywords: argan stand; argan plantation; argan oil; chemical composition; physicochemical parameters

Citation: Sabiri, C.; Tazi, B.; Maata, N.; Rahim, S.; Taki, H.; Bennamara, A.; Saad, L.; Derouiche, A. Nutritional Assessment and Comparison of the Composition of Oil Extracted from Argan Nuts Collected from a Plantation and Two Natural Forest Stands of ARGAN Trees. *Forests* **2023**, *14*, 180. https://doi.org/10.3390/f14020180

Academic Editors: Mohammed S. Lamhamedi, Damase P. Khasa and Steeve Pépin

Received: 27 November 2022
Revised: 9 January 2023
Accepted: 11 January 2023
Published: 18 January 2023

1. Introduction

The forests of the argan tree, or *Argania spinosa* (L.) Skeels, a species endemic to Morocco, currently extend only to the arid and semi-arid zones of southwestern Morocco. Argan trees cover an area of 828,000 hectares, and they are the second-most important forest species in Morocco after the holm oak [1]. During the Tertiary period, when it first appeared, the argan tree is thought to have spread over a large part of Morocco [2]. In the Quaternary period, it would have been pushed back to the southwest by the glacial invasion. This probably explains the presence of two small areas of argan trees in the upper Grou valley to the southeast of Rabat and the northwestern foothills of the Beni-Snassen, near Oujda [3]. *Argania spinosa* trees adjust their physiological status and leaf attributes to environmental conditions, allowing plants to thrive under a dry climate [4]. All parts of the argan tree are usable and present great economic, medicinal, and therapeutic interests [5] thanks to extracts from its organs (fruit and leaves) [6]. Despite the economic, social, and environmental importance of this tree, little interest has been given to its installation outside its biotope in Morocco and abroad [7]. The failure of the extension, planting, and reforestation operations of the argan tree is due to major obstacles such as rainfall, humidity, cold, and different traits related to root architecture [8]. Pre-inoculation with an effective strain of an endomycorrhizal fungus (*Glomus intraradices* Schenck & Smith) provides a clear advantage for the growth of argan tree seedlings. This technique deserves to be considered in the current practice of nurseries, as it is already practiced in some countries for other woody species [9,10].

Argan oil is known for its richness in unsaturated fatty acids and its minor compounds, polyphenols, tocopherols, sterols, etc., which give it its antioxidant properties. The quantification of these minor compounds is important and their determination is necessary, as their variation can be accompanied by nutritional health modifications. Indeed, some compounds, such as polyunsaturated fatty acids (essential fatty acids) or vitamin E (tocopherols), are responsible for the nutritional interest in argan oil. This particular biochemical profile of argan oil has often led to its being considered as a new functional food or health food [11]. Argan oil is composed of triaglycerols, of which 80% comprises unsaturated fatty acids, which are known for their role in primary and secondary prevention of cardiovascular disease [12]. These are dominated by oleic acid at 48%, followed by linoleic acid at approximately 32%. The saturated fatty acids in argan oil are palmitic acid (approximately 13%) and stearic acids (approximately 5%). The unsaponifiable fraction of argan oil contains equal amounts (20%) of sterols and triterpenes. Schottenol and spinasterol are the two main sterols, and these molecules seem to have protective properties for the epidermis. Finally, α-tocopherol (vitamin E) plays the role of a regulator in the immune system and in inflammation [12]. In addition, the polyphenols contained in argan oil play a promoting role in the prevention of and therapy for diseases with underlying inflammatory conditions, including cancer, neurodegenerative diseases, obesity, diabetes, type II diabetes, and cardiovascular disease [13]. Research conducted on argan oil attributes nutritional, medicinal (pharmacological), and cosmetic benefits to it. Its consumption regulates the lipid profile [14], protects against cardiovascular diseases, atherosclerosis, and certain cancers [15–17], and also increases androgenic activity [18]. Argan oil has been used traditionally for many centuries. The development of processes to extract oil by mechanical cold presses from non-torrified almonds has motivated some laboratories to integrate it into cosmetic products, such as soaps, shampoos, and creams [17,19]. It is used to nourish hair, prevent hair loss, and fortify dull and brittle hair [20]. It is also used for the treatment of chapped, dry, or dehydrated skin and acne. Long-term use reduces the speed of wrinkle appearance and helps with the healing of scars [21,22]. The quality of the oil is linked to the concentration of many trace metals. Some metals are known to increase the rate of oil oxidation. The assessment of elemental concentration in vegetable oil is vital because of their elicit toxic effect on the human health if consumed in excessive quantities. The main objective of this study is to evaluate and compare the composition of argan oil from an argan tree plantation outside its natural area with those from two natural forest stands grown in southwestern Morocco.

2. Materials and Methods

2.1. Plant Material

The first argan oil originated from "Tafingoult", a small rural commune in Taroudant Province of the Souss-Massa region (Morocco), at the following location (latitude: 30°46′0″ N; longitude: −8°22′60″ W). It is characterized by a hot and cold semi-arid climate (Köppen-Geiger climate classification) [23]. The soil of "Tafingoult" belongs to the category of silty soils of the textural class known as "sandy silt", and the most abundant species is *Argania spinosa*, followed by *Tetraclinis articulata* and *Quercus rotundifolia*.

The second argan oil originated from "Smimou", a town in Essaouira Province of the Marrakech-Safi region (Morocco) at the following location (latitude: 31°12′49″ N; longitude: −9°42′21″ W). It is characterized by a hot and cold semi-arid climate (Köppen-Geiger climate classification) [24]. The soil of "Smimou" belongs to the category of silty soils of textural class known as "silt", and the most abundant species is *Argania spinosa*, followed by *Tetraclinis articulata*.

The planted argan trees in Casablanca were planted in 2008. The argan plants used were grown in the "Ounagha forest nursery" in Essaouira in polyethylene bags until they were suitable for transplantation (1 year). The argan tree plantation is located in an urban area of Casablanca, in "Hay Laymoune", and in the vicinity of the city in "Dar Bouazza" (see Figures S1–S5).

2.2. Sampling and Processing of Argan Nuts

Three types of argan oil were extracted by cold mechanical pressure from unroasted argan kernels. These are the argan oils from the fruits of the plantation in "Casablanca" (AOC) and the two other oils from forest trees in the prefectures of "Taroudant" (AOT) and "Essaouira" (AOE). The three samples were obtained from the dried fruits of the 2018 collection, when the planted trees were 8 years old, and the wholes were treated in parallel under the same conditions of pulping, crushing, and extracting [25]. The argan fruits were collected in October 2018 and dried in the traditional way by exposure to the sun, then depulped. The nuts were crushed and the almonds were directly extracted in May 2019. The extraction was performed using 446 g of kernels from each provenance (Figure 1). The obtained oils were stored after decantation in sterile, hermetically sealed, coated bottles to avoid the impact of external factors. The first analyses were carried out during the week preceding the extraction. All of the results were obtained 3 weeks after the extraction of the samples. All of the analyses were carried out at the Official Laboratory of Chemical Analysis and Research of Casablanca, "LOARC", and the National School of Agriculture Meknes (ENAM).

(a) (b) (c)

Figure 1. Sampling and extraction of argan oil from the plantation and two natural forest stands of argan trees. (**a**) Argan nuts from a plantation in Casablanca. (**b**) Composite samples of argan kernels collected from the three sampling sites (Essaouira, Casablanca, and Taroudant). (**c**) Argan oil extracted by cold mechanical pressure.

2.3. Physico-Chemical Parameters

The physicochemical parameters used for classifying the three argan oils (Casablanca, Taroudant, and Essaouira) were the acidity, the peroxide value, and the spectrophotometric characteristics.

2.3.1. Acidity

Acidity is an indicator used to determine the content of free fatty acids resulting from the hydrolysis of triglycerides. It is expressed as a percentage of oleic acid and detected by ISO 660 [26]. In this experiment, 10 g of argan oil was added to a mixture of solvents of ethanol (50 mL) and diethyl ether (v:v) previously neutralized by potassium hydroxide (KOH 0.1 N) in the presence of phenolphthalein. The fatty acids were titrated with a potassium solution (0.1 N) until the appearance of pink coloration. The volume of the added potassium solution was noted, and then the acidity was expressed as a percentage of free oleic acid according to the following formula:

$$\text{Acidity \%} = (N \times V \times M)/(\text{Wt of sample (g)} \times 10) \quad (1)$$

where:

N: normality of standard KOH solution used for titration (0.1 mole/liter);
V: volume (mL) of standard KOH solution used for titration;
M: molecular weight of oleic acid (282 g/mole);
Wt: weight (g) of the sample.

2.3.2. Peroxide Value

The peroxide value of each type of oil was determined according to the protocol ISO 3960 [27]. A 5 g sample of argan oil was solubilized in a mixture of acetic acid (30 mL) and isooctane (20 mL), and 1 mL of saturated potassium iodide (KI) solution was then added to the mixture. After a one-minute reaction, 100 mL of distilled water was added with 1 mL of starch solution. The obtained mixture was then titrated with a 0.01% sodium thiosulfate solution until a black color was obtained. The added volume was noted, and the peroxide value was calculated using the formula:

$$\text{Peroxide Value in meq } O_2/\text{kg} = (V \times 1000 \times N)/(\text{Wt of sample (g)}) \quad (2)$$

where:

V: volume (mL) of sodium thiosulfate used for the determination;
N: normality of sodium thiosulfate solution ($Na_2S_2O_3$);
Wt: weight (g) of the sample.

2.3.3. Spectrophotometric Characteristics

An analytical method was used to determine abnormal oxidized compounds in virgin argan oil, according to the International Oleic Council standard [28]. A 0.25 g sample was adjusted to 25 mL with cyclohexane, and after homogenization, the absorbance was measured at 232 nm, 266 nm, 270 nm, and 274 nm using double beam UV–Vis Spectrophotometer (6850 UV/Vis. Spectrophotometer-JENWAY). The variation of the specific extinction (Δk) was calculated using the formula:

$$\Delta K = \Delta K = K270\text{ nm} - [(K274\text{ nm} + K266\text{ nm})/2] \quad (3)$$

where:

K270 nm: absorbance at 270 nm.
K274 + K266 nm: absorbance at 270 nm plus or minus 4 nm (K 274 + K266).

2.4. Chemical Composition

2.4.1. Fatty Acids

A test sample containing between 0.3 and 0.4 g of argan oil was introduced into a glass screw tube with 4 mL of methanolic potassium hydroxide solution. After stirring to esterify the mixture, extraction of the methyl esters was carried out by adding 4 mL of the isooctane solution. In order to increase the density of water and to separate the two phases (water and oil), 1 g of sodium bisulfate monohydrate was added to the mixture, which was agitated at the vortex and left to rest for 4 min. A 0.5 mL intake of the mixture was then diluted by hexane (up to 10 times) in a vial. The vial was finally placed in the in the autosampler rack of gas chromatography (GC) for the injection of the solution [29]. The analysis of fatty acid methyl esters was processed by gas chromatography (GC; HP 6890, Agilent Technologies, Santa Clara, CA, USA) equipped with a flame ionization detector and a capillary column (Carbowax 20M, 30 m × 0.32 mm, 0.25 μm thickness, Agilent Technologies, Santa Clara, CA, USA). The nitrogen flow rate was 2.5 mL/min. The temperature was programmed from 140 to 240 °C at 10 °C/min. The injected quantity was 1 μL. Fatty acids were identified by comparing their retention times to those of the standards.

2.4.2. Sterol Composition

The fat, with the addition of α-cholestanol (250 μL) as internal standard, was saponified with potassium hydroxide in ethanolic solution (25 mL). Then, the unsaponifiable matter was extracted with diethyl ether (200 mL). The sterol fraction was separated from the unsaponifiable extract by silica gel plate chromatography (TLC) at about 2 cm; then, the sterols recovered from the silica gel were analyzed by capillary column gas chromatography (Varian 3800 instrument) equipped with a VF-1 ms column and using helium (20 to 35 cm/s); hydrogen (30 to 50 cm/s) as carrier gas. The column temperature was isothermal at 260 ± 5 °C, and the temperature of the injector and detector was 280 °C–300 °C. The injected quantity was 1 μL for each analysis [30].

2.4.3. Tocopherol Composition

For this step, 0.25 g of the sample was dissolved in 25 mL of isooctane. The separation of the different tocopherols was achieved by high performance liquid chromatography (HPLC). The tocopherol content was determined by HPLC using Shimadzu instruments equipped with a C18-Varian column (250 mm × 4.6 mm × 5 μm). Detection was performed using a fluorescence detector (excitation wavelength 290 nm, detection wavelength 330 nm). The tocopherols content was determined using calibration factors determined from standard solutions [31].

2.4.4. Heavy Metals

To measure the totality of metals, a total mineralization of the sample to be analyzed was recommended. For the mineralization assay, 0.3 g of oil was mixed with 7 mL of nitric acid (HNO_3) and 1 mL of hydrogen peroxide (H_2O_2). The mixture was placed in the microwave for 1 h. After cooling, the content was poured into a 50 mL flask and adjusted with distilled water. Finally, the determination of metals was achieved by Atomic Absorption Spectrometry (AAS) using double beam UV–Vis Spectrophotometer (6850 UV/Vis. Spectrophotometer-JENWAY) [32]. The concentration of lead, mercury, and cadmium in the samples was calculated using the following formula:

$$\text{Pb, Cd (mg/Kg)} = \frac{(\text{Sample concentration (mg/L)} - \text{ blanc concentration (mg/L)})}{(20 \times \text{ weight of the sample (g)})} \quad (4)$$

$$\text{Hg (mg/Kg)} = \frac{(\text{Sample concentration (mg/L)} - \text{ blanc concentration (mg/L)})}{(50 \times \text{ weight of the sample (g)})} \quad (5)$$

2.5. Statistical Analyses

Statistical analyses of the different variables were performed using the analysis of variance. The hypothesis of the normality of the error terms of all the variables was checked. Mean comparisons were made using the Bonferroni test at a 5% significance level. All statistical analyses were carried out using SPSS Statistics Software (IBM SPSS Statistics 25.0) [33]. Each reported value is the mean of duplicate or triplicate samples prepared for each type of argan oil. The results of the ANOVA and post hoc Bonferroni tests were considered statistically significant if $p < 0.05$.

3. Results

3.1. Physico-Chemical Parameters

3.1.1. Acidity

The obtained results revealed that Casablanca argan oil had a high acidity level of 0.37% compared to that of the Essaouira and Taroudant oils, which had levels of 0.25% and 0.2%, respectively, without reaching the significance level (Table 1).

Table 1. Comparison of physico-chemical parameters of argan oil from a plantation (Casablanca) and two natural forest stands (Taroudant and Essaouira) of argan trees.

Samples	% Acidity [1]	Peroxide Value $meqO_2/Kg$	OD at 232 nm [1] (K232 nm)	OD at 270 nm [1] (K270 nm)	Variation of Specific Extinction at 270 nm (ΔK)
Casablanca	0.37 ± 0.02 [a]	0	1.37 ± 0.06 [b]	0.19 ± 0.00 [b]	0.00 ± 0.00 [a]
Taroudant	0.25 ± 0.04 [a]	0	1.70 ± 0.00 [a]	0.20 ± 0.00 [b]	0.00 ± 0.00 [a]
Essaouira	0.2 ± 0.03 [a]	0	1.50 ± 0.01 [a,b]	0.34 ± 0.00 [a]	0.00 ± 0.00 [a]

[1] Mean values ± standard deviation (n = 2). Values in the same column with different letters are significantly different at $p < 0.05$, according to the Bonferroni test.

3.1.2. Peroxide Value

The results of the analyses of the three oils (Casablanca, Essaouira, and Taroudant) (Table 1) show a definitive absence of hydroperoxides (0 $meqO_2/kg$).

3.1.3. Spectrophotometric Characteristics

Casablanca oil recorded ultraviolet absorbance values that were close to those of the other two oils studied (Table 1). The absorbance values obtained at 232 nm and 270 nm in the argan oils of Casablanca, Taroudant, and Essaouira were in the order of 1.37 and 0.19 (ΔK = 0.00), 1.702 and 0.198 (ΔK = 0.00), and 1.501 and 0.348 (ΔK = 0.00), respectively.

3.2. *Chemical Composition*

3.2.1. Fatty Acids

Analysis of the fatty acid profile results indicated that the differences in the percentages of certain fatty acids, including C-14:0, C-16:1, C-17:0, C-17:1, and C-18:0, were not statistically significant (Table 2). The comparison of the results of AOC analysis with those of AOT and AOE revealed that AOC was significantly less rich in oleic acid compared to AOT and AOE (42.66%, 47.55%, and 47.14%, respectively). In addition, AOC was significantly richer in linoleic acid than AOT and AOE (36.95%, 31.65%, and 34.28%, respectively). For linolenic acid, AOC was significantly richer, with a percentage of 0.11% compared to 0.08% for AOT and AOE. Trans fatty acids were absent in all of the studied samples (Table 2).

Table 2. Comparison of cis and trans fatty acid concentration (%) of argan oil from argan trees grown on a plantation (Casablanca: AOC) and in two natural forest stands (Taroudant: AOT and Essaouira: AOE).

Fatty Acids [1]	Casablanca %	Taroudant %	Essaouira %
Myristic acid (C-14:0)	0.13 ± 0.00 [a]	0.14%± 0.00 [a]	0.14 ± 0.00 [a]
Palmitic acid (C-16:0)	12.97 ± 0.00 [b]	13.42 ± 0.06 [a]	12.35 ± 0.04 [c]
Palmitoleic acid (C-16:1)	0.10 ± 0.00 [a]	0.11 ± 0.00 [a]	0.12 ± 0.00 [a]
Margaric acid (C-17:0)	0.09 ± 0.00 [a]	0.08 ± 0.00 [a]	0.07 ± 0.00 [a]
Heptadecenoic acid (C-17:1)	0.02 ± 0.00 [a]	0.02%± 0.00 [a]	0.01% ± 0.00 [a]
Stearic acid (C-18:0)	6.16 ± 0.06 [a]	6.11 ± 0.04 [a]	5.04 ± 0.01 [b]
Oleic acid (C-18:1 (w-9))	42.66 ± 0.11 [b]	47.55 ± 0.12 [a]	47.14 ± 0.00 [a]
Linoleic acid (C-18:2 (w-6))	36.95 ± 0.18 [a]	31.65 ± 0.18 [b]	34.28 ± 0.01 [c]
Linolenic acid (C-18:3 (w-3))	0.11 ± 0.00 [a]	0.08 ± 0.00 [b]	0.08 ± 0.00 [b]
Arachidic acid (C-20:0)	0.42 ± 0.00 [a]	0.40 ± 0.00 [a]	0.34 ± 0.00 [c]
Gadoleic (C-20:1)	0.40 ± 0.00 [b]	0.43 ± 0.00 [a]	0.43 ± 0.00 [a]
Trans fatty acids	0.00 [a]	0.00 [a]	0.00 [a]
Elaidic acid (C_{18}:1T)	0.00 [a]	0.00 [a]	0.00 [a]
Linolelaidic acid (C_{18}:2T)	0.00 [a]	0.00 [a]	0.00 [a]

[1] Mean values ± standard deviation (n = 2). Values in the same column with different letters are significantly different at $p < 0.05$, according to the Bonferroni test.

3.2.2. Sterol Composition

The sterol assay results presented in Table 3 show that the difference between the percentages of sterol derivatives, especially campesterol, spinasterol, schottenol, and others, in the three studied oils was statistically non-significant, and that AOC and AOE were significantly richer in cholesterol than AOT (0.08 ± 0.001%, 0.09 ± 0.01 and % 0.2± 0.05%, respectively). For delta-7-aventiserol, AOC was significantly richer than AOE (4± 0.08%, 3.55 ± 0.04%). The total sterol content in AOC compared to AOT constituted almost identical percentages, with a non-significant difference (143.17 ± 6.56 and 143.33 ± 4.09, respectively) (Table 3).

Table 3. Comparison of sterol concentration (%) of argan oil from argan trees grown on a plantation (Casablanca: AOC) and in two natural forest stands (Taroudant: AOT and Essaouira: AOE).

Sterol Composition [1]	Casablanca %	Taroudant %	Essaouira %
Cholesterol	0.08 ± 0.00 [a]	0.2± 0.05 [b]	0.09 ± 0.01 [a]
Campesterol	0.19 ± 0.02 [a]	0.20 ± 0.02 [a]	0.18 ± 0.02 [a]
Spinasterol	33.67 ± 0.44 [a]	33.79 ± 0.17 [a]	33.14 ± 0.49 [a]
Schottenol	39.97 ± 0.85 [a]	39.84 ± 0.65 [a]	40.65 ± 0.39 [a]
Delta-7-Aventiserol	4± 0.08 [a]	3.86 ± 0.09 [a, b]	3.55 ± 0.04 [b]
Others	9.98 [a]	10.01 [a]	10.91 [a]
Total sterols (mg/100 g)	143.33 ± 6.56 [b]	143.17 ± 4.09 [b]	151.74 ± 4.09 [a]

[1] Mean values ± standard deviation (n = 2). Values in the same column with different letters are significantly different at $p < 0.05$, according to the Bonferroni test.

3.2.3. Tocopherol Composition

The results of the tocopherol assay presented in Table 4 show that AOC was significantly richer in gamma-tocopherols, with 90.53% ± 0.22 compared to AOT and AOE (87.15 ± 0.06% and 87.88 ± 0.06%, respectively). On the other hand, AOC and AOE were significantly richer in total tocopherols than AOT (758.47 ± 5.65, 749.43 ± 0.04 and 662.80 ± 3.06) (Table 4).

Table 4. Comparison of tocopherol concentration (%) of argan oil from argan trees grown on a plantation (Casablanca: AOC) and in two natural forest stands (Taroudant: AOT and Essaouira: AOE).

Tocopherols Composition [1]	Casablanca	Taroudant	Essaouira
Alpha-tocopherols %	4.24 ± 0.05 [a]	7.04 ± 0.11 [a]	**4.49 ± 0.21 [a]**
Gamma-tocopherols %	90.53 ± 0.22 [a]	87.15 ± 0.06 [b]	**87.88 ± 0.06 [c]**
Delta-tocopherols %	5.23± 0.17 [a]	5.81 ± 0.17 [a]	**7.64 ± 0.14 [a]**
Total tocopherols (mg/kg)	758.47 ± 5.65 [a]	662.80 ± 3.06 [b]	749.43± 0.04 [a]

[1] Mean values ± standard deviation (n = 2). Values in the same column with different letters are significantly different at p <0.05, according to the Bonferroni test.

3.2.4. Heavy Metals

The results showed similar concentrations for the three oils studied. Indeed, mercury and cadmium were represented by a rate lower than 0.015 mg/kg, and lead by a concentration lower than 0.07 mg/kg.

4. Discussion

This study demonstrated that the free acidity of the analyzed samples remained below 0.8%, revealing that the three samples should be classified as extra virgin argan oil according to the Moroccan standard N.M. 08.5.090 [34]. These results were in accordance with those reported in previous studies [35,36], which estimated an acidity ranging from 0.12% to 0.64% [35].

In terms of peroxide value, the three argan oils (Casablanca, Taroudant, and Essaouira) were free of hydroperoxides (0 meqO_2/kg), and, therefore, did not contain primary oxidation products due to the presence of antioxidants contained in the fruit of the argan tree, which enriches the oil with these substances during its extraction [21]. Thus, the analyzed argan oils (Casablanca, Taroudant, and Essaouira) complied with the Moroccan standard NM. 08.5.090, which sets a value of 15 meq O_2/kg of oil for extra virgin argan oil [37]. These results are superior to those obtained by Gharby et al. (2011), whose values ranged from 0.03 to 0.93 meq O_2/kg oil, and Azizi et al. (2022), who found values that varied between 1.07 and 3.30 meq O_2/kg. This may be due to our control of the different stages of processing of our products, from the harvesting of the fruits to the obtaining and conservation of the oil, as well as during the analysis of the different parameters.

The determination of the specific extinction of argan oil allowed us to highlight its quality and to confirm the results given by the peroxide value determination. The absorbance results of the three oils at 270 nm conformed with the Moroccan standard (NM.08.5.090), since these values were all lower than (or equal to) 0.35 in the results reported by Rahmani [37]. The specific extinction of argan oil was also determined by Gharby et al. in 2011 [35], for a cosmetic oil from Taroudant. The results revealed K270 values that generally varied between 0.02 and 0.16. The specific extinction of argan oil was also determined by Hilali [38] on samples taken from 21 trees. This study showed values of K270 that generally varied between 0.228 and 0.426.

Moroccan standard NM. 08.5.090 for argan oil has not yet set a value for extinction at 232. Regarding extra virgin olive oil, it specifies this for K232 nm values below 2.50 [39]. By extrapolation, we deduced that the absorbance of our samples at 232 nm was low. Indeed, the lowest value determined for AOC was 1.37 $\pm$ 0.06. The maximum value was recorded for AOT (1.70 $\pm$ 0.00) and the average value for AOE (1.50 $\pm$ 0.00). The values of the specific extinction variation (ΔK) for the three studied samples remain within the limits set by the Moroccan standard (the standard recommends a ΔK that is always $\leq$0.01) [37]. According to the obtained values of absorbance in the ultraviolet, and comparing our results with the Moroccan standard, we can classify our oils in the category of "Extra virgin oil" according to the Moroccan standard N.M. 08.5.090.

As far as contaminants are concerned, the results obtained in this study show that cadmium and mercury contamination was very limited, since the levels recorded were lower than 0.015 mg/kg, which does not exceed 0.002 mg/kg [40]. All samples contained less than 0.07 mg/kg of lead, and were, therefore, in compliance with European regulations, which tolerate a value less than or equal to 0.1 mg/kg.

It was noted that the three samples analyzed showed approximately identical percentages of saturated fatty acids and some monounsaturated fatty acids, such as C16:1, C17:1, and C20:1, with the predominance of palmitic and stearic acid. These results are similar to the results reported in the study carried out by Khallouki in 2003 [41]. AOC was significantly richer in linoleic acid (36.95% $\pm$ 0.18%) and linolenic acid (0.11% $\pm$ 0.00%) compared to AOT and AOE, which had, respectively, 31.65% $\pm$ 0.18% and 34.28% $\pm$ 0.01% for linoleic acid and 0.08% for linolenic acid. A comparison of our results to those published by Khallouki reveals that the composition of AOC in linolenic acid (36.95% $\pm$ 0.18%) is similar to that found by Khallouki (36%) [41]. The percentage of linolenic acid in AOC (0.11% $\pm$ 0.002%) found in this study was higher than the results reported by Khallouki, who did not detect the presence of linolenic acid, and higher than the result reported by Gharby et al. in 2021, who had worked on cosmetic argan oil prepared from regurgitated nuts before and after deodorization (0.10 $\pm$ 0.01 and 0.07 $\pm$ 0.01, respectively) [42]. The comparison of our results with those of Gharby showed that AOC was richer with linoleic acid (36.95% $\pm$ 0.180%) compared to the levels (32.1 $\pm$ 0.15 and 31.36 $\pm$ 0.20) reported by Gharby [42]. AOC was also richer in linoleic acid than the three argan oils that were analyzed by Zarrouk (33.71 $\pm$ 0.73, 33.38 $\pm$ 0.40 and 31.99 $\pm$ 0.31) [43]. Argan oil from a plantation in Casablanca was less rich in oleic acid than the other oils analyzed. It would have compensated for the lack of oleic acid by a higher percentage of linoleic acid. The variations observed in the fatty acid composition of argan oil could be attributed to various factors, including the geographical origin and the effect of climate, especially rainfall [44,45]. The rest of the fatty acids analyzed in AOC presented rates close to those found in argan oils from the other two regions, Taroudant and Essaouira (AOE and AOT). After extraction, the three samples did not undergo any treatment, which explains the absence of trans fatty acids; thus, they were classified as pure oils. Indeed, the presence of trans fatty acids in argan oil suitable for consumption is an indicator of the presence of refined oil. The Moroccan standard tolerates a percentage of trans fats less than or equal to 0.05% [34]. The most dominant sterols in the three analyzed samples were schottenol (AOC: 39.97%, AOT: 39.84%, and AOE: 40.65%) and spinasterol (AOC: 33.67%, AOT: 33.79%, and AOE: 33.14%). This is in line with the studies established by Khallouki in 2003 [41] and Zarrouk in 2019 [43], which also detected a dominance of these components. AOE contains a total sterol content that exceeds that present in the other two analyzed oils, AOT and AOE.

The γ-tocopherol is the most represented vitamin in argan oil tocopherols. Its concentration was significantly higher in AOC compared to AOT and AOE (90.53% $\pm$ 0.22%, 87.15% $\pm$ 0.06 and 87.88% $\pm$ 0.06, respectively). These results are in accordance with the results reported by Azizi et al. (2022), who indicated that the γ-tocopherols were the most abundant, followed by α-tocopherols and, finally, the δ-tocopherols. On the other hand, the total tocopherol content in AOC was statistically higher (758.47 $\pm$ 5.65 mg/kg of oil), compared to that of AOT (662.80 $\pm$ 3.06 mg/kg of oil). The total tocopherol content of AOC exceeded that reported by Khallouki et al., 2003 (629 mg/kg oil) [41] and Gharby et al., 2021 (75 $\pm$ 5.5 and 60 $\pm$ 3.5 mg/100 g of oil) [42].

5. Conclusions and Research Needs

The chemical and phytochemical proprieties of argan oil (*Argania spinosa* L. skeels) from three regions of Morocco (an argan plantation in the region of Casablanca and two spontaneous cultures) were analyzed and compared. It was found that the argan trees, although they were grown outside their biotope in the region of Casablanca, produced high-quality oil which was richer in linoleic and linolenic acid, as well as in certain unsaponifiable compounds such as sterols and tocopherols. These properties constitute an added value for its use in the nutritional and pharmaco-cosmetic field. The findings of this study will be useful for selecting the domestication of the argan tree as a real opportunity to mitigate of the pressure on this resource, to develop it, and to conserve it, which will have a positive impact on development both in the socio-economic and ecological fields. However, in order to better understand the effect of genotype environment-interaction on the yield and quality of argan oil, it is necessary to produce experimental designs of the same argan genotypes in different reforestation sites. The production and multiplication of the selected genotypes requires the development of clonal or varietal forestry specific to the argan tree.

Supplementary Materials: The following supporting information can be downloaded at: https://www.mdpi.com/article/10.3390/f14020180/s1, Figure S1: Photo of the three argan trees taken from the terrace in the background tramway of Casablanca; Figure S2: Argan tree N°1; Figure S3: Argan tree N°2 with tramway of Casablanca in the background in left; Figure S4: Argan tree N°3 with tramway of Casablanca in the background in right; Figure S5: Branch of argan tree planted in Casablanca with argan fruits, and the tramway of Casablanca in the background.

Author Contributions: Conceptualization, A.D.; methodology, A.D., L.S., N.M. and C.S.; software, C.S.; validation, A.D., B.T., S.R., H.T. and A.B.; formal analysis, C.S. and A.D.; resources, A.D., L.S. and N.M.; data curation, C.S.; writing—original draft preparation, C.S.; writing—review and editing, C.S. and A.D.; visualization, C.S., A.D. and B.T.; supervision, A.D. and B.T.; project administration, A.D. All authors have read and agreed to the published version of the manuscript.

Funding: This research received no external funding.

Data Availability Statement: Not applicable.

Acknowledgments: We would like to thank Ben Abdenbi Fatouma, Naih Aicha and Mbarak Mohamadi for their active participation in the supply of the argan oil and Derouiche Yazid, UNC Charlotte USA, for his proofreading of this document.

Conflicts of Interest: The authors declare no conflict of interest.

References

1. Kabouss, A.E.; Charrouf, Z.; Oumzil, H.; Faid, M.; Lamnaouer, D.; Miyata, Y.Y.; Miyahara, K. Characterization of flavonoids of Argan tree leaves (*Argania spinosa* (L.) Skeels, Sapotaceae) and study of their antimicrobial activity. *Actes Inst. Agron. Vet.* **2001**, *21*, 157–162.
2. Boukhobza, M.; Pichon-Prum, N. Argan tree, an economic and medicinal resource for Morocco. *Phytotherapy* **1988**, *27*, 21–26.
3. Mousadik, A.E.; Petit, R.J. Chloroplast DNA phylogeography of the argan tree of Morocco. *Mol. Ecol.* **1966**, *5*, 547–555. [CrossRef] [PubMed]
4. Díaz-Barradas, M.C.; Zunzunegui, M.; Ain-Lhout, F.; Jáuregui, J.; Boutaleb, S.; Álvarez-Cansino, L.; Esquivias, M.P. The specific form of extinction in ultraviolet light. *Plant Soil* **2010**, *337*, 217–231. [CrossRef]
5. Charrouf, Z.; Guillaume, D. Secondary metabolites from *Argania spinosa* (L.). *Skeels. Phytochem. Rev.* **2002**, *1*, 345–354. [CrossRef]
6. Pumareda, L.; Henry, F.; Charrouf, Z.; Pauly, G.; Flaconne, G. Valorization of argan leaves: Environmental impact. *BFT* **2006**, *287*, 35–44.
7. Nouaïm, R. *L'arganier au Maroc: Entre Mythes et Réalités*; L'HARMATTAN: Paris, France, 2005; p. 230.
8. Msanda, F.; Aboudi, A.E.; Peltier, J.P. Biodiversity and biogeography of the Moroccan argan forest. *Cah. Agric.* **2005**, *14*, 357–364.
9. Nouaïm, R.; Chaussod, R. Mycorrhizal dependency of two clones of micro propagated Argan tree (*Argania spinosa*): I. Growth and biomass production. *Agrofor. Syst.* **1994**, *27*, 53–65. [CrossRef]
10. Nouaim, R.; Chaussod, R. Role of mycorrhizae in the hydrous and mineral nutrition of plants, especially woody plants in arid zones. Mycorrhization of forest plants in arid and semi-arid environments and the fight against desertification in the Mediterranean Rim. *Cah. Options Méditerr.* **1997**, *20*, 9–26.
11. Kamal, R.; Kharbach, M.; Eljemli, M.; Bouklouze, A.; Cherrah, Y.; Aloui, K. In vivo anti-inflammatory activity of argan oil. *Phytothérapie* **2019**, *18*, 255–261. [CrossRef]
12. Lewis, E.D.; Meydani, S.N.; Wu, D. Regulatory role of vitamin E in the immune system and inflammation. *IUBMB Life* **2019**, *71*, 487–494. [CrossRef] [PubMed]
13. Yahfoufi, N.; Alsadi, N.; Jambi, M.; Matar, C. The Immunomodulatory and Anti- Inflammatory Role of Polyphenols. *Nutrients* **2018**, *10*, 1618. [CrossRef] [PubMed]
14. Derouiche, A.; Cherki, M.; Drissi, A.; Bamou, Y.; Messal, M.E.; Oudghiri, A.I.; Lecerf, J.M.; Adlouni, A. Nutritional intervention study with argan oil in man: Effects on lipids and apolipoproteins. *Ann. Nutr. Metab.* **2005**, *49*, 196–201. [CrossRef] [PubMed]
15. Cherki, M.; Derouiche, A.; Drissi, A.; Messal, M.E.; Bamou, Y.; Oudghiri, A.I.; Lecerf, J.M.; Adlouni, A. Consumption of argan oil may have an antiatherogenic effect by improving paraoxonase activities and antioxidant status: Intervention study in healthy man. *Nutr. Metab. Cardiovasc.* **2005**, *15*, 352–360. [CrossRef]
16. Derouiche, A.; Saile, R.; Drissi, A.; Cherki, M.; Bamou, Y.; Oudghiri, A.I.; Lecef, J.M.; Adlouni, A. We-P14:437 Effect of tow naturals products oils, Argan and Olive, in the nutritional prevention on atherosclerosis in man. *Atheroscler. Suppl.* **2006**, *7*, 443. [CrossRef]
17. Adlouni, A. L'huile d'argan, de la nutrition à la santé. *Phytotherapy* **2010**, *8*, 89–97. [CrossRef]
18. Derouiche, A.; Jafri, A.; Driouch, I.; Khasmi, M.E.; Adlouni, A.; Benajiba, N.; Bamou, Y.; Sail, R.; Benouhoud, M. Effect of argan and olive oils consumption on the hormonal profile of androgens among Moroccan healty adult adult men. *NPC* **2013**, *8*, 51–53. [PubMed]
19. Guillaume, D.; Charrouf, Z. Argan oil. *Altern. Med. Rev.* **2001**, *16*, 275–276.
20. Charrouf, Z.; Guillaume, D. Phenols and polyphenols from Argania spinosa. *Am. J. Food Technol.* **2007**, *2*, 679–683. [CrossRef]
21. Charrouf, Z.; Guillaume, D. Huile d'argan une production devenue adulte. *Les Technol. Lab.* **2007**, *2*, 4–7.
22. Boucetta, K.Q.; Charrouf, Z.; Aguenaou, H.; Derouiche, A.; Bensouda, Y. The effect of dietary and/or cosmetic argan oil on postmenopausal skin elasticity. *Clin. Interv. Aging* **2015**, *10*, 339–349. [PubMed]
23. DB-City: Toutes Les Infos sur Les Pays, Régions, Villes et Villages. Available online: https://fr.db-city.com/fr.db-city.com/Maroc--Souss-Massa--Taroudannt--Tafingoult (accessed on 2 January 2023).
24. DB-City: Toutes Les Infos sur les Pays, Régions, Villes et Villages. Available online: https://fr.db-city.com/fr.db-city.com/Maroc--Marrakech-Safi--Essaouira--Smimou (accessed on 2 January 2023).
25. Gharby, S.; Charrouf, Z. Argan Oil: Chemical Composition, Extraction Process, and Quality Control. *Front. Nutr.* **2022**, *8*, 1251. [CrossRef] [PubMed]
26. *Standard Method ISO 660*; Animal and Vegetable Fats and Oils-Determination of Acid Value and Acidity. ISO: Geneva, France, 1996.
27. *ISO 3960*; Corps Gras D'origines Animale et Végétale-Détermination de L'indice de Peroxyde-Détermination Avec Point D'arrêt Iodométrique. DE NORMALISATION, Organisation Internationale: Geneva, France, 2007.
28. *Standard Method ISO 3656: 2002*; Animal and Vegetable Fats and Oils—Determination of Ultraviolet Absorbance Expressed as Specific uv Extinction. ISO: Geneva, France, 2002.
29. *Standard Method ISO 12966-2*; Corps Gras D'origines Animale et Végétale—Chromatographie en Phase Gazeuse des Esters Méthyliques D'acides Gras—Partie 2: Préparation des Esters Méthyliques D'acides Gras. ISO: Geneva, France, 2011.
30. *IOC/T.20/Doc. No. 30*; Determination of the Sterol Composition and Content and Alcoholic Compounds by Capillary Gas Chromatography. IOC (International Olive Council): Geneva, France, 2011.
31. *Standard Method ISO 9936*; Animal and Vegetable Fats and Oils—Determination of Tocopherol and Tocotrienol Contents by High-Performance Liquid Chromatography. ISO: Geneva, France, 1997.

32. Méthodes Standard UICPA pour l'analyse des Huiles, Graisses et Produits Dérivés. In *Determination of Lead*, 7th ed.; 1er Supplément, 1988; p.2.632.
33. IBM Corp. *IBM SPSS Statistics for Windows*; Version 25.0; IBM Corp.: Armonk, NY, USA, 2017.
34. *Moroccan Standard NM 08.5.090*; Animal and Vegetable Fats and Oils—Argan Oils Specifications. SNIMA (Service de Normalisation Industrielle Marocaine): Rabat, Morocco, 2003.
35. Gharby, S.; Harhar, H.; Badr, K.; El Monfalouti, H.; Haddad, A.; Charrouf, Z. Analyse chimique et sensorielle de l'huile d'argane. *Les Technologies de Laboratoire* **2011**, *6*, 13–23.
36. Azizi, S.-E.; Dalli, M.; Mzabri, I.; Berrichi, A.; Gseyra, N. Chemical characterization of oils produced by some native and introduced genotypes of argan tree in eastern Morocco using HPLC-DAD/GC-MS, and the evaluation of their physicochemical parameters. *OCL* **2022**, *29*, 14. [CrossRef]
37. Rahmani, M. Composition chimique de l'huile d'argane «vierge». *Cah. Agric.* **2005**, *14*, 461–465.
38. Hilali, M.; Charrouf, Z.; Soulhi, A.E.A.; Hachimi, L.; Guillaume, D. Influence of origin and extraction method on argan oil physico-chemical characteristics and composition. *J. Agric. Food Chem.* **2005**, *53*, 2081–2087. [CrossRef]
39. *IOC/T.15/NC No. 3/Rev. 7*; Trade Standard Applying to Olive Oils and Olive Pomace Oils. IOC (International Olive Council): Geneva, France, 2013.
40. Lapera, L.; Lococo, F.; Mavrogeni, E.; Giuffrida, D.; Dugo, G. Determination of copper (II), lead (II), cadmium (II) and zinc (II) in virgin olive oils produced in Sicily and Apulia by derivative potentiometric stripping analysis. *Ital. J. Food Sci.* **2002**, *14*, 389–399.
41. Khallouki, F.; Younos, C.; Soulimani, R.; Oster, T.; Charrouf, Z.; Spiegelhalder, B.; Bartsch, H.; Owen, R.W. Consumption of argan oil (Morocco) with its unique profile of fatty acids, tocopherols, squalene, sterols and phenolic compounds should confer valuable cancer chemopreventive effects. *Eur. J. Cancer Prev.* **2003**, *12*, 67–75. [CrossRef]
42. Gharby, S.; Guillaume, D.; Elibrahimi, M.; Charrouf, Z. Physico-Chemical Properties and Sensory Analysis of Deodorized Argan Oil. *ACS Food Sci. Technol.* **2021**, *1*, 275–281. [CrossRef]
43. Zarrouk, A.; Martine, L.; Grégoire, S.; Nury, T.; Meddeb, W.; Camus, E.; Badreddine, A.; Durand, P.; Namsi, A.; Yammine, A.; et al. Profile of Fatty Acids, Tocopherols, Phytosterols and Polyphenols in Mediterranean Oils (Argan Oils, Olive Oils, Milk Thistle Seed Oils and Nigella Seed Oil) and Evaluation of their Antioxidant and Cytoprotective Activities. *Curr. Pharm. Des.* **2019**, *25*, 1791–1805. [CrossRef]
44. Fellat-Zarrouck, K. Étude de Corps Gras d'Origine Marocaine: Huile d'Olive, Adultérations Huile de Sardine, Sardina Pilchardus Huile d'Argan, Argania Spinosa. Doctoral Dissertation, Aix-Marseille 1, AV ESCAD NORMANDIE NIEMEN 13013 MARSEILLE, 1987.
45. Maurin, R.; Fellat-Zarrouck, K.; Kirs, M. Huile d'argan, Argania spinosa du Maroc, Triglycérides. In Proceedings of the Euro-Lipid Congres, Angers, France, 6–9 June 1989.

forests

MDPI

Article

Diversity of Endomycorrhizal Fungi in Argan Forest Stands: Implications for the Success of Reforestation Programs

Matike Ganoudi [1,*], Imane Ouallal [2], Abdelali El Mekkaoui [3], Mounir Majid [4], Mohammed Ibriz [5] and Driss Iraqi [1]

1 Biotechnology Research Unit, Center of Agricultural Research of Rabat National Institute of Agricultural Research, Avenue Ennasr, BP 415 Rabat Principal, Rabat 10090, Morocco; driss.iraqi@inra.ma
2 Laboratory of Natural Resources and Sustainable Development, Ibn Tofail University, Kenitra 14000, Morocco; imaneouallal@gmail.com
3 Environment and Conservation of Natural Resources, Research Unit, Regional Center of Agricultural Research of Rabat, National Institute of Agricultural Research, Avenue Ennasr, BP 415 Rabat Principal, Rabat 10090, Morocco; abdelali.elmekkaoui@inra.ma
4 Biotechnology and Biotransformations Laboratory, Food Science and Nutrition Department, Hassan II Institute of Agronomy and Veterinary Medicine, P.O. Box 6202, Rabat 10112, Morocco; m.mounir@iav.ac.ma
5 Department of Biology, Genetics and Biometrics Laboratory, Faculty of Sciences, BP 133, Kenitra 14000, Morocco; mohammed.ibriz@uit.ac.ma
* Correspondence: matike.ganoudi@inra.ma; Tel.: +212-621123831

Abstract: Over the last few decades, argan trees (*Argania spinosa* L.) skeels have faced harsh ecological conditions and anthropogenic pressure, leading to a dramatic decline in surface and density of cultivation. Nowadays, most techniques used to regenerate argan trees have failed. Arbuscular mycorrhizal fungi (AMF) are root symbionts that increase plant resistance to biotic and abiotic stresses during transplantation. The exploration of these symbiotic fungi from different soils of argan stands is the starting point for the selection and production of high-performance organisms adapted to the reforestation sites. The objective of this study is to investigate the composition of the AMF community associated with the argan tree rhizosphere. Forty adult argan trees were sampled in eight forest sites representative of the distribution and genetic diversity of argan forest stands. Five sub-samples of rhizospheric soil were taken around each tree. Our results revealed the presence of different AMF structures (i.e., hyphae, vesicles/and arbuscules) in root samples. Based on morphological characterization, six genera of AMF spores were identified with a dominance of the genera *Septoglomus* (34%). In addition, soil organic matter and phosphorus concentrations showed a highly significant correlation with AMF spore density. The chi-square test showed a highly significant dependence of the distribution of genera on the site conditions of forest stands. These AMF could be tested and used during the inoculation of argan seedlings in forest nurseries for the success of restoration and reforestation programs, as well as for the development and sustainable improvement of this agroforestry system.

Keywords: *Argania spinosa*; rhizosphere; arbuscular mycorrhizal fungi (AMF); diversity; reforestation

Citation: Ganoudi, M.; Ouallal, I.; El Mekkaoui, A.; Majid, M.; Ibriz, M.; Iraqi, D. Diversity of Endomycorrhizal Fungi in Argan Forest Stands: Implications for the Success of Reforestation Programs. *Forests* **2023**, *14*, 1649. https://doi.org/10.3390/f14081649

Academic Editors: Timothy A. Martin, Mohammed S. Lamhamedi, Damase P. Khasa and Steeve Pépin

Received: 30 March 2023
Revised: 28 May 2023
Accepted: 3 June 2023
Published: 15 August 2023

1. Introduction

The argan tree is an endemic tree of Morocco belonging to the Sapotaceae family. It is a thermo-xerophilic species that grows in the center-west of Morocco between Agadir and Essaouira, Argan stands occupy an area of 871,210 hectares [1], and it is ranked as the second forest tree species in this country. The argan tree is considered a valuable species because it provides several useful parts. The wood is used as fuel (mainly as charcoal). The argan cake is the seed waste produced during the extraction process, and it is traditionally used for skin care and livestock nutrition. Different biological activities of argan cake have been cited, essentially as antioxidant, chemoprotective, anti-inflammatory, and antimicrobial [2]. Argan oil is highly used for the manufacture of cosmetic products such as oil, soap, shampoo, and cosmetic creams, as well as for human consumption [3]. On the other hand, argan forests protect the soil against erosion, prevent desertification, and preserve ecological balance and biodiversity [4]. However, drought and strong anthropogenic pressure hamper the natural regeneration of the species and limit its range. To this end, several artificial regeneration programs have been implemented to improve and rehabilitate the degraded areas of the argan tree. To remedy this problem, it is necessary to improve the techniques of plant production in vitro and in nurseries. The results of ecological surveys carried out in the Sousse Massa region have shown that the argan tree, similar to the majority of forest species, forms endomycorrhizal associations. It was also noted during these surveys a predominance and wide distribution of the genera *Glomus* [5]. Arbuscular mycorrhizal fungi (AMF) have received a great deal of interest over the past twenty years, due to their favorable effects, mainly on the uptake of water and nutrients, particularly the weakly mobile phosphorus ion from the soil [6]. Inoculation of plants with AMF can not only facilitate their establishment but also improve the physicochemical and biological properties of the soil [7–9]. Argan reforestation requires young tree seedlings to be raised in nurseries and quickly adapt to the dry climate of the native range of these trees. Mycorrhizal inoculation significantly increases the growth and health of young argan seedlings, thereby increasing their fitness and survival after planting [10,11].

According to previous results [12], a local variant of an AMF species is more beneficial than a foreign species for its host. Therefore, it is essential to explore autochthonous AMF to select high-performance symbionts adapted to reforestation sites. Hence, the search for diversified, potent fungi with a broad spectrum of applications in reforestation programs is crucial. In this context, the objective of this work is to determine the diversity of communities of AMF associated with the argan tree in different edapho-climatic conditions in the northeast, northwest, and center-west of Morocco.

2. Materials and Methods

2.1. Sampling Sites in Argan Ecosystems

Eight sites of argan trees located in the northeast region (Aklim/Benisnassen), northwest region (Rabat/Oued Cheratt), and center-west region (Agadir, Ait Melloul, Essaouira, Ouled Taima, Talouine, and Taroudant), were prospected. These stations represent different eco-geographical and climatic conditions in Moroccan argan ecosystems (Figure 1).

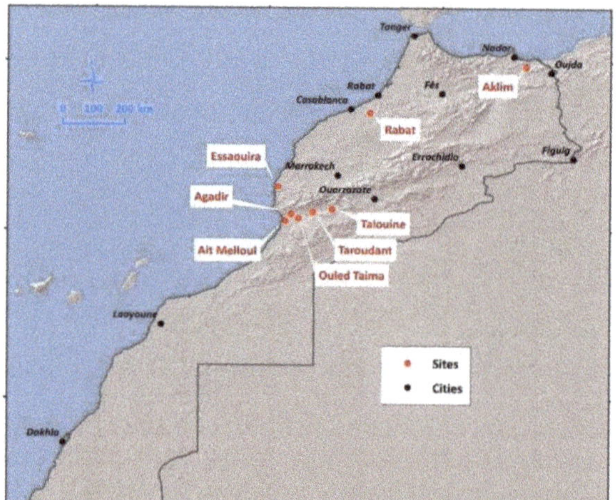

Figure 1. Location of the sampling sites.

2.2. Collection of Soil and Root Samples

Samples were collected from the eight argan stands in February, during the peak of the rainy season, when AMF populations often reach their maximum sporulation (Kaushal, 2000). First, we removed the roots growing beside the argan trees. Argan roots were distinctive for their hardness and pale red color.

Forty adult argan trees, representative of the general appearance of the tree at the eight stations, were sampled. To collect samples, five sub-samples of rhizospheric environments were taken around each tree at depths between 15 and 30 cm, where the finest roots likely to be mycorrhizal can be found. Care was taken not to damage the rootlets of the argan trees that were harvested. The five samples of "rhizospheric soils" from each argan tree were mixed in a plastic container to obtain a composite average sample (2 kg) that was representative of the rhizosphere of the sampled tree. Thus, five composite samples were taken at each station.

At the same time, samples of fine roots, likely to be mycorrhizal and more easily observable under a microscope, were collected from each argan tree in a bottle containing a solution of glycerol, ethanol, and distilled water (1 v/1 v/1 v). All samples were numbered, darkened, and transferred to the laboratory, where they were stored in a refrigerator at 4 °C for further analyses.

2.3. Physical and Chemical Analyses of Sampled Soil

The main physicochemical properties of the soil samples were analyzed according to the following conventional methods: The properties measured were pH in water at 1:1, and available phosphorus was determined by a colorimetric method as detailed in Olsen and Somers [13]. Total nitrogen was determined using the Kjeldahl method, and organic matter was determined using the Walkley and Black method [14,15]. This method consists of oxidizing the organic matter in the soil with a mixture of potassium dichromate in a hot sulfuric medium until the release of CO_2. The analysis of sodium and potassium was determined by photometry according to the extraction method adapted for ammonium acetate.

2.4. AMF Root Colonization

Root colonization was estimated following the staining method described by Phillips and Hayman [16]. Approximately 3 g of fine roots, cut into 0.5–1.0 cm pieces, were first cleared in 10% KOH and placed in a water bath at 90 °C for 1 h. They were rinsed with deionized water, neutralized with 1% HCl for a few minutes, and bleached with a solution of H_2O_2 for 10 min at 90 °C in the water bath. Finally, they were stained with 2% ink (Parker blue ink, USA) in 1% HCl at 70 °C for 30 min. The roots were subsequently cut into 1 cm lengths and mounted on slides for microscopic observations.

The evaluation of mycorrhizal colonization is performed on batches of 30 stained root fragments and mounted in three groups of ten between a slide and coverslip in lactic acid. The samples are then carefully examined under a microscope at a magnification of 100× to note the mycorrhizal structures. Occasionally, a magnification of 400× is used to better distinguish between the arbuscular and vesicular structures. The parameters noted include the frequency (the number of mycorrhizal root fragments) and intensity (proportion of colonized cortex estimated in relation to the entire root system) percentages of mycorrhization of the root system, as well as the contents of arbuscules and vesicles manifested inside the root cortex. These parameters are determined according to the method of (Trouvelot et al. [17].

2.5. Extraction of AMF Spores from Rhizosphere Soil Mixtures and Identification of Spores

The "wet sieving and decanting" method described by Gerdemann and Nicolson [18] and adapted by Cranenbrouck et al. [19] was used to isolate the spores of AMF. About 100 g of composite representative soil from each site was submerged in 1 L of water in a beaker and stirred for 1 min with a spatula. The supernatant was passed through four superimposed decreasing meshes sieves (500, 250, 106, and 50 µm). This operation was repeated twice. This combination is suitable for collecting AMF spores. The upper sieve (500 μm) retains soil and root debris as well as mycorrhizal roots, while the other sieves (250, 106, and 38 μm) retain spores belonging to all the AMF genera. The contents of the 80 and 50 μm sieves are recovered with water (approximately 40 mL) in two beakers, which are then transferred into four tubes and centrifuged for 5 min at 5000 rpm. The supernatant, containing light debris and dead spores, is quickly removed, and the remaining volume (about 20 mL) is kept for the next step. The soil material was resuspended in a 1.4 M sucrose solution and centrifuged at 1000 rpm for 4 min. The spores are washed very carefully with water through a fine sieve (50 μm) to remove the remaining sucrose solution. These spores are then collected in a small Petri dish (50 mm) for observation under a microscope. The estimate of the number of spores in each soil is made by counting the spores in 1 mL of supernatant and extrapolating the total volume to 100 mL. The spores are collected one by one. The fungal spores recovered were first presented under a binocular magnifying glass (×40), then under a microscope (×100). Spores were considered viable if they had clear contents and an intact wall. They have been classified according mainly to their morphological characteristics (size, color, shape, consistency, wall structure, attachment of the suspensory hypha and ornamentation, and lipid reserves) based on the INVAM website (http://invam.caf.wvu.edu, accessed on 20 June 2022).

2.5.1. Spores' Density

Spore density was determined by direct counting under a binocular loupe. The total number of spores present in 100 g of soil was calculated as the average of five samples.

2.5.2. Species Richness

Genus richness was determined by calculating the number of morphotypes observed per site sample.

2.5.3. Occurrence Rate

The distribution of AMF genera between the studied sites was calculated by the percentage of sites where each genus was detected (the percentage of sites including a particular genus).

2.5.4. Shannon and Pielou Index

The Shannon index [20] expresses diversity by taking into account the number of genera and the abundance of individuals within each of these genera. It is calculated as follows:

$$H' = -\sum_{i=1}^{S} pi \ \log_2(pi)$$

pi = proportional abundance or percent importance of genera: pi = ni/N; S = Total number of genera

ni = number of individuals of genera in the sample

N = total number of individuals of all genera in the sample.

The Pielou index [21], which is the regularity of the distribution of species, is calculated as follows:

$E = H'/\log_2 (S)$

S = Total number of genera

H' = Shannon index

2.6. Statistical Analyses

A one-way ANOVA followed by a Tukey honestly significant difference post-hoc test ($p \leq 0.05$) was used to test the arcsine-transformed percentages of argan root colonization. The normal distribution of residuals was checked before analysis. A Pearson correlation was conducted at $p \leq 0.05$ to assess the relationships between mycorrhizal colonization and available P. A factor analysis of correspondence was performed on the AMF community composition by sites of sampling using the XLSTAT software.

All statistical analyses were performed using the SAS software database.

3. Results

3.1. Soil Physical Parameters

The results of the physicochemical properties of argan soil at the eight studied sites (Table 1) showed that the Taliouin and Ouled Taima soils are rich in clay (32% and 22%, respectively), while the Aklim soil s are rich in silt (75%), and the Ait Melloul soils are rich in sand (79%). The pH of rhizospheric samples taken from the various argan tree sites is alkaline, ranging from 7.2 for the Rabat site to 8.2 for the Taroudant site. Available phosphorus in the sampled soils is consistently low at all sites, and potassium content reaches 1971.2 ppm in the Aklim soil. Soils from Aklim and Agadir have the highest organic matter and total carbon contents.

Table 1. Physical and chemical properties of soil samples.

Sites	% Clay	% Silt	% Sand	pH Water	pH KCl	P_2O_5 ppm	K_2O ppm	% SOM	% SOC
Agadir	20.6 ± 0.4 d	36.4 ± 0.2 d	43.6 ± 0.3 d	7.5 ± 0.1 bc	7.2 ± 0.1 b	21.1 ± 15.9 c	588.8 ± 0.4 b	4.3 ± 0.1 a	1.8 ± 0.0 b
Ait Melloul	4.9 ± 0.1 f	15.3 ± 0.2 h	78.9 ± 0.1 a	7.3 ± 0.2 c	7.3 ± 0.1 ab	47.2 ± 0.0 b	360.2 ± 0.1 f	2.1 ± 0.1 bc	1.3 ± 0.0 d
Aklim	5.7 ± 0.1 f	74.8 ± 0.2 a	19.2 ± 0.2 g	7.6 ± 0.1 bc	7.5 ± 0.a	14.1 ± 0.1 cd	1971.2 ± 0.1 a	4.4 ± 0.1 a	2 ± 0.1 b
Essaouira	9.0 ± 0.1 e	44.7 ± 0.4 b	44.4 ± 0.4 d	7.8 ± 0.0 ab	7.1 ± 0.1 b	4.2 ± 0.0 d	161.5 ± 0.1 g	4.2 ± 0.1 a	0.8 ± 0.0 e
Ouled Taima	21.6 ± 0.2 c	29.5 ± 0.3	48.1 ± 0.1 c	7.4 ± 0.1 bc	7.2 ± 0.1 b	30.4 ± 0.1 cb	404.1 ± 0.1 e	0.9 ± 0.0 d	0.6 ± 0.0 f
Rabat	20.1 ± 0.2 d	24.7 ± 0.1 f	36.8 ± 0.1 e	7.2 ± 0.1 bc	7.3 ± 0.1 ab	29.1 ± 0.1 c	134.4 ± 0.1 h	1.9 ± 0.1 c	2.8 ± 0.0 a
Taliouin	32 ± 0.5 a	43.1 ± 0.5 c	24.4 ± 0.3 f	7.7 ± 0.1 bc	7.2 ± 0.1 b	69.0 ± 0.1 a	555.7 ± 0.1 c	2.03 ± 0.1 c	1.2 ± 0.0 d
Taroudant	20.9 ± 0.2 b	23.2 ± 0.2 g	53.8 ± 0.5 b	8.2 ± 0.0 a	7.1 ± 0.1 b	47.2 ± 0.1 b	495.1 ± 0.1 d	2.3 ± 0.0 b	1.3 ± 0.1 c

Mean values within the same column followed by different letters differ significantly ($p < 0.05$) using Tukey's test HSD at 5%. Data are presented as mean ± standard deviation ($n = 3$). SOM: soil organic matter, SOC: soil organic carbon.

3.2. Root Colonization and Spore Density of AMF

Roots from eight sites were examined, and different structures were observed, including arbuscules, vesicles, and extracellular hyphae (Figure 2), in all roots from the sites examined. The results show a significant difference between sites. The frequency of mycorrhization was high at all the sites studied, with values ranging from 80% to 100%. Similarly, mycorrhization intensity varied significantly between sites, with the highest level observed in Aklim (37% ± 6.27) and the lowest level in the Taroudant site (20% ± 4.01). The percentage of vesicle colonization was significantly higher at the Taliouin site (53% ± 10.97) compared to other sampling sites. Conversely, no significant difference in arbuscular colonization was observed between sites (Table 2). Examination of the rhizospheric soils of *Argania spinosa* showed that the average spore density varied from site to site, with the highest at the Essaouira site (120 spores per 100 g of soil) and the lowest at the Taroudant site (40 spores per 100 g of soil) (Table 2).

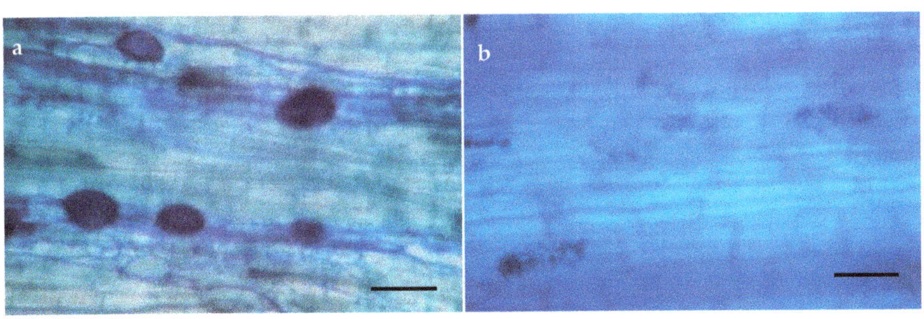

Figure 2. The different structures observed in the roots collected from the argan tree: (**a**) vesicles and hyphae; (**b**) Arbuscules. 1 cm = 100 µm.

Table 2. Mycorrhizal Frequency, intensity, arbuscular, and vesicular contents of argan roots.

Sites	% F	% I	% V	% A	SD/100 g of Soil
Agadir	100.0 ± 0.0 a	22.3 ± 2.3 b	35.3 ± 4.2 ab	52.2 ± 6.4 a	101 a
Ait Melloul	72.0 ± 3.0 a	26.1 ± 4.5 ab	42.7 ± 4.0 a	50.2 ± 9.9 a	60 b
Aklim	98.0 ± 1.6 ab	36.6 ± 6.2 a	34.5 ± 3.8 abc	52.0 ± 5.7 a	98 ab
Essaouira	99.6 ± 0.3 bc	30.8 ± 2.7 a	31.5 ± 2.7 abc	51.8 ± 6.1 a	120 a
Ouled Taima	84.0 ± 5.5 c	25.7 ± 2.4 ab	41.1 ± 2.9 a	49.8 ± 3.6 a	57 bc
Rabat	84.0 ± 3.2 c	22.9 ± 2.0 b	21.5 ± 3.7 bcd	52.4 ± 1.6 a	46 bc
Taliouin	80.0 ± 4.4 cd	27.3 ± 10.3 ab	52.8 ± 10.9 a	38.2 ± 12.2 a	50 bc
Taroudant	88.0 ± 6.0 d	19.5 ± 4.0 b	48.7 ± 7.6 a	40.3 ± 6.3 a	40 c

Mean values in the same column followed by different letters differ significantly ($p < 0.05$) using the Tukey HSD test at 5%, data presented as mean ± standard deviation (n = 5). %F: mycorrhizal frequency, %I: mycorrhizal intensity, %V: vesicular, %A: arbuscular, and SD: spore density.

3.3. AMF Community Composition in Soils

Our research revealed that the isolated spores belonged to the three family glomale orders, Glomeraceae, Gigasporaceae, and Acaulosporaceae. The analysis of the morpho-anatomical characteristics of this spore community revealed the presence of six genera: Septoglumus, Acaulospora, Rhizophagus, Glomus, Funneloformis, and Racocetra (Figure 3).

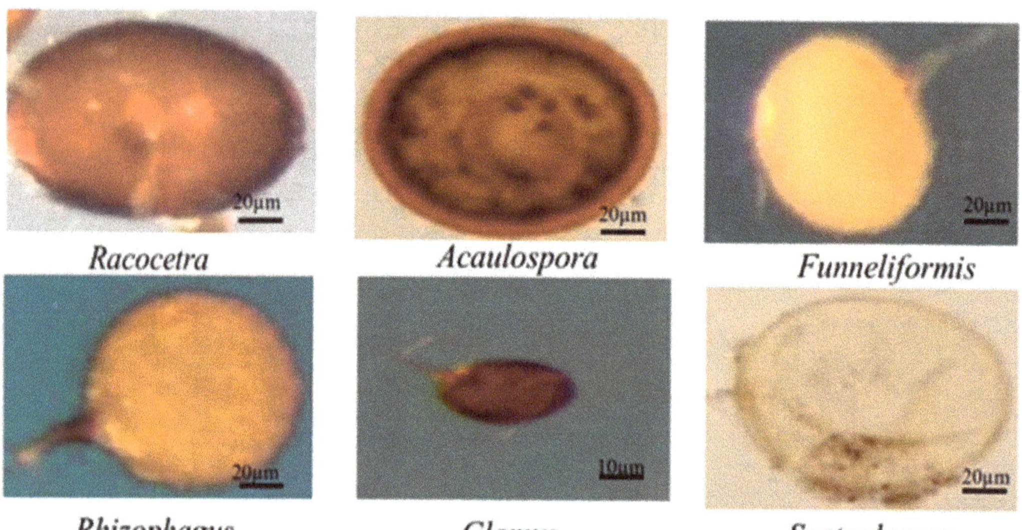

Figure 3. Types of arbuscular mycorrhizal fungi isolated from the argan tree rhizosphere.

The analysis of the relative abundance of AMF genera presents in soil samples from different sites revealed the presence of only one genus at Ait Melloul (Septoglomus) and Aklim (Rhizophagus) sites, the presence of two genera at Agadir and Taroudant (Septoglomus, Rhizophagus), Rabat (Rhizophagus, Racocetra), and Taliouin (Septoglomus, Funneliformis) sites, and the presence of three genera (Septoglomus, Glomus, Acaulospora) at Essaouira site. Regarding the occurrence rate of the genera, Septoglomus was present in 34% of the surveyed sites, followed by Rhizophagus in 22%, while the other genera were present in 11% of the studied sites (Figure 4).

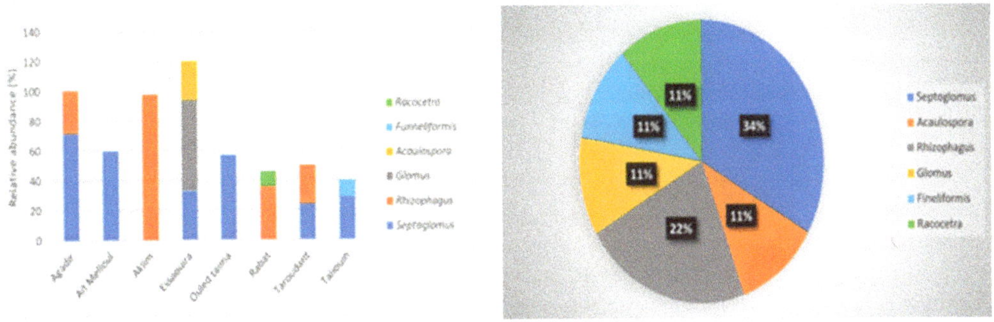

(a) (b)

Figure 4. (**a**) Relative abundance of arbuscular mycorrhizal fungi genera presents in soil samples from different sites, (**b**) occurrence rate of each genus found under the argan tree rhizosphere.

Factorial analysis of correspondences (FAC) showed a highly significant correlation between the sites and genera of AMF ($p < 0.0001$). Furthermore, it subdivided the sites into two classes: Class 1 (Taliouin, Oueled Taima, Ait Melloul, Agadir, Aklim, Taroudant, and Rabat) and Class 2 (Essaouira) (Figure 5).

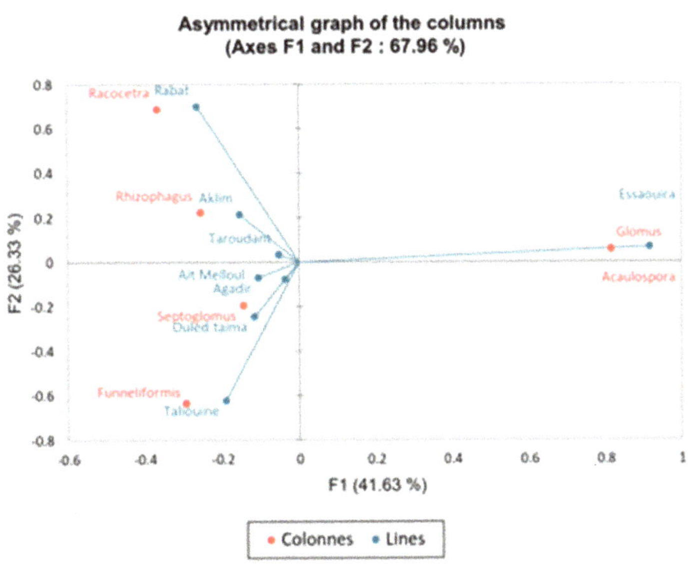

Figure 5. Factorial analysis of correspondences (FAC) of arbuscular mycorrhizal fungi community composition according to sampling sites.

The relationship between physicochemical soil properties and AMF spore density is represented in Table 3. In fact, a positive and highly significant correlation was observed between spore density (SD) and soil organic matter ($r = 0.8802$, $p \leq 0.01$). On the other hand, a highly significant inverse correlation was observed between spore density and P_2O_5 ($r = -0.8688$, $p \leq 0.01$).

Table 3. Pearson correlation coefficients between soil physicochemical parameters and arbuscular mycorrhizal fungi spore density (SD) of argan rhizosphere soil samples.

	%Clay	%Silt	%Sand	pH Water	pH KCl	P_2O_5 ppm	K_2O ppm	%SOM	%SOC
SD	−0.58754 0.1256	0.54890 0.1589	−0.12613 0.7660	−0.05867 0.8902	−0.10489 0.8048	−0.86885 0.0051 **	0.29033 0.4855	0.88022 0.0039 **	−0.11987 0.7774

Pearson Correlation Coefficient, N = 8 Prob > |r| under H0: Rho = 0; SOM: Soil Organic Matter, SOC: Soil Organic Carbon. Effects marked with ** are significant (** $p \leq 0.01$).

Calculation of the Shannon diversity index showed a specific diversity with different abundances between the sites studied. Indeed, the Taliouin and Ouled Taima sites showed the highest diversity index compared to the other sites. Similarly, both sites had the highest Pielou index (E). The Pielou index (E) showed a highly variable distribution of genera between sites, indicating an imbalance in the AMF community present in the eight sites studied (Figure 6).

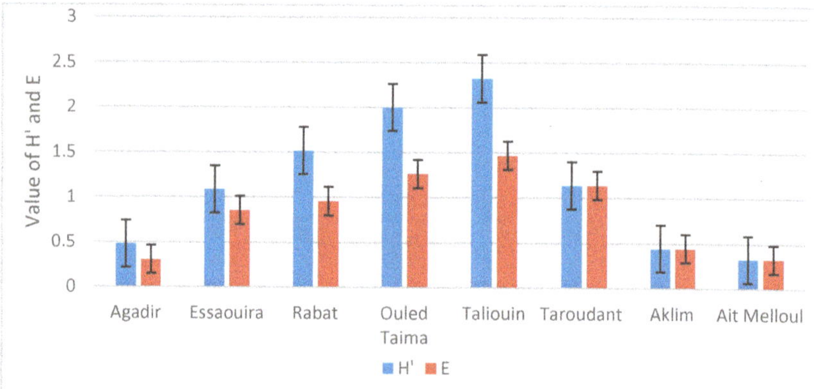

Figure 6. Shannon index (H') and the Pielou (E) index of the eight stations, standard deviation (n = 5).

4. Discussion

Knowledge of AMF diversity in forest areas is an important biological parameter for the assessment of environmental disturbances. The use of indigenous soil mycorrhizae has advantages in quick acclimatization and ecosystem restoration, such as their good tolerance to local conditions and low ecological risk [22]. Indeed, the study will provide a basis for AMF utilization in the success of argan reforestation programs in the region.

According to studies conducted in the surveyed areas, the roots of the argan tree carry endomycorrhizal structures: vesicles, arbuscular, and hyphae. This is explained by several authors who confirm the dependence of this tree on AMF [22–24]. The quantified mycorrhization frequencies were always high (80%–100%) in all the argan plantations studied, which confirms that *Argania spinosa* is a highly mycotrophic species. The highest mycorrhizal frequencies were noted at the sites of Agadir (100%), Essaouira (99.6%), and Aklim (98%). The mycorrhizal intensity varied between 36% for the Aklim site and 19% for the Taroudant site. In addition, our study showed a relationship between the number of spores and the mycorrhizal intensity, which was confirmed by some authors [25,26]. According to [27,28], the sporulation may depend on the type of AMF, soil characteristics, and climatic conditions. The contents of arbuscules and vesicles change from one site to another.

The variability of the parameters of root colonization by AMF may be explained by differences in the physicochemical properties of the soils and the climatological conditions of the sites studied. Root colonization variability is strongly dependent on plant fitness, and in some situations, a reduced root biomass owing to dryness might result in a higher frequency of AMF colonization [29].

The results of the available phosphorus content of the soil samples collected from the different sites were low. Indeed, the negative correlation recorded between the levels of root colonization (Frequency and intensity) and the concentration of available phosphorus confirmed the adaptation of AMF to poor soils in P [30–32]. These results are also consistent with those reported by several authors [10,24], who state that the frequencies of mycorrhization are high in soils with low levels of total phosphorus in argan. However, this relationship does not seem to be valid on other sites.

The results of AMF isolation from the argan rhizosphere of the studied sites showed variations in the abundance and number of morphotypes. The highest density was observed in the Essaouira plain site (120 spores per 100 g of soil). This number is much lower than that reported by several authors: 207 spores per 100 g [24], 188 spores per 100 g [10], and 1128 spores per 100 g [33]. Red. [34] reported 900 à 2080 spores per 100 g, which is also higher than our finding. These results also remain low compared to those found in the rhizosphere of certain Moroccan species, such as *Tetraclinis articulata* [35], *Cupressus atlantica* [36], *Olea europaea* [37], *Date palm* [38]. The highest spore densities (120, 101, and 98 spores per 100 g of dry soil) were observed in the sites of Essaouira, Agadir, and Aklim, respectively, which had high organic matter content (4.2, 4.3, and 4.4%). Indeed, it has been noted that the addition of organic amendments (manure, compost, and crop residues) can stimulate the multiplication of AMF [39]. Furthermore, the differences recorded may be due to the physicochemical and microbiological properties of soils in arid and semi-arid areas [40], microclimate fluctuations [41], vegetation cover, sampling season [42], spore formation, and deterioration of germination [43].

The preliminary identification, based only on the morphological criteria of the spores, showed the presence of six genera of AMF (*Septoglumus, Acaulospora, Rhizophagus, Glomus, Funneloformis*, and *Racocetra*) belonging to three families (*Glomaceae, Gigasporaceae*, and *Acaulosporaceae*). These results are very important compared to those of Ouallal et al. [24], who reported only two genera (*Glomus and Scutellospora*) of arbuscular mycorrhizal fungi in the nine Moroccan argan trees that they studied. In addition, Elmaati et al. [33] revealed the dominance of the *Rhisophagus* and *Glomus* genera in five argan sites. However, Sellal et al. [10] identified five genera (*Glomus, Scutellospora, Entrophospora, Pacispora*, and *Gigaspora*) from spores isolated in the rhizosphere of argan trees from seven sites in the regions of Essaouira, Agadir, Taroudant, and Tiznit.

The results of the occurrence rate revealed the dominance of the *Septoglomus* genera in the sampling sites, with a distribution rate of 34%, followed by *Rhizophagus* with 22%, while the other genera were present in 11% of the studied sites. The dominance of the genera *Septoglomus* in several different environments indicates high plasticity and high adaptation to different impacts of biotic or abiotic origin [44]. According to the authors, the dominance of *Glomus* is due to its ability to produce more spores in less time than other genera, such as *Gigaspora* and *Acaulospora*. This dominance of the *Glomus* genera has also been reported in several studies on the argan tree rhizosphere [10,24,33].

According to Chagnon et al. [45], *Acaulospora* is considered to be stress tolerant. In fact, it has frequently been reported in harsh climatic circumstances, acidic soils (i.e., pH 3.6–4.20; Morton, 2017), and high-altitude locations, at 2500 m [46]. In our investigation, this genus was found in alkaline soils (pH 7.81) at 150 m above sea level. This shows the capacity of the genus to thrive in a variety of different environments. *Acaulospora* generally predominates in dry forests but not in wet forests [47].

In general, *Glomus, Rhizophagus, Septoglomus, Funneliformis*, and *Acaulospora* were already identified in the rhizosphere of the argan tree in semi-arid areas.

The Shannon Diversity and Pielou index indicated that the AMF community was highly diverse with respect to abundance and species distribution; this study was consistent with results found by Ouallal et al. [24], who showed a similar diversity between study sites. The indices used here are among the most commonly used in AMF studies [48].

5. Conclusions

To our knowledge, our results report for the first time the composition of the arbuscular mycorrhizal fungi community in several forest sites representative of the genetic diversity and ecological distribution of argan forest stands: the northeast region (Aklim/Bnisnassen), the northwest (Rabat/Oued cheratt), and the center-west (Agadir, Ait Melloul, Essaouira, Ouled Taima, Taliouin, and Taroudant). The results revealed a strong correlation between the density of AMF spores and the physicochemical properties of the soils. The results also showed an abundance and a diversification of AMF associated with the species. In all sites, we identified six genera of AMF (*Septoglomus*, *Rhizophagus*, *Glomus*, *Acaulospora*, *Fineloformus*, and *Racocetra*). This richness highlighted in the rhizosphere of the argan tree is very important; it gives useful information for the selection of an effective inoculum according to the different pedoclimatic zones. This diversity encourages further research into the selection of native AMF species that show good adaptation to environmental conditions and climatic variation between sites. It is well established that AMF provides vigorous and robust plants that withstand transplanting and weather conditions. Indeed, the mycorrhization of plants in Moroccan nurseries with native inoculum before their transfer to the field is an essential step for the success of plantations in reforestation programs and the restoration of degraded ecosystems.

Author Contributions: Conceptualization, M.G. and D.I.; methodology, M.G.; software M.G.; validation, D.I.; formal analysis, M.G. and I.O.; investigation, M.G.; resources, M.G.; data curation M.G. and I.O.; writing—original draft preparation M.G.; writing—review and editing, I.O.; visualization, M.G.; supervision, D.I.; project administration M.G., I.O., A.E.M., M.M., M.I. and D.I.; funding acquisition, D.I. All authors have read and agreed to the published version of the manuscript.

Funding: This research and the publication fees were supported by National Institute of Agronomic Research Avenue Ennasr, BP 415 Rabat Principal, Rabat 10090, Morocco.

Data Availability Statement: Publicly available sources of the data used in this study are described in the article. However, other relevant data or information can be requested from the authors on reasonable grounds.

Acknowledgments: This study was carried out as part of the INRA research project. We thank the technicians and researchers of the Biotechnology Laboratory for their valuable advice. We would also like to thank the Soil Analysis Laboratory of the Environment and Conservation of Natural Resources Research Unit, INRA, in Rabat, for their help in analyzing the soil sample, and Fatima Gaboun (INRA) for her support with statistical analysis.

Conflicts of Interest: The authors declare no conflict of interest.

References

1. Moukrim, S.; Lahssini, S.; Mharzi Alaoui, H.; Rifai, N.; Arahou, M.; Rhazi, L. Modélisation de la distribution spatiale des espèces endémiques pour leur conservation: Cas de l'*Argania spinosa* (L.) Skeels. *Rev. DÉcologie Terre Vie* **2018**, *73*, 153–166. [CrossRef]
2. Mechqoq, H.; El Yaagoubi, M.; El Hamdaoui, A.; Momchilova, S.; Guedes da Silva Almeida, J.R.; Msanda, F.; El Aouad, N. Ethnobotany, Phytochemistry and Biological Properties of Argan Tree (*Argania spinosa* (L.) Skeels) (Sapotaceae)—A Review. *J. Ethnopharmacol.* **2021**, *281*, 114528. [CrossRef] [PubMed]
3. Chaussod, R.; Adlouni, A.; et Christon, R. L'arganier et l'huile d'argane au Maroc: Vers la mutation d'un système agroforestier traditionnel? Enjeux et contribution de la recherche. *Cah. Agric.* **2005**, *14*, 351–356.
4. Krajnc, B.; Bontempo, L.; Luis Araus, J.; Giovanetti, M.; Alegria, C.; Lauteri, M.; Augusti, A.; Atti, N.; Smeti, S.; Taous, F.; et al. Selective Methods to Investigate Authenticity and Geographical Origin of Mediterranean Food Products. *Food Rev. Int.* **2021**, *37*, 656–682. [CrossRef]
5. Bousselmane, F.; Kenny, L.; et Achouri, M. Effet des mycorhizes à vésicules et arbuscules sur la croissance et la nutrition de l'arganier (*Argania spinosa* L.). *Rev. Maroc. Des. Sci. Agron. Et. Vétérinaires* **2022**, *22*, 193–198.
6. Smith, S.E.; Read, D.J. *Mycorrhizal Symbiosis*, 3rd ed.; Academic Press: London, UK, 2008.
7. Del Mar Alguacil, M.; Torrecillas, E.; Kohler, J.; Roldán, A. A Molecular Approach to Ascertain the Success of "in Situ" AM Fungi Inoculation in the Revegetation of a Semiarid, Degraded Land. *Sci. Total Environ.* **2011**, *409*, 2874–2880. [CrossRef]
8. Chen, M.; Arato, M.; Borghi, L.; Nouri, E.; Reinhardt, D. Beneficial Services of Arbuscular Mycorrhizal Fungi—From Ecology to Application. *Front. Plant Sci.* **2018**, *9*, 1270. [CrossRef] [PubMed]

9. Chifetete, V.W.; Dames, J.F. Mycorrhizal Interventions for Sustainable Potato Production in Africa. *Front. Sustain. Food Syst.* **2020**, *4*, 593053. [CrossRef]
10. Sellal, Z.; Ouazzani Touhami, A.; Chliyeh, M.; Dahmani, J.; Benkirane, R.; Douira, A. Arbuscular Mycorrhizal Fungi Species Associated with Rhizosphere of *Argania spinosa* (L.) Skeels in Morocco. *Int. J. Pure Appl. Biosci.* **2016**, *4*, 82–99. [CrossRef]
11. Ganoudi, M.; Calonne-Salmon, M.; Ibriz, M.; Declerck, S. In Vitro Mycorrhization of *Argania spinosa* L. Using Germinated Seeds. *Symbiosis* **2021**, *85*, 57–68. [CrossRef]
12. Requena, N.; Jimenez, I.; Toro, M.; Barea, J.M. Interactions between Plant-growth-promoting Rhizobacteria (PGPR), Arbuscular Mycorrhizal Fungi and Rhizobium Spp. in the Rhizosphere of Anthyllis Cytisoides, a Model Legume for Revegetation in Mediterranean Semi-arid Ecosystems. *New Phytol.* **1997**, *136*, 667–677. [CrossRef]
13. Olsen, S.R.; Sommers, L.E. Phosphorus. In *Methods of Soil Analysis: Part 2*; Page, A.L., Ed.; Chemical and Microbiological Properties; ASA and SSSA: Madison, WI, USA, 1982; pp. 403–430.
14. El Mekkaoui, A.; Moussadek, R.; Mrabet, R.; Douaik, A.; El Haddadi, R.; Bouhlal, O.; Elomari, M.; Ganoudi, M.; Zouahri, A.; Chakiri, S. Effects of Tillage Systems on the Physical Properties of Soils in a Semi-Arid Region of Morocco. *Agriculture* **2023**, *13*, 683. [CrossRef]
15. El Mekkaoui, A.; Moussadek, R.; Mrabet, R.; Chakiri, S.; Douaik, A.; Ghanimi, A.; Zouahri, A. The Conservation Agriculture in Northwest of Morocco (Merchouch Area): The Impact of No-till Systems on Physical Properties of Soils in Semi-Arid Climate. *E3S Web Conf.* **2021**, *234*, 37. [CrossRef]
16. Phillips, J.M.; Hayman, D.S. Improved Procedures for Clearing Roots and Staining Parasitic and Vesicular-Arbuscular Mycorrhizal Fungi for Rapid Assessment of Infection. *Trans. Br. Mycol. Soc.* **1970**, *55*, 158-IN18. [CrossRef]
17. Trouvelot, A. Mesure Du Taux de Mycorhization VA d'un Système Radiculaire. Recherche de Méthodes d'estimation Ayant Une Signification Fonctionnelle. In Proceedings of the 1st European Symposium on Mycorrhizae, Dijon, France, 1–5 July 1985; pp. 217–221.
18. Gerdemann, J.W.; Nicolson, T.H. Spores of Mycorrhizal Endogone Species Extracted from Soil by Wet Sieving and Decanting. *Trans. Br. Mycol. Soc.* **1963**, *46*, 235–244. [CrossRef]
19. Cranenbrouck, S.; Voets, L.; Bivort, C.; Renard, L.; Strullu, D.-G.; Declerck, S. Methodologies for in Vitro Cultivation of Arbuscular Mycorrhizal Fungi with Root Organs. In *In Vitro Culture of Mycorrhizas*; Declerck, S., Fortin, J.A., Strullu, D.-G., Eds.; Soil Biology; Springer: Berlin/Heidelberg, Germany, 2005; Volume 4, pp. 341–375. ISBN 978-3-540-24027-3.
20. Shannon, C.E. A Mathematical Theory of Communication. *Bell Syst. Tech. J.* **1948**, *27*, 44–53. [CrossRef]
21. Pielou, E.C. The measurement of diversity in different types of biological collections. *J. Theor. Biol.* **1966**, *13*, 131–144. [CrossRef]
22. Bossou, L.-D.R.; Houngnandan, H.B.; Adandonon, A.; Zoundji, C.; Houngnandan, P. DiversitÉ des champignons mycorhiziens arbusculaires associés à la culture du maïs (*Zea mays* L.) au Bénin. *Int. J. Biol. Chem. Sci.* **2019**, *13*, 597. [CrossRef]
23. El Mrabet, S.; Ouahmane, L.; El Mousadik, A.; Msanda, F.; Abbas, Y. The Effectiveness of bucular Mycorrhizal Inoculation and Bio-Compost Addition for Enhancing Reforestation with Argania Spinosa in Morocco Available. *Open J. For.* **2014**, *4*, 14–23.
24. Ouallal, I.; Abbas, Y.; Ech-cheddadi, S.; Ouajdi, M.; Ouhadach, M.; El Yacoubi, H.; Kerdouh, B.; El Goumi, Y.; Rochdi, A. Diversité des champignons endomycorhiziens de l'arganier et potentiel mycorhizogène des sols rhizosphériques des arganeraies du Sud-Ouest marocain. *Bois For. Trop.* **2019**, *338*, 73. [CrossRef]
25. Jensen, A.; Jakobsen, I. The Occrrence of Vesicular-Arbuscular Mycorrhiza in Barley and Wheat Grown in Some Danish Soils with Different Fertilizer Treatments. *Plant Soil.* **1980**, *55*, 403–414. [CrossRef]
26. Khaddari, A.E.; Gabardi, S.E.; Touhami, A.O.; Aoujdad, J.; Ouajdi, M.; Dahmani, J. diversity of endomycorrhizal fungi in 456 the thuya rhizosphere, sefrou region (middle eastern atlas, morocco). *Plant Cell. Biotechnol. Mol. Biol.* **2019**, *20*, 1143–1159.
27. Walker, C.; Mize, C.W. Populations of Endogonaceous Fungi at Two Locations in Central Iowa. *Can. J. Bot.* **1982**, *60*, 2518–2529. Available online: https://cdnsciencepub.com/doi/abs/10.1139/b82-305 (accessed on 16 March 2023). [CrossRef]
28. Diagne, O.; Ingleby, K. Ácologie des Champignons Mycorhiziens Arbusculaires Infectant Acacia Raddiana. In *Un Arbre au Désert*; Grouzis, M., Le Floc'h, É., Eds.; IRD Éditions: Marseille, France, 2003; pp. 205–228. ISBN 978-2-7099-1522-9.
29. Heikham, E.; Rupam, K.; Bhoopander, G. Arbuscular mycorrhizal fungi in alleviation of salt stress: A review. *Ann. Bot.* **2009**, *104*, 1263–1280.
30. Garbaye, J. *La Symbiose Mycorhizienne, une Association Entre les Plantes et les Champignons*; Editions Quae: Versailles, France, 2013; 251 Synthèses; ISBN 9782759219643. Available online: Http://Digital.Casalini.It/9782759219643 (accessed on 12 August 2022).
31. Mohammad, M.J.; Malkawi, H.I.; Shibli, R. Effects of Arbuscular Mycorrhizal Fungi and Phosphorus Fertilization on Growth and Nutrient Uptake of Barley Grown on Soils with Different Levels of Salts. *J. Plant Nutr.* **2003**, *26*, 125–137. [CrossRef]
32. Kachkouch, W.; Touati, J.; Ouazzani Touhami, A.; Filali-Maltouf, A.; El Modafar, C.; Moukhli, A.; Oukabli, A.; Benkirane, R.; Douira, A. Diversity of arbuscular mycorrhizal fungi in the rhizosphere of Olea europaea in three regions of Morocco (Tafilalt, Zagora and Taounate). *Int. J. Pure App. Biosci.* **2014**, *2*, 178–195.
33. Elmaati, Y.; Msanda, F.; Mousadik, A.E.; Elhamdaoui, A.; El, S.; Ouahmane, L. contribution to the characterization of mycorrhizae in the south west of morocco and their effect on growth parameters of Argania spinosa. *Am. J. Innov. Res. Appl. Sci.* **2015**, *1*, 235–243.

34. Kenny, L.; Galiana, A.; Bellefontaine, R. *Projet UE/MEDA/ADS. Appui a L'amelioration de la Situation de L'emploi de la Femme Rurale et Gestion Durable de L'arganeraie dans le Sud-ouest du Maroc—Theme 2: Multiplication Vegetative et Symbioses Racinaires de L'arganier: Optimisation des Agrosystemes a Base D'arganier*; Agence du Developpement Social (Maroc): Rabat, Maroc; Agropolis (France): Montpellier, France, 2009; Rapport final; pp. 1–71.
35. Abbas, Y.; Ducousso, M.; Abourouh, M.; Azcón, R.; Duponnois, R. Diversity of Arbuscular Mycorrhizal Fungi in Tetraclinis Articulata (Vahl) Masters Woodlands in Morocco. *Ann. For. Sci.* **2006**, *63*, 285–291. [CrossRef]
36. Ouahmane, L.; Hafidi, M.; Thioulouse, J.; Ducousso, M.; Kisa, M.; Prin, M.; Galiana, A.; Boumezzough, A.; Duponnois, R. Improvement of Cupressus Atlantica Gaussen Growth by Inoculation with Native Arbuscular Mycorrhizal Fungi. *J. Appl. Microbiol.* **2006**, *103*, 1364–5072. [CrossRef] [PubMed]
37. Sghir, F.; Chliyeh, M.; Kachkouch, W.; Khouader, M.; Touhami, A.O.; Benkirane, R.; Douira, A. Mycorrhizal status of Olea europaea spp. oleaster in Morocco. *J. Appl. Biosci.* **2013**, *61*, 4478–4489. [CrossRef]
38. Khirani, S.; Boutaj, H.; Modafar, C.E.; Khelil, A.O.E. Arbuscular mycorrhizal fungi associated with date palm in Ouargla region (southeastern algeria). *Plant Cell. Biotechnol. Mol. Biol.* **2020**, *21*, 15–28.
39. Gosling, P.; Hodge, A.; Goodlass, G.; Bending, G.D. Arbuscular Mycorrhizal Fungi and Organic Farming. *Agric. Ecosyst. Environ.* **2006**, *113*, 17–35. [CrossRef]
40. Houngnandan, P.; Yemadje, R.G.H.; Kane, A.; Boeckx, P.; et Van Cleemput, O. Les glomales indigènes de la forêt claire à Isoberlinia doka (Craibet Stapf) à Wari-Maro au centre du Bénin. *Tropicultura* **2009**, *27*, 83–87.
41. Dalpeé, Y. Inventaire de la flore endomycorhizienne des rivages et dunes maritimes du Québec, du Nouveau-Brunswick et de la Nouvelle-Écosse. *Le Nat. Can.* **1989**, *116*, 219–236.
42. Gemma, J.N.; Koske, R.E.; Carreiro, M. Seasonal dynamics of selected species of VA mycorrhizal fungi in a sand dune. *Mycol. Res.* **1989**, *92*, 317–321. [CrossRef]
43. Smith, T.F. The effect of season and crop rotation on the abundance of spores of vesicular-arbuscular (VA) mycorrhizal endophytes. *Plant Soil* **1980**, *57*, 475–479. [CrossRef]
44. Zangaro, W.; Moreira, M. Micorrizas arbusculares nos biomas Floresta Atlân- tica e Floresta de Araucária. In *Micorrizas 30 anos de Pesquisa no Brasil*; Siqueira, J.O., de Souza, F.A., Cardoso, E.J.B.N., Tsai, S.M., Eds.; UFLA: Lavras, Brazil, 2010; pp. 279–310.
45. Chagnon, P.L.; Bradley, R.L.; Maherali, H.; Klironomos, J.N. A trait-based framework to understand life history of mycorrhizal fungi. *Trends Plant. Sci.* **2013**, *18*, 484–491. [CrossRef] [PubMed]
46. Loján, P.; Senés-Guerrero, C.; Suárez, J.P.; Kromann, P.; Schüßler, A.; Declerck, S. Potato field-inoculation in Ecuador with Rhizophagus irregularis: No impact on growth performance and associated arbuscular mycorrhizal fungal communities. *Symbiosis* **2018**, *73*, 45–56. [CrossRef]
47. Da Silva, I.R.; de Mello CM, A.; Neto RA, F.; da Silva DK, A.; de Melo, A.L.; Oehl, F.; Maia, L.C. Diversity of arbuscular mycorrhizal fungi along an environmental gradient in the Brazilian semi-arid. *Appl. Soil. Ecol.* **2014**, *84*, 166–175. [CrossRef]
48. Prreira, C.M.R.; Goto, B.T.; Silva DKA, D.A.; Ferreira, A.C.A.; de Souza, F.A.; de Silva, G.A.; da Maia, L.C.; Oehl, F. Acaulospora Reducta Sp. Nov. and A. Excavata Two Glomeromycota Fungi with Pitted Spores from Brazil. *Mycotaxon* **2016**, *130*, 983–995. [CrossRef]

MDPI
St. Alban-Anlage 66
4052 Basel
Switzerland
www.mdpi.com

Forests Editorial Office
E-mail: forests@mdpi.com
www.mdpi.com/journal/forests

www.ingramcontent.com/pod-product-compliance
Lightning Source LLC
Chambersburg PA
CBHW040312270926
51850CB00007BC/1109
9783036587684